AF251877

Stochastic Integral and Differential Equations in Mathematical Modelling

Stochastic Integral and Differential Equations in Mathematical Modelling

Santanu Saha Ray

National Institute of Technology, Rourkela, India

World Scientific

NEW JERSEY · LONDON · SINGAPORE · BEIJING · SHANGHAI · HONG KONG · TAIPEI · CHENNAI · TOKYO

Published by

World Scientific Publishing Europe Ltd.

57 Shelton Street, Covent Garden, London WC2H 9HE

Head office: 5 Toh Tuck Link, Singapore 596224

USA office: 27 Warren Street, Suite 401-402, Hackensack, NJ 07601

Library of Congress Cataloging-in-Publication Data
Names: Saha Ray, Santanu, author.
Title: Stochastic integral and differential equations in mathematical modelling /
    Santanu Saha Ray, National Institute of Technology, Rourkela, India.
Description: New Jersey : World Scientific, [2023] | Includes bibliographical references and index.
Identifiers: LCCN 2022045798 | ISBN 9781800613577 (hardcover) |
    ISBN 9781800613584 (ebook) | ISBN 9781800613591 (ebook other)
Subjects: LCSH: Stochastic models. | Stochastic analysis--Mathematical models. |
    Stochastic integral equations. | Stochastic differential equations.
Classification: LCC QA274.2 .S24 2023 | DDC 519.2/2--dc23/eng20230111
LC record available at https://lccn.loc.gov/2022045798

British Library Cataloguing-in-Publication Data
A catalogue record for this book is available from the British Library.

For any available supplementary material, please visit
https://www.worldscientific.com/worldscibooks/10.1142/Q0401#t=suppl

Desk Editors: Sanjay Varadharajan/Adam Binnie/Shi Ying Koe

Typeset by Stallion Press
Email: enquiries@stallionpress.com

Preface

The theory of deterministic chaos has enjoyed during the last three decades a rapidly increasing audience of mathematicians, physicists, engineers, biologists, economists, etc. However, this type of "chaos" can be understood only as quasi-chaos in which all states of a system can be predicted and reproduced by experiments. Meanwhile, many experiments in natural sciences have brought about hard evidence of stochastic effects. The best-known example is perhaps the Brownian motion where pollen submerged in a fluid experiences collisions with the molecules of the fluid and thus exhibits random motions. The study of stochasticity was initiated in the early years of the 1900s. Einstein, Smoluchowsky and Langevin wrote pioneering investigations. This work was later resumed and extended by Ornstein and Uhlenbeck.

This research monograph concerns the analysis of discrete-time approximations for stochastic differential equations (SDEs) driven by Wiener processes. The first chapter of the book provides a theoretical basis for working with SDEs and stochastic processes.

In the present dissertation, various analytical methods like Kudryashov method, improved sub-equation method, Jacobi elliptic function (JEF) expansion method and extended auxiliary equation method have been utilised for getting analytical solutions for stochastic differential equations, viz. Wick-type stochastic Zakharov–Kuznetsov (ZK) equation, Wick-type stochastic Kudryashov–Sinelshchikov equation, Wick-type stochastic modified Boussinesq equations, Wick-type stochastic Kersten–Krasil'shchik coupled KdV-mKdV equations and Wick-type stochastic nonlinear Schrödinger equations have been presented by using various analytical methods.

Wavelet methodologies such as hybrid Legendre Block-Pulse functions, second-kind Chebyshev wavelets, Bernstein polynomials and two-dimensional second-kind Chebyshev wavelets have been used to solve the stochastic integral equations. Furthermore, by applying wavelet methods, the approximate solutions of the stochastic Volterra–Fredholm integral equation, stochastic mixed Volterra–Fredholm integral equation, multidimensional stochastic integral equations, fractional stochastic Itô–Volterra integral equation and nonlinear fractional stochastic Itô–Volterra integral equation have been discussed in this work.

Also, semi-implicit Euler–Maruyama scheme and Chebyshev spectral collocation have been applied to solve stochastic Fisher equation and stochastic FitzHugh–Nagumo equation. These equations have a lot of applications in physical phenomena. The Fisher equation is one of the reaction-diffusion equations and is widely used in the study of biological invasion, and the FitzHugh–Nagumo model is one of the classical standard models in neuroscience.

Also, numerical methods, viz. Euler–Maruyama method, order 1.5 strong Taylor method, split-step forward Euler–Maruyama method, derivative-free Milstein method and higher-order approximation scheme, have been successfully employed to fractional differential stochastic point kinetics equation for obtaining mean neutron population.

About the Author

 **Santanu Saha Ray** is currently a Professor and former Head of the Department of Mathematics, National Institute of Technology, Rourkela, India. Dr. Saha Ray completed his Ph.D. in 2008 from Jadavpur University, Kolkata, India. He received his M.C.A. (Master of Computer Applications) degree in 2001 from the Indian Institute of Engineering Science and Technology (IIEST), erstwhile Bengal Engineering College, Shibpur, India. He completed a Master's degree in applied mathematics at Calcutta University, Kolkata, India, in 1998 and a bachelor's (honours) degree in mathematics at St. Xavier's College (currently known as St. Xavier's University, Kolkata), Kolkata, India, in 1996. He was elected Fellow of the Institute of Mathematics and Its Applications, United Kingdom, in 2018.

Stanford University, USA, together with the publishing house Elsevier and SciTech Strategies, has released a report on the top 2% best scientists in the world in various fields. Dr. Saha Ray is enlisted in the world's top 2% Scientists List of Stanford University released recently and published in the open access science journal *PLOS* (*Public Library of Science*). In India, Dr. Saha Ray is in No. 1 position in the field of "Numerical and Computational Mathematics" and his corresponding world ranking is 107. He is at the top of the most coveted list of "world ranking of top 2% scientists" from India as per a subject-wise analysis conducted by a team of scientists at Stanford University, led by Dr. John PA Ioannidis. Stanford University's new list of the top 2% scientists in the world in various fields includes over 1000 scientists from India. Recently, he was duly elected for Full Membership in Sigma Xi, the Scientific Research Honour Society, USA.

Dr. Saha Ray has more than 21 years of teaching experience at the undergraduate and postgraduate levels in glorious institutes like the National Institute of Technology, Rourkela, and two renowned private engineering institutes in Kolkata, West Bengal. He has more than 20 years of research experience in various fields of Applied Mathematics. He has published many peer-reviewed research papers in numerous fields and various international SCI journals of repute. For a detailed citation overview, the reader may be referred to Scopus. To date, he has more than 222 research papers published in journals of international repute, including more than 193 SCI journal papers. He was awarded an IOP Publishing Top Cited Author Award 2018 from India in the field of physics published across the whole IOP Publishing portfolio in the past three years (2015–2017), using citations recorded in Web of Science.

He has solely authored a book entitled *Graph Theory with Algorithms and Its Applications: in Applied Science and Technology* published by Springer. A solely authored book entitled *Fractional Calculus with Applications for Nuclear Reactor Dynamics* has been published by the CRC Press of the Taylor & Francis Group. Another solely authored book entitled *Numerical Analysis with Algorithms and Programming* has been also published in the CRC press of Taylor & Francis group, USA. Another three books entitled *Wavelet Methods for Solving Partial Differential Equations and Fractional Differential Equations*, *Generalized Fractional Order Differential Equations Arising in Physical Models* and *Novel Methods for Solving Linear and Nonlinear Integral Equations* have been published by the CRC press of Taylor & Francis group, USA. Recently, his book entitled *Nonlinear Differential Equations in Physics* has been launched by Springer Nature.

Currently, he is acting as editor-in-chief for the Springer Scopus international journal entitled *International Journal of Applied and Computational Mathematics*. He is also an associate editor of a Springer SCIE international journal *Mathematical Sciences* and reviewer of several journals of Elsevier, Springer and Taylor & Francis. He had also been the lead guest editor in the International SCI journals of Hindawi Publishing Corporation, USA.

He has contributed papers on several topics, such as fractional calculus, mathematical modelling, mathematical physics, stochastic modelling, integral equations, and wavelet methods. He is a member of the Society for Industrial and Applied Mathematics (SIAM) and the American Mathematical Society (AMS).

He was the principal investigator of the two *Board of Research in Nuclear Sciences* research project, with grants from Bhabha Atomic Research Centre, Mumbai, India. He was also the principal investigator of a research project financed by the *SERB, Department of Science and Technology*, Government of India, and a research project financed by the *National Board for Higher Mathematics*, Department of Atomic Energy, Government of India, respectively. Currently, he has been acting as the principal investigator of a research project financed by the *SERB, Department of Science and Technology*, Government of India. Another research project has been recently approved by the *National Board for Higher Mathematics*, Department of Atomic Energy, Government of India.

Under his sole supervision, five research scholars had been awarded Ph.D. from NIT Rourkela.

It is not out of place to mention that he had attended the workshop organised by the West Bengal University of Technology (WBUT) on "Review of Engineering Degree Curriculum of Mathematics" held at the National Institute of Technical Teacher's Training and Research (NITTTR) Kolkata, from July 26–30, 2004. In that Workshop Curriculum of Mathematics at the undergraduate level as well as postgraduate level of the West Bengal University of Technology was revised.

He was invited to deliver a lecture in a Short Term Training Programme on "Mathematical Modelling" organised by the Department of Science, National Institute of Technical Teacher's Training and Research (NITTTR), Kolkata, from December 4–8, 2006.

It is worth noting to mention that he was invited to deliver a lecture in a workshop on "Fractional Order systems" organised by the Instrumentation and Electronics Engineering Department, Jadavpur University, Salt Lake Campus, Kolkata, sponsored by IEEE Kolkata

Chapter, DRDL Hyderabad, and BRNS (DAE) Mumbai, from March 28–29, 2008.

Moreover, he was also invited to deliver a lecture in a Short Term Training Programme on "Mathematical Modelling" organised by the Department of Science, National Institute of Technical Teacher's Training and Research (NITTTR), Kolkata, from November 24–28, 2008.

He was invited as an expert speaker in the One Week Online Short Term Training Programme (STTP) on Fractional Calculus: Foundations to Frontiers (FCFF 2020) during December 28, 2020 to January 1, 2021 held in Applied Mathematics and Humanities Department, SVNIT, Surat.

He was also invited as an expert speaker and session chairman in the Conference on "Fractional Calculus: Analysis and Applications" during August 20–21, 2021 held in the Center for Advanced Study, Department of Mathematics, Savitribhai Phule Pune University, Pune.

He also delivered an invited talk at the Second Online Conference on Nonlinear Dynamics and Complexity, which took place at ISEP, Porto, Portugal, October 4–6, 2021.

He delivered an invited talk at the "International Conference on Fractional Calculus, 2022" organised by the School of Mathematics and Statistics, University of Hyderabad, India, during January 18–19, 2022.

He delivered an invited talk at the 3rd International Conference on Recent Developments in Engineering and Technology (ICRDET-2022) which was held in the virtual mode under the auspices of the RTU, Kota, and Anand-ICE (India), during February 25–26, 2022.

Additionally, he was invited to present an invited lecture for the CONVERGENCE-2022 Webinar Series, which took place at SASTRA Deemed University on June 25, 2022.

He was convener of "Symposium on Recent Trends and Emerging Applications of Mathematical Sciences (SRTEAMS-2013)" sponsored by CSIR, New Delhi; DST, Govt. of India and INSA, New Delhi, held from May 16–17, 2013. He was chairman as well convener of "National Conference on Recent Advances in Mathematics and its Applications (NCRAMA-2018)" sponsored by NBHM, DAE, Govt. of India; SERB DST, Govt. of India and CSIR, Govt. of India, held at NIT Rourkela from December 7–8, 2018. Also, he was chairman as well convener of "1st International Conference on Applied Analysis, Computation and Mathematical Modelling in Engineering (AACMME-2021)" held at NIT Rourkela from February 24–26, 2021. He served as the editor of a proceedings book published by Springer titled *Applied Analysis, Computation and Mathematical Modelling in Engineering: Select Proceedings of AACMME 2021*. Additionally, he served as the guest editor of the special issue of AACMME-2021 for the Elsevier Journal *Journal of Computational and Applied Mathematics*. He recently organised a short-term training programme (STTP) titled "Emerging Applications of Mathematics and Statistics in Engineering Science and Technology (EAMSEST-2022)" at NIT Rourkela in hybrid mode from May 9–15, 2022, sponsored by SERB and NBHM, for which he served as chairman and convener.

He had attended as an invited participant in "International Conference on Mathematical Modeling in Physical Sciences" held in Madrid, Spain, from August 28–31, 2014. Also, he attended as an invited participant in "Global Conference on Applied Physics and Mathematics" held in Rome, Italy, from July 25–27, 2016. Travel grant was sponsored by SERB, DST, Government of India.

He also served as an expert member of the Board of Studies in the Department of Mathematics of NIT Arunachal Pradesh from 2019 and in the Department of Mathematics of Veer Surendra Sai University of Technology, Burla, Odisha, from 2022, and as an expert examiner of Ph.D. defences and other examinations for various universities. He was also an expert member of the faculty recruitment interview board. He also served as an adjudicator for several national and international theses from various national and international universities/institutions.

Contents

List of Figures

List of Tables

Chapter 1

Introduction and Preliminaries of Stochastic Calculus

1.1 Origins of Stochastic Calculus

In 1827, the (then already) famous Scottish botanist Robert Brown
observed a rather curious phenomenon. Brown was interested in the
tiny particles found inside grains of pollen, which he studied by sus-
pending them in water and observing them under his microscope.
Remarkably enough, it appeared that the particles were constantly
jittering around in the fluid. At first, Brown thought that the particles were alive, but he was able to rule out this hypothesis after he
observed the same phenomenon when using glass powder, and a large
number of other inorganic substances, instead of the pollen particles.
A satisfactory explanation of Brown's observation was not provided
until the publication of Einstein's famous 1905 paper.

The first mathematical characterisation of Brownian motion was
proposed by Einstein in 1905. Einstein attempted to deduce the mass
of water molecules from observable quantities such as the temperature and viscosity of the water, and stochastic differential equations
were introduced which have been used for modelling such types of
dynamic phenomena, where the exact dynamics of the system are
uncertain. One motivation for studying such equations is that various physical phenomena can be modelled as random processes and
when such a phenomenon enters a physical system, the model of
stochastic differential equations can be obtained.

The best-known stochastic process to which stochastic calculus is applied is the Wiener process (named in honour of Norbert Wiener), which is used for modelling Brownian motion as described by Louis Bachelier in 1900 and by Albert Einstein in 1905 and other physical diffusion processes in space of particles subject to random forces. Since the 1970s, the Wiener process has been widely applied in financial mathematics and economics to model the evolution in time of stock prices and bond interest rates.

1.2   Motivation and Objectives

Stochastic mathematical models have received increasing attention for their ability of representing intrinsic uncertainty in complex systems, e.g., representing various scales in particle simulations at molecular and mesoscopic scales, as well as extrinsic uncertainty, e.g., stochastic external forces, stochastic initial conditions or stochastic boundary conditions. The modelling of systems by differential equations usually requires that the parameters involved be completely known. Such models often originate from problems in physics or economics where there is insufficient information on parameter values. In some cases, the parameter values may depend in a complicated way on the microscopic properties of the medium. In addition, the parameter values may fluctuate due to some external or internal "noise", which is random — or at least appears so to one. One important class of stochastic mathematical models is stochastic partial differential equations (SPDEs), which can be seen as deterministic partial differential equations (PDEs) with finite or infinite dimensional stochastic processes. As a purely mathematical construction, white noise can be a good model for rapid random fluctuations.

The stochastic integral equation also has numerous applications to the problems in reactor dynamics, chemical kinetics, fluid dynamics, quantum physics, population growth, financial applications, turbulence and systems theory. The applications of wavelet theory in numerical methods for solving differential equations are more than a decade old. In the early nineties, people were very optimistic because it seemed that many pertinent properties of wavelets would automatically lead to an efficient numerical method for differential equations. As a powerful tool, wavelets have been extensively used in signal processing, image processing, pattern recognition, computer graphics and many other areas. Wavelets permit the accurate representation of a variety of functions and operators. The wavelet methods can also be applied for stochastic integral equations with Itô stochastic integral.

1.3 Framework of Stochastic Calculus

1.3.1 *Function and distribution space*

There are (at least) two ways of constructing the classical Wiener–Itô chaos expansion:

(A) by Hermite polynomials,
(B) by multiple Itô integrals.

1.3.1.1 *Hermite functions for Wiener–Itô chaos expansion*

The Hermite polynomials $h_n(x)$ are defined by

$$h_n(x) = (-1)^n e^{\frac{1}{2}x^2} \frac{d^n}{dx^n}\left(e^{-\frac{1}{2}x^2}\right); \quad n = 0, 1, 2, \ldots \tag{1.1}$$

Thus, the first Hermite polynomials are

$$h_0(x) = 1, \quad h_1(x) = x, \quad h_2(x) = x^2 - 1, \quad h_3(x) = x^3 - 3x,$$
$$h_4(x) = x^4 - 6x^2 + 3, \quad h_5(x) = x - 10x^3 + 15x \ldots \tag{1.2}$$

1.3.1.2 *Orthogonality of Hermite polynomials*

Using integration by parts, for $k \neq l$, it can be shown that

$$\int_{-\infty}^{+\infty} h_k(x)h_l(x)e^{-x^2}\,dx = 0,$$

and for $k \neq l$,

$$\int_{-\infty}^{+\infty} h_k(x)^2 e^{-x^2}\,dx = \sqrt{\pi}2^k k!.$$

In other words, the Hermite polynomials are orthogonal for the Gaussian distribution with mean 0 and variance $\frac{1}{2}$. Thus, an equivalent formulation of the fact that Hermite polynomials are an orthogonal basis for $L^2(\mathbb{R}, w(x)dx)$ consists in introducing Hermite functions and in saying that the Hermite functions are an orthonormal basis for $L^2(\mathbb{R})$.

In this section, $(S(\mathbb{R}^d))$ and $(S(\mathbb{R}^d))^*$ are the Hida test function and distribution space on $\mathbb{R}^d$. Let the Hermite function $\xi_n(x)$ be defined by

$$\xi_n(x) = e^{-(1/2)x^2} h_n(\sqrt{2}x)/((n-1)!\pi)^{1/2}, \quad n \geq 1, \tag{1.3}$$

where $h_n(x)$ is the Hermite polynomial.

Then, the set $\{\xi_n(x)\}_{n \geq 1}$ represents an orthogonal basis for $L^2(\mathbb{R})$. Let $\alpha = (\alpha_1, \ldots, \alpha_d)$ denote d-dimensional multi-indices with $\alpha_1, \ldots, \alpha_d \in \mathbb{N}$. It follows the family of tensor products $\xi_\alpha = \xi_{(\alpha_1, \ldots, \alpha_d)} = \xi_{\alpha_1} \otimes \cdots \otimes \xi_{\alpha_d} (\alpha \in \mathbb{N}^d)$ which forms an orthogonal basis for $L^2(\mathbb{R}^d)$.

Let $\alpha^{(i)} = (\alpha_1^{(i)}, \ldots, \alpha_d^{(i)})$ be the ith multi-index number in some fixed ordering of all d-dimensional multi-indices $\alpha = (\alpha_1, \ldots, \alpha_d) \in \mathbb{N}^d$. Suppose, it possesses the property that $i < j$ implies

$$\alpha_1^{(i)} + \cdots + \alpha_d^{(i)} \leq (\alpha_1^{(j)} + \cdots + \alpha_d^{(j)}).$$

Now,

$$\eta_i = \xi_{\alpha_1^{(i)}} \otimes \cdots \otimes \xi_{\alpha_d^{(i)}}, \quad i \geq 1.$$

Consider the multi-indices (of arbitrary length) as elements of $(\mathbb{N}_0^{\mathbb{N}})_c$ for all sequences $\alpha = (\alpha_1, \alpha_2, \ldots) \in \mathcal{J}$, where $\mathcal{J} = (\mathbb{N}_0^{\mathbb{N}})_c$ with elements $\alpha_i \in \mathbb{N}_0$ (where $\mathbb{N}_0 = \mathbb{N} \cup \{0\}\})$ and with compact support, with only finitely many $\alpha_i \neq 0$. Then, for $\alpha \in \mathcal{J}$, it can be defined as

$$H_\alpha(\omega) = \prod_{i=1}^{\infty} h_{\alpha_i}(\langle \omega, \eta_i \rangle), \omega \in (S(\mathbb{R}^d))^*.$$

1.3.2 *Kondratiev stochastic test function space and stochastic distribution space*

The theory of generalised functions of infinitely many variables with special spaces of test and generalised functions and with the pairing generated by the Gaussian measure was developed by Kondratiev [1–3], see also [4–6] (afterwards the said spaces are called the Kondratiev spaces), and independently by Hida [7,8] (the corresponding spaces are called the Hida spaces).

The construction of the Kondratiev spaces of test functions has been considered with orthogonal bases given by a generating function $\gamma(\lambda)h(x;\alpha(\lambda))$, where h satisfies assumptions accepted in Ref. [9].

Let $\mathcal{H}_p, p \in \mathbb{Z}_+ := \mathbb{N} \cup \{0\}$ be a family of real separable Hilbert spaces [10] such that

- for each $p \in \mathbb{Z}_+$ $\mathcal{H}_{p+1}$ is densely and continuously embedded into $\mathcal{H}_p$ (it is convenient to suppose that for each $p \in \mathbb{Z}_+$ $\| \cdot \|_{\mathcal{H}_{p+1}} \geq \| \cdot \|_{\mathcal{H}_p}$, the general case can be reduced to this one [11]);
- the embeddings $\mathcal{H}_2 \hookrightarrow \mathcal{H}_1$ and $\mathcal{H}_3 \hookrightarrow \mathcal{H}_2$ are quasinuclear, i.e., the corresponding embedding operators are of Hilbert–Schmidt type.

A chain (a rigging of $\mathcal{H}_0$) has been considered

$$\mathcal{N}' \supset \cdots \supset \mathcal{H}_{-p} \supset \cdots \supset \mathcal{H}_0 \supset \cdots \supset \mathcal{H}_p \supset \cdots \supset \mathcal{N}, \qquad (1.4)$$

where $\mathcal{N} = pr \lim_{p \in \mathbb{Z}_+} \mathcal{H}_p$ is the projective limit of the sequences of spaces $\{\mathcal{H}_p\}_{p \in \mathbb{Z}_+}$ (it means that $\mathcal{N} = \cap_{p \in \mathbb{Z}_+} \mathcal{H}_p$ with a topology of the projective limit — the weaker topology such that for each $p \in \mathbb{Z}_+$ the embedding $\mathcal{N}$ into $\mathcal{H}_p$ is continuous, see, e.g., Refs. [11,12] for details) and $\mathcal{H}_{-p}$ are Hilbert spaces dual of $\mathcal{H}_p$ with respect to $\mathcal{H}_0$, $\mathcal{N}' = \cup_{p \in \mathbb{Z}_+} \mathcal{H}_{-p}$ (often it can be convenient to introduce on $\mathcal{N}'$ a topology of inductive limit — the strongest topology such that for each $p \in \mathbb{Z}_+$ the embedding of $\mathcal{H}_{-p}$ into $\mathcal{N}'$ is continuous, in this case, one writes $\mathcal{N}' = ind \lim_{p \in \mathbb{Z}_+} \mathcal{H}_{-p}$). In some versions of the white noise analysis, it can be necessary to assume in addition to the chain (1.4) is nuclear (or, which is the same that the space $\mathcal{N}$ is nuclear); it means that for each $p \in \mathbb{Z}_+$ there exists $p' \in \mathbb{N}$ such that the embedding $\mathcal{H}_{p'} \hookrightarrow \mathcal{H}_p$ is quasinuclear.

Using properties of holomorphic functions [13] and the kernel theorem [11,12], it is shown in Ref. [10] that for each $x \in Q$ there exists an expansion

$$h(x;\lambda) = \sum_{n=0}^{\infty} \frac{1}{n!} \langle h_n(x), \lambda^{\otimes n} \rangle,$$

$$h_n(x) \in \mathcal{H}_{-2,\mathbb{C}}^{\hat{\otimes} n},$$

$$\lambda \in B_x := \{\lambda \in \mathcal{H}_{-2,\mathbb{C}} \cap U_0 : |\lambda|_2 < R_x, R_x > 0\}, \qquad \lambda^{\otimes 0} := 1, \quad (1.5)$$

here and in the following denoted by $\langle \cdot, \cdot \rangle$ the dual pairings in tensor power of the complexification of chain (1.4) and by $|\cdot|_p$ the norms in tensor powers of $\mathcal{H}_{p,\mathbb{C}}$, $p \in \mathbb{Z}$, $\mathcal{H}_{-2,\mathbb{C}}^{\hat{\otimes}0} := \mathbb{C}$. Note that the series (1.5) converges uniformly in every closed ball B_x. Assume in addition that
$B := \cap_{x \in Q} B_x$ is a nonempty open set.
Let $K > 1$ be some constant, $p \in \mathbb{N}\backslash\{1\}$, $q \in \mathbb{N}$.

Definition 1.3.2.1. A Hilbert space of formal series

$$(\mathcal{H}_p)_q := \left\{ f(x) = \sum_{n=0}^{\infty} \left\langle h_n(x), f^{(n)} \right\rangle, f^{(n)} \in \mathcal{H}_{p,\mathbb{C}}^{\hat{\otimes}n}, \right. \tag{1.6}$$
$$\left. x \in Q : \|f\|_{(\mathcal{H}_p)_q}^2 := \sum_{n=0}^{\infty} (n!)^2 K^{qn} |f^{(n)}|_p^2 < \infty \right\},$$

with corresponding to $\|\cdot\|_{(\mathcal{H}_p)_q}$ scalar product is called the Kondratiev space of test functions.

Together with spaces $(\mathcal{H}_p)_q$, one can consider the spaces of test functions $(\mathcal{H}_p) := pr \lim_{q \in \mathbb{N}} (\mathcal{H}_p)_q$ and $(\mathcal{N}) := pr \lim_{q \in \mathbb{N}, p \in \mathbb{N}\backslash\{1\}} (\mathcal{H}_p)_q$; these spaces also are called the Kondratiev ones.

Definition 1.3.2.2.

(a) Kondratiev stochastic test function space
Let N be a natural number. For $0 \leq \rho \leq 1$, let

$$(S)_\rho^N = (S)_\rho^{m;N}$$

consists of those

$$f = \sum_\alpha c_\alpha H_\alpha \in \mathbb{L}^2(\mu_m) = \bigoplus_{k=1}^{N} L^2(\mu_m) \quad \text{with } c_\alpha \in \mathbb{R}^N,$$

such that

$$\|f\|_{\rho,k}^2 = \sum_{\alpha \in J} c_\alpha^2 (\alpha!)^{1+\rho} (2\mathbb{N})^{k\alpha} < \infty \quad \text{for all } k \in \mathbb{N}_0,$$

where $(2\mathbb{N})^{k\alpha} = (2\cdot1)^{k\alpha_1}(2\cdot2)^{k\alpha_2}\ldots(2\cdot m)^{k\alpha_m}$, if Index $(\alpha) = m$.

Let

$$(S)_{\rho,k} := \{f : \|f\|_{\rho,k} < \infty\},$$

and define

$$(S)_\rho := \bigcap_{k \in \mathbb{N}_0} (S)_{\rho,k},$$

endowed with projective topology [14].
(b) Kondratiev stochastic distribution space
Let N be a natural number. For $0 \le \rho \le 1$,

$$(S)_{-\rho}^N = (S)_{-\rho}^{m;N}$$

consists of those

$$F = \sum_\alpha b_\alpha H_\alpha \quad \text{with } b_\alpha \in \mathbb{R}^N,$$

such that

$$\|F\|_{-\rho,-k}^2 = \sum_{\alpha \in J} b_\alpha^2 (\alpha!)^{1-\rho} (2\mathbb{N})^{-k\alpha} < \infty \quad \text{for all } k \in \mathbb{N}_0.$$

Set

$$(S)_{-\rho,-k} := \{F : \|F\|_{-\rho,-k} < \infty\},$$

and define

$$(S)_{-\rho} := \bigcup_{k \in \mathbb{N}_0} (S)_{-\rho,-k},$$

endowed with inductive topology [14].
The family of semi-norms $\|f\|_{\rho,k}$; $k \in \mathbb{N}$ advances to a topology on $(S)_\rho^N$, and $(S)_{-\rho}^N$ can be represented as the dual of $(S)_\rho^N$ by the following action:

$$\langle F, f \rangle = \sum_\alpha \langle b_\alpha, c_\alpha \rangle \alpha!,$$

if $F = \sum_\alpha b_\alpha H_\alpha \in (S)_{-\rho}^N$; $f = \sum_\alpha c_\alpha H_\alpha \in (S)_\rho^N$ and $\langle b_\alpha, c_\alpha \rangle$ is the usual inner product in $\mathbb{R}^N$.

1.3.3 *Hida stochastic test function space and distribution space*

Definition 1.3.3.1.

(a) The Hida test function space $(S)^N$ consists of

$$f = \sum_\alpha c_\alpha H_\alpha \in L^2(\mu_m) \quad \text{with } c_\alpha \in \mathbb{R}^N,$$

such that

$$\sup_\alpha \{c_\alpha^2 \alpha! (2\mathbb{N})^{k\alpha}\} < \infty \quad \text{for all } k < \infty.$$

(b) The Hida distribution space $(S)^{*,N}$ consists of

$$F = \sum_\alpha b_\alpha H_\alpha \quad \text{with } b_\alpha \in \mathbb{R}^N,$$

such that

$$\sup_\alpha \{b_\alpha^2 \alpha! (2\mathbb{N})^{-q\alpha}\} < \infty \quad \text{for some } q < \infty.$$

1.3.4 *Wick product*

The Wick product $F \lozenge G$ is

$$F = \sum_\alpha a_\alpha H_\alpha,$$

$$\text{and } G = \sum_\alpha b_\alpha H_\alpha \in (S)_{-1}^{m;N} \text{ with } a_\alpha, b_\alpha \in \mathbb{R}^N$$

is defined as

$$F \lozenge G = \sum_{\alpha,\beta} (a_\alpha, b_\beta) H_{\alpha+\beta}.$$

1.3.5 *The Hermite transform*

The Hermite transform or $\mathcal{H}$-transform transforms Wick products into ordinary products and convergence in $(S)_{-1}$ into bounded, pointwise convergence in a certain neighbourhood of zero in $\mathbb{C}^N$ (the set of all sequences of complex numbers).

Definition 1.3.5.1. Let $F = \sum_\alpha b_\alpha H_\alpha \in (S)_{-1}^N$ with $b_\alpha \in \mathbb{R}^N$. Then, the Hermite transform of F is defined as

$$\mathcal{H}F(z) = \tilde{F}(z) = \sum_\alpha b_\alpha z^\alpha \in \mathbb{C}^N, \text{ if convergent,}$$

where $z = (z_1, z_2, \dots) \in \mathbb{C}^N$ and $z^\alpha = z_1^{\alpha_1} z_2^{\alpha_2}, \dots, z_n^{\alpha_n}, \dots$, if $\alpha = (\alpha_1, \alpha_2, \dots) \in \mathcal{J}$, where $z_j^0 = 1$.

From Definition 1.4.1.1.3, for $X, Y \in (S)_{-1}^N$,

$$\mathcal{H}(X \Diamond Y(z)) = \tilde{X}(z)\tilde{Y}(z). \tag{1.7}$$

The right-hand side of Eq. (1.7) is the complex bilinear product between two elements of $\mathbb{C}^N$ and is defined by

$$(z_1^1, \dots, z_n^1)(z_1^2, \dots, z_n^2) = \sum_{k=1}^n z_k^1 z_k^2, \quad \text{where } z_k^i \in \mathbb{C}.$$

Let $X = \sum_\alpha a_\alpha H_\alpha \in (S)_{-1}^N$ and $c_0 = X(0) \in \mathbb{R}^N$ be called expectation of X, viz., $E(X)$. Let us assume $g : V \to \mathbb{C}^m$ is an analytic function, where V is a neighbour of $E(X)$. Assume that the Taylor series of g about $E(X)$ contains coefficients in $\mathbb{R}^m$. Then, the Wick version is defined as

$$g^\Diamond(X) = \mathcal{H}^{-1}(g \circ \tilde{X}) \in (S)_{-1}^m.$$

In other words, if g has the Taylor series expansion

$$g(z) = \sum a_\alpha (z - E(X))^\alpha, \quad \text{where } a_\alpha \in \mathbb{R}^m, \text{ then}$$

$$g^\Diamond(z) = \sum a_\alpha (z - E(X))^{\Diamond \alpha} \in (S)_{-1}^m.$$

Suppose that modelling consideration leads to consider a stochastic partial differential equation to express as

$$A(t, x, \partial_t, \nabla_x, U, \omega) = 0, \tag{1.8}$$

where A is some given function, $U = U(t, x, \omega)$ is an unknown (generalised) stochastic process and the operators are given as

$$\partial_t \equiv \frac{\partial}{\partial t}, \nabla_x \equiv \left(\frac{\partial}{\partial x_1}, \dots, \frac{\partial}{\partial x_d} \right) \text{ when } x = (x_1, \dots, x_d) \in \mathbb{R}^d.$$

The Wick-type SPDE is of the form

$$A^{\Diamond}(t, x, \partial_t, \nabla_x, U, \omega) = 0. \tag{1.9}$$

Equation (1.9) undergoes Hermite transform so that the Wick products are converted into ordinary products and the equation can be written as

$$\tilde{A}(t, x, \partial_t, \nabla_x, \Phi, z_1, z_2, \dots) = 0, \tag{1.10}$$

where $\Phi = \mathcal{H}(U)$ is the Hermite transform of U and $z_1, z_2, \dots$ are complex numbers.

Suppose that a solution $u = u(t, x, z)$ of the equation can be obtained as $\tilde{A}(t, x, \partial_t, \nabla_x, \Phi, z_1, z_2, \dots) = 0$ for each $z = (z_1, z_2, \dots) \in \mathbb{K}_q(r)$ for some q, r where $\mathbb{K}_q(r) = \{ z = (z_1, z_2, \dots) \in \mathbb{C}^N$ and $\sum_{\alpha \neq 0} |z^{\alpha}|^2 (2\mathbb{N})^{q\alpha} < r^2 \}$.

1.3.6 *Inverse Hermite transform*

Theorem 1.3.6.1 ([15]). *Suppose $u(t, x, z)$ be a solution (in the usual strong, pointwise sense) of Eq. (1.10) for (t, x) in some bounded open set $G \subset \mathbb{R} \times \mathbb{R}^d$, and for all $z \in \mathbb{K}_q(r)$, for some q, r. Moreover, suppose that $u(t, x, z)$ and all its partial derivatives, which are involved in Eq. (1.10), are bounded for $(t, x, z) \in G \times \mathbb{K}_q(r)$, continuous with respect to $(t, x) \in G$ for all $z \in \mathbb{K}_q(r)$ and analytic with respect to $z \in \mathbb{K}_q(r)$, for all $(t, x) \in G$. Then, there exists $U(t, x) \in (S)_{-1}$ such that $u(t, x, z) = (\Phi(t, x)(z))$ for all $(t, x, z) \in G \times \mathbb{K}_q(r)$ and $U(t, x)$ solves (in the strong sense in $(S)_{-1}$) Eq. (1.8) in $(S)_{-1}$.*

Proposition 1.3.6.2. *For any domain* $\mathfrak{D}$, $L^2(\mathfrak{D})$ *is a real Hilbert space with inner product*

$$\langle u, v \rangle_{L^2(\mathfrak{D})} := \int_{\mathfrak{D}} u(\boldsymbol{x})v(\boldsymbol{x})d\boldsymbol{x}.$$

Definition 1.3.6.3 (trace class). For a separable Hilbert space H, a non-negative definite operator $L \in \mathfrak{L}(H)$ is of trace class if $\operatorname{Tr} L < \infty$, where the trace is defined as $\operatorname{Tr} L := \sum_{j=1}^{\infty} \langle L\phi_j, \phi_j \rangle$, for an orthonormal basis $\{\phi_j : j \in N\}$ of H.

1.3.7 *Probability space* $(\Omega, \mathcal{F}, \mathbb{P})$

The probability space consists of

(a) the sample space Ω, the set of all possible outcomes,
(b) the set of all events $\mathcal{F}$, it is a collection of subsets of Ω,
(c) the probability measure $\mathbb{P}$, i.e., a function from events to probabilities.

Definition 1.3.7.1 (probability measure). A measure $\mathbb{P}$ on $(\Omega, \mathcal{F})$ is a probability measure if it has unit total mass, $\mathbb{P}(\Omega) = 1$.

Definition 1.3.7.2 (almost surely). An event F happens almost surely (a.s. or $\mathbb{P}$-a.s.) with respect to a probability measure $\mathbb{P}$ if $\mathbb{P}(F) = 1$.

1.3.8 *Filtration*

Definition 1.3.8.1 ([16]). Let $(\Omega, \mathcal{F}, \mathbb{P})$ be a probability space.

(a) A filtration $\{\mathcal{F}_t : t \geq 0\}$ is a family of sub σ-algebras of $\mathcal{F}$ that are increasing: i.e., $\mathcal{F}_s$ is a sub σ-algebra of $\mathcal{F}_t$ for $s \leq t$. Each $(\Omega, \mathcal{F}_\mathbf{t}, \mathbb{P})$ is a measure space and assume it is complete.
(b) A filtered probability space is a quadruple $(\Omega, \mathcal{F}, \mathcal{F}_t, \mathbb{P})$, where $(\Omega, \mathcal{F}, \mathbb{P})$ is a probability space and $\{\mathcal{F}_t : t \geq 0\}$ is a filtration of $\mathcal{F}$.

Stochastic processes that conform to the notion of time described by the filtration $\mathcal{F}_t$ are known as adapted processes.

Definition 1.3.8.2. Let $(\Omega, \mathcal{F}, \mathcal{F}_t, \mathbb{P})$ be a filtered probability space. A stochastic process $\{X(t) : t \in [0, T]\}$ is $\mathcal{F}_t$-adapted if the random variable $X(t)$ is $\mathcal{F}_t$-measurable for all $t \in [0, T]$.

Definition 1.3.8.3. A real-valued stochastic process $B(t)$, $t \in [0, T]$ is called Brownian motion, if

(i) $B(0) = 0$ a.s.,
(ii) the process has independent increments for $0 \leq t_0 \leq t_1 \leq \cdots \leq t_n \leq T$,
(iii) $B(t) - B(s) \sim N(0, t - s)$ for $0 \leq s \leq t$,
(iv) $B(t)$ is continuous as a function of t.

Corollary 1.3.8.4. *Let H be a separable Hilbert space and X be an H-valued Gaussian with $\mu = E[X]$. Then, $X \in L^2(\Omega, H)$ and the covariance operator $\mathcal{C}$ of X is a well-defined trace class operator on X. It follows $X \sim N(\mu, \mathcal{C})$.*

Definition 1.3.8.5. Let $\{\mathcal{N}_t\}_{t \geq 0}$ be an increasing family of σ-algebras of subsets of Ω. A process $g(t, \omega) : [0, \infty) \times \Omega \to \mathbb{R}^n$ is called $\mathcal{N}_t$-adapted if for each $t \geq 0$ the function $\omega \to g(t, \omega)$ is $\mathcal{N}_t$-measurable.

Definition 1.3.8.6. Let $\mathcal{V} = \mathcal{V}(S, T)$ be the class of functions $f(t, \omega) : [0, \infty) \times \Omega \to \mathbf{R}$ such that

(i) The function $(t, \omega) \to f(t, \omega)$ is $\mathcal{B} \times \mathcal{F}$-measurable, where $\mathcal{B}$ denotes the Borel algebra on $[0, \infty)$ and $\mathcal{F}$ is the σ-algebra on Ω.
(ii) f is adapted to $\mathcal{F}_t$, where $\mathcal{F}_t$ is the σ-algebra generated by the random variables $B(s)$, $s \leq t$.
(iii) $E(\int_S^T f^2(t, \omega) dt) < \infty$.

Definition 1.3.8.7 (the Itô integral [17]). Let $f \in \mathcal{V}(S, T)$. Then, the Itô integral of f is defined by

$$\int_S^T f(t, \omega) dB_t(\omega) = \lim_{n \to \infty} \int_S^T \varphi_n(t, \omega) dB_t(\omega) \quad (\text{lim in } L^2(P)),$$

where φ_n is a sequence of elementary functions such that

$$E\left(\int_S^T (f(t, \omega) - \varphi_n(t, \omega))^2 dt\right) \to 0 \quad \text{as } n \to \infty.$$

1.3.9 *Q-Wiener process*

Assumption 1.3.9.1 ([17]). Let U be a separable Hilbert space with norm $\|\cdot\|_U$ and inner product $\langle\cdot,\cdot\rangle_U$. $Q \in \mathcal{L}(U)$ is non-negative definite and symmetric, where $\mathbb{L}(U)$ is the set of bounded linear operators $L : U \to U$. Furthermore, Q has an orthonormal basis $\{x_j : j \in \mathbb{N}\}$ of eigenfunctions with corresponding eigenvalues $q_j \geq 0$ such that $\sum_{j \in \mathbb{N}} q_j < \infty$ (i.e., Q is of trace class).

Let $(\Omega, \mathcal{F}, \mathcal{F}_t, \mathbb{P})$ be a filtered probability space (Definition 1.3.7.1). From Definition 1.3.8.3 and the Gaussian distribution $N(0, Q)$ on Hilbert space (Corollary 1.3.8.4). The Q-Wiener process is defined as follows.

Definition 1.3.9.2. A U-valued stochastic process $\{W(t) : t \geq 0\}$ is a Q-Wiener process if

(a) $W(0) = 0$ a.s.,
(b) $W(t)$ is continuous as a function $\mathbb{R}^+ \to U$, for each $\omega \in \Omega$,
(c) $W(t)$ is $\mathcal{F}_t$-adapted and $W(t) - W(s)$ is independent of $\mathcal{F}_s$ for $s < t$, and
(d) $W(t) - W(s) \sim N(0, (t-s)Q)$ for all $0 \leq s \leq t$.

Theorem 1.3.9.3. *Let Q satisfy Assumption 1.3.9.1. Then, $W(t)$ is a Q-Wiener process if and only if*

$$W(t) = \sum_{j=1}^{\infty} \sqrt{q_j}\chi_j\beta_j(t) \quad \text{a.s.,} \tag{1.11}$$

where $\beta_j(t)$ are iid $\mathcal{F}_t$-Brownian motions and the series converges in $L^2(\Omega, U)$. Moreover, Eq. (1.11) converges in $L^2(\Omega, C([0, T], U))$ for any $T > 0$.

Proof. It may be referred to Ref. [16]. $\square$

Corollary 1.3.9.4. *From Eq. (1.11),*

$$Cov(\langle W(t), \chi_j\rangle_U, \langle W(t), \chi_k\rangle_U) = tq_j\delta_{jk}. \tag{1.12}$$

Hence, $W(t) \sim N(0, tQ)$.

Consider a bounded domain D and let $H = U = L^2(D)$ and Q satisfy Assumption 1.3.9.1. Then, for a kernel $q \in L^2(D \times D)$, in terms of the Q-Wiener process,

$$Cov(W(t, \boldsymbol{x}), W(t, \boldsymbol{y})) = tq(\boldsymbol{x}, \boldsymbol{y}), \quad \text{for } (\boldsymbol{x}, \boldsymbol{y}) \in D \times D,$$

and $W(1, \boldsymbol{x})$ is mean-zero Gaussian random field with covariance $q(\boldsymbol{x}, \boldsymbol{y})$.

1.3.10   *Cylindrical Wiener process*

In case of $Q = I$, which is not a trace class on an infinite dimensional space U (as $q_j = 1$ for all j) so that Eq. (1.11) does not converge in $L^2(\Omega, U)$. To extend the definition of a Q-Wiener process, cylindrical Wiener process is introduced.

Definition 1.3.10.1 (cylindrical Wiener process). Let U be a separable Hilbert space. The cylindrical Wiener process (also called space-time white noise) is the U-valued stochastic process $W(t)$ defined by

$$W(t) = \sum_{j=1}^{\infty} \chi_j \beta_j(t),$$

where $\{\chi_j\}$ is any orthonormal basis of U and $\beta_j(t)$ are iid $\mathcal{F}_t$-Brownian motions.

1.3.11   *Real-valued Toeplitz and circulant matrix*

Definition 1.3.11.1 (Toeplitz matrix). An $N \times N$ real-valued matrix $\boldsymbol{C} = (c_{ij})$ is Toeplitz if $c_{ij} = c_{i-j}$ for some real numbers $c_{1-N}, \ldots, c_{N-1}$. It can be written as

$$\begin{pmatrix} c_0 & c_{-1} & \cdots & c_{2-N} & c_{1-N} \\ c_1 & c_0 & c_{-1} & \ddots & c_{2-N} \\ \vdots & \ddots & \ddots & \ddots & \vdots \\ c_{N-2} & \ddots & c_1 & c_0 & c_{-1} \\ c_{N-1} & c_{N-2} & \cdots & c_1 & c_0 \end{pmatrix},$$

and the entries of C are constant on each diagonal. An $N \times N$ Toeplitz matrix is uniquely defined by the vector $\boldsymbol{c} = [c_{1-N}, \ldots, c_{-1}, c_0, c_1, \ldots, c_{N-1}]^T \in \mathbb{R}^{2N-1}$, containing the entries of the first row and column. A symmetric Toeplitz matrix has $c_{i-j} = c_{j-i}$ and is defined by its first column $\boldsymbol{c}_1 = [c_0, c_1, \ldots, c_{N-1}]^T \in \mathbb{R}^N$.

Definition 1.3.11.2 (circulant matrix). An $N \times N$ real-valued Toeplitz matrix $C = (c_{ij})$ is circulant if each column is a circular shift of the elements of the preceding column (so that the last entry becomes the first entry). That is,

$$\begin{pmatrix} c_0 & c_{N-1} & \cdots & c_2 & c_1 \\ c_1 & c_0 & c_{N-1} & \ddots & c_2 \\ \vdots & \ddots & \ddots & \ddots & \vdots \\ c_{N-2} & \ddots & c_1 & c_0 & c_{N-1} \\ c_{N-1} & c_{N-2} & \cdots & c_1 & c_0 \end{pmatrix},$$

which is uniquely determined by the first column $\boldsymbol{c}_1 = [c_0, c_1, \ldots, c_{N-1}]^T \in \mathbb{R}^N$. Let $c_{ij} = c_{i-j}$ for $i \leq j \leq i$ and $c_{ij} = c_{i-j+N}$ for $i+1 \leq j \leq N$. Symmetric circulant matrices also have

$$c_{N-j} = c_j, \quad \text{for } j = 1, 2, \ldots, N-1,$$

and hence at most $\lfloor N/2 \rfloor + 1$ distinct entries.

1.3.11.1 *Circular embedding*

Definition 1.3.11.3. An $N \times N$ symmetric Toeplitz matrix $\mathbb{C}$ can always be embedded inside a larger symmetric circulant matrix $\tilde{C}$. This is known as circulant embedding.

Definition 1.3.11.4 (minimal circulant extension). Given a symmetric Toeplitz matrix $C \in \mathbb{R}^{N \times N}$ with first column $\boldsymbol{c}_1 = [c_0, c_1, \ldots, c_{N-1}]^T \in \mathbb{R}^N$, the minimal circulant extension is the circulant matrix $\tilde{C} \in \mathbb{R}^{2N' \times N'}$ with the first column $\tilde{\boldsymbol{c}}_1 = [c_0, c_1, \ldots, c_{N'}, c_{N'-1}, \ldots, c_1]^T \in \mathbb{R}^{2N'}$, where $N' = N - 1$.

1.3.12   *Fractional Brownian motion process*

Fractional Brownian motion (FBM) [18] has a number of nice properties, one of which is "self-similarity". A process $\{X(t), t \in \mathbb{R}\}$ is self-similar with index $H > 0$ if for any $a > 0$, the process $\{X(at), t \in \mathbb{R}\}$ has the same finite-dimensional distributions as $\{a^H X(t), t \in \mathbb{R}\}$. Thus, like a fractal, there is scaling, but it is not the trajectories of the process that scale but the probability distribution, the "odds". This is why this type of scaling is sometimes called "statistical self-similarity" or, more precisely, "statistical self-affinity".

The FBM process is based on the following three properties:

(i)   the process is Gaussian with zero mean,
(ii) it has stationary increments,
(iii) it is self-similar with index H, $0 < H < 1$.

Fractional Brownian motion reduces to Brownian motion when $H = \frac{1}{2}$, but in contrast to Brownian motion, it has dependent increments when $H \neq \frac{1}{2}$.

The stochastic representation of FBM [4,19] is

$$B^H(t) := \frac{1}{\Gamma(H + \frac{1}{2})} \left(\int_{-\infty}^{0} \left[(t - s)^{H - \frac{1}{2}} - (-s)^{H - \frac{1}{2}} \right] dB(s) \right.$$
$$\left. + \int_{0}^{t} (t - s)^{H - \frac{1}{2}} dB(s) \right), \tag{1.13}$$

where Γ represents the Gamma function $\Gamma(\alpha) := \int_0^{\infty} x^{\alpha - 1} \exp(-x) dx$ and $0 < H < 1$ is called the Hurst parameter.

To obtain a fractional Brownian motion, the key idea is to fractionally integrate white noise process W. The fractional integration operator of order s is defined as [20]

$$D^{(-s)} f(x) = \frac{1}{\Gamma(s)} \int_0^x (x - \xi)^{s-1} f(\xi) d\xi. \tag{1.14}$$

As it is known that fractionally integrating white noise of order $H + 1/2$ is equivalent to the definition of fractional Brownian motion (1.15). Using the definition of fractional integration (1.16) with

$s = H + \frac{1}{2}$, this yields

$$B_H(t) = \frac{1}{\Gamma(H + \frac{1}{2})} \int_0^t (t - s)^{H - \frac{1}{2}} W(s) ds$$

$$= \frac{1}{\Gamma(H + \frac{1}{2})} \int_0^t (t - s)^{H - \frac{1}{2}} dB(s).$$

1.4 Summary

The purpose of this chapter is to introduce and review some useful definitions and properties of stochastic calculus. Mainly, the definitions and properties of stochastic test function and distribution space, Wiener process, Brownian motion and FBM have been discussed. The definitions of Brownian motion and fractional Brownian motion play a pivotal role in forming analytical and numerical solutions of various equations which will be discussed in depth in the following chapters.

Chapter 2

Analytical Solutions of Stochastic Differential Equations

2.1 Introduction

In many practical applications regarding the field of science and engineering, the physical systems are modelled by stochastic nonlinear PDEs (SNPDEs). Because, in many of the cases exact solutions are very difficult or even impossible to obtain for SNPDEs, the approximate analytical solutions are particularly important for the study of dynamic systems for analysing their physical nature. In the case of approximate analytical solutions, the success of a certain approximation method depends on the nonlinearities that occur in the studied problem, and thus a general algorithm for the construction of such approximate solutions does not exist in the general cases.

The Zakharov–Kuznetsov (ZK) equation is a well-studied canonical two-dimensional extension of the KdV equation [21]. It has been derived formally in a long-wave, weakly nonlinear regime from the Euler–Poisson system [22]. This equation has been used to model waves on shallow water surfaces [23]. In 1974, ZK [24] derived an equation which describes weakly nonlinear ion-acoustic waves in strongly magnetised lossless plasma composed of cold ions and hot isothermal electrons. The investigation of exact travelling wave solutions to these nonlinear partial differential equations (NPDEs) has also been observed as a field of great interest to many mathematicians and physicists because of its significant role in understanding the behaviour of nonlinear physical phenomena. As a result, numerous techniques for obtaining travelling wave solutions have been developed over the last three decades [25].

An excellent deal of attention has been intended by the researchers towards solving the Wick-type stochastic differential equations such as the stochastic time-fractional KdV equation [26–28], stochastic KdV equation [29,30] and stochastic ZK equations [22,31–33] due to their travelling wave nature and various applications in quantum theory.

In 2010, Kudryashov and Sinelshchikov obtained a nonlinear PDE for describing the pressure waves in a mixture of liquid and gas bubbles taking into consideration the viscosity of liquid and the heat transfer [34]. The Kudryashov–Sinelshchikov equation is the generalisation of the Korteweg–de Vries (KdV) and the Burgers–Korteweg–de Vries (BKdV) equation.

The dynamics of bubbly liquids is of exceptional sensible value in lots of fields of utility ranging from underwater acoustics to therapeutic and diagnostic scientific applications in blood waft. Gas bubbles immersed in a homogeneous fluid equip it with some microstructure during which the behaviour of the liquid is vastly modified. The propagation of waves through a bubbly liquid can also be described in the simplest case by the Korteweg–de Vries (KdV) equation. A more general evolution equation of the third order has been derived by Kudryashov and Sinelshchikov taking into account viscosity and heat transfer [35].

The Boussinesq-type equations are formally derived by integrating the three-dimensional Euler equations through the water depth using a polynomial approximation of the vertical profile of the velocity field, thereby reducing the three-dimensional problem to an equivalent two-dimensional problem that is relatively more efficient to solve numerically. In recent years, efforts have been made by a number of researchers to extend the range of applicability of the Boussinesq system to deeper water by improving the dispersion characteristics of the equation [36,37].

In recent years, modified Boussinesq equations (MBE) have attracted scientists in all aspects of wave dynamics due to their academic values as well as wide practical applications, especially for simulating wave propagation in coastal zones. Traditionally, classical Boussinesq equations (BE) adopted the depth-integrated velocities to be key variables associated with the surface elevation to form a set of two equations to describe the wave motion for a single-layer fluid [38]. This equation is approximately described as the propagation of waves in certain nonlinear dispersive systems. They are considered as intermediate long-wave equations since they represent an intermediate dynamic, in complexity and completeness, situated between the complete dynamic of the full initial equations describing any wave number and any amplitude, and some strong long waves and small amplitude limits [39].

There are many powerful mathematical techniques for finding exact solutions of nonlinear differential equations [40–44]. In the sub-equation method, the solutions are written in the form of a polynomial and the degree of this polynomial is obtained by considering the homogeneous balance principle. This method has been implemented not only in obtaining exact solutions for integer order differential equations but also in obtaining solutions of various differential equations of fractional order.

The Kersten–Krasil'shchik coupled KdV-mKdV equations arise as the classical part of one of the super extensions of the KdV equation and also can be considered as a sort of coupling between the KdV (with respect to u) and the mKdV equations (with respect to v) [45]. Its complete integrability was shown by Kersten and Krasil'shchik using the existence of infinite series of symmetries [46], and its singular analysis and Lax pair were given by Kalkanli, Sakovich and Yurdusen using the Painlevé test and prolongation technique [47]. The KdV equation is one of the most popular soliton equations and was originally derived by Korteweg and de Vries in the 19th century as water waves equations. It is a useful approximation in many studies when one wishes to include a simple nonlinearity and a simple dispersive effect. Some of these studies are ion-acoustic and magneto hydro dynamic waves in plasma, longitudinal dispersive waves in elastic rods, pressure waves in liquid–gas bubble mixtures, rotating flow down a tube and thermally excited photon packets in low-temperature nonlinear crystals [48,49].

A prominent expansion method that has emerged is the Jacobi elliptic function (JEF) expansion method [50]. The JEF expansion method is reliable and can be applied to many other nonlinear evolution equations. The main idea of this method is to take full advantage of the elliptic equation that JEFs satisfy and use its solutions in

Jacobi elliptic function method. It is interesting that some methods are special cases of the JEF method [51].

Wei *et al.* proposed the JEF expansion method for obtaining solutions of nonlinear stochastic equations [52]. Ghany obtained solutions of the Wick-type stochastic fractional KdV equations by the fractional Riccati equation and modified fractional sub-equation method [53,54]. Chen and Li applied the modified Riccati equation method and obtained solutions of stochastic mKdV equations [55]. Saha Ray and Singh applied the Kudryashov method for obtaining the white noise solution of Wick-type stochastic ZK equation [56]. Saha Ray and Singh also obtained white noise solutions for Wick-type Kudryashov–Sinelschikov equation and Wick-type Boussinesq equation by using improved sub-equation method [57,58].

The generalised nonlinear Schrödinger equation (NLSE) is a generic model that is very important in nonlinear optics, where it describes the full spatiotemporal optical solitons or light bullets [59]. Variable coefficient NLSE [60] is one of the most fundamental equations of quantum mechanics and has many physical applications, e.g., in gravitation, movement of trapped ions and coherent state studies. The NLSE is the continuum limit of a number of discrete systems including the discrete self-trapping equation in a special case, the Ablowitz–Ladik equation and under certain approximations the Davydov equations [61].

Zayed and Alurrfi [62,63] applied the extended auxiliary equation method (EAEM) for obtaining exact solutions of nonlinear Schrödinger-type equations. Chen and Xie [60] obtained periodic-like solutions of variable coefficient and Wick-type stochastic NLSE. Singh and Saha Ray [64] obtained exact solutions for the Wick-type stochastic Kersten–Krasil'shchik coupled KdV-mKdV equations by using the JEF expansion method. Özgil *et al.* [65] obtained exact solutions of perturbed NLSE with Kerr law nonlinearity.

In this chapter, the analytical solutions for dynamical systems which are modelled by stochastic partial differential equations (SPDEs) have been presented. Stochastic differential equations have been frequently encountered in science and engineering and have also been intensely studied and researched in recent years. In this chapter, various SPDEs arising in the field of physics and mathematics have been studied. Then, their analytical solutions have been obtained using various analytical techniques, such as the Kudryashov method, improved sub-equation method, Jacobian elliptic function expansion method and extended auxiliary equation method.

2.2 Outline of Present Study

In this chapter, Hermite transform has been used for converting the Wick-type SDEs to deterministic PDEs and has shown the use of inverse Hermite transformation in obtaining stochastic solutions in the white noise space for the same.

In Section 2.4, the exact solution of Wick-type stochastic ZK equation has been obtained by using the Kudryashov method. Hermite transform has been used for transforming the Wick-type stochastic ZK equation to deterministic PDE. Also, inverse Hermite transform has been applied for obtaining a set of stochastic solutions in the white noise space.

In Section 2.5, Wick-type stochastic Kudryashov–Sinelshchikov (KS) equation has been solved by using improved sub-equation method. Hermite transform has been used for transforming the Wick-type stochastic KS equation to deterministic PDE. Also, inverse Hermite transform has been applied for a set of stochastic solutions in the white noise space.

In Section 2.6, Wick-type stochastic fractional MBE has been solved by using improved sub-equation method. Hermite transform has been used for transforming the Wick-type stochastic modified Boussinesq equation to deterministic PDE. Also, inverse Hermite transform has been applied for a set of stochastic solutions in the white noise space.

In Section 2.7, Wick-type stochastic Kersten–Krasil'shchik coupled KdV-mKdV equations have been solved by using the JEF expansion method. Hermite transform has been used for transforming the Wick-type stochastic Kersten–Krasil'shchik coupled KdV-mKdV equation into deterministic PDE. Also, inverse Hermite transform has been applied for a set of stochastic solutions in the white noise space.

In Section 2.8, Wick-type stochastic NLSE has been solved by using the extended auxiliary equation method. Hermite transform has been used for transforming the Wick-type stochastic nonlinear Schrödinger equation into deterministic PDE. Also, the stochastic solutions have been obtained by applying the inverse Hermite transform.

2.2.1   *Wick-type stochastic ZK equation*

Consider the Wick-type stochastic ZK equation as follows:

$$U_t + a(t)U \Diamond U_x + b(t)U_{xxx} + c(t)U_{xyy}$$
$$= W(t) \Diamond R^{\Diamond}(t, U, U_x, U_{xxx}, U_{xyy}), \tag{2.1}$$

which is the perturbation of the ZK equation of the form

$$u_t + a(t)uu_x + b(t)u_{xxx} + c(t)u_{xyy} = 0, \tag{2.2}$$

by random force $W(t) \Diamond R^{\Diamond}(t, U, U_x, U_{xxx}, U_{xyy})$. Here, $a(t), b(t)$ and $c(t)$ are functions of t, $W(t) = \frac{dB(t)}{dt}$ defines the Gaussian white noise and $B(t)$ is a BM, and "$\Diamond$" denotes the Wick product as defined by Holden *et al.* [15].

Set $R(t, u, u_x, u_{xxx}, u_{xyy}) = \alpha uu_x - \beta u_{xxx} - \gamma u_{xyy}$. Then, $R(t, u, u_x, u_{xxx}, u_{xyy})$ is a function of u, u_x, u_{xxx} and u_{xyy} for some constants α, β, γ. $R^{\Diamond}$ is the Wick-type representation of the functional R.

The ZK is a natural multidimensional extension of the KdV equation, quite different from the well-known Kadomtsev–Petviashvili (KP) equation though [66]. The ZK equation is not completely integrable but it has a Hamiltonian structure and possesses two invariants. On the other hand, the KdV and the Kadomtsev–Petviashvili equations are integrable [67].

The well-studied model for the weakly nonlinear waves in shallow water is the Korteweg–de Vries (KdV) equation which has been described as

$$u_t + 6uu_x + u_{xxx} = 0, \quad x, t \in \mathbb{R}. \tag{2.3}$$

The KdV equation is a completely integrable Hamiltonian system and its soliton solution [68], found analytically for equation in the form (2.3), is given by

$$u(x,t) = \frac{c}{2}\sec h^2\left(\frac{1}{2}\sqrt{c}(x - x_0 - ct)\right). \tag{2.4}$$

This is a localised wave solution with negative amplitude and is also of the well-known bell-shaped form. The nature of the solution can be described as follows:

- These waves are called solitons. Solitons are localised waves that keep their shape as they travel (in contrast to dispersive wave packets).
- Since they originate from a nonlinear PDE, the principle of super-position does not apply.
- Solitons have a surprising property. Two solitons can pass through each other without any effect on their shape (but they experience a phase shift). This is not obvious given the nonlinear nature of the KdV equation.

This model also arises in several other physical contexts, e.g., plasma physics, stratified internal waves and ion-acoustic waves are a few to mention [69–71]. The KP equation

$$(u_t + u_{xxx} + uu_x)_x + \varepsilon u_{yy} = 0, \quad (x,y) \in \mathbb{R}^2,\ t \in \mathbb{R}, \tag{2.5}$$

devised by Kadomtsev and Petviashvili [72], describes the propagation of weakly nonlinear long waves on the surface of a fluid. On the other hand, the ZK equation

$$u_t + u_{xxx} + u_{xyy} + uu_x = 0, \quad (x,y) \in \mathbb{R}^2,\ t \in \mathbb{R}, \tag{2.6}$$

developed by ZK in [24], governs the behaviour of weakly nonlinear ion-acoustic waves in a plasma comprising cold ions and hot isothermal electrons in the presence of a uniform magnetic field [73,74].

2.2.2 *Wick-type stochastic Kudryashov–Sinelshchikov equation*

Consider the Wick-type stochastic KS as follows:

$$U_t + \gamma(t)U\Diamond U_x + U_{xxx} - \varepsilon(t)(U\Diamond U_{xx})$$
$$- \kappa(t)U_x\Diamond U_{xx} - v(t)U_{xx} - \delta(t)(U\Diamond U_x)_x$$
$$= W(t)\Diamond R^{\Diamond}(t, U, U_x, U_{xx}, U_{xxx}), \tag{2.7}$$

which is the perturbation of the KS equation of the form [75]

$$u_t + \gamma u u_x + u_{xxx} - \varepsilon(u u_{xx})_x - k u_x u_{xx} - v u_{xx} - \delta(u u_x) = 0, \quad (2.8)$$

by random force $W(t) \lozenge R^{\lozenge}(t, U, U_x, U_{xxx}, U_{xyy})$. Here, $\gamma(t), \varepsilon(t),$ $\kappa(t), \nu(t)$ and $\delta(t)$ are arbitrary functions of t, $W(t) = \frac{dB(t)}{dt}$ defines the Gaussian white noise and $B(t)$ is a BM, and "$\lozenge$" denotes the Wick product as defined by Holden *et al.* [15].
$R(t, u, u_x, u_{xx}, u_{xxx}) = -a u u_x - b u_{xxx} + c u_x u_{xx} + d u u_{xxx} + e u_{xx} +$ $f u_x^2 + g u u_{xx}$ is a function of u, u_x, u_{xx} and u_{xxx} for some constants a, b, c, d, e, f, g. $R^{\lozenge}$ is the Wick-type representation of the functional R.

2.2.3   *Wick-type stochastic modified Boussinesq equation*

Consider the Wick-type stochastic MBE as follows:

$$U_{tt} - U_{xx} + P(t) \lozenge U \lozenge U_x^2 + Q(t) \lozenge U^2 \lozenge U_{xx} + R(t) \lozenge U_{xxxx} = 0, \quad (2.9)$$

which is the perturbation of the MBE of the form [39]

$$u_{tt} - u_{xx} + P(t) u u_x^2 + Q(t) u^2 u_{xx} + R(t) u_{xxxx} = 0. \qquad (2.10)$$

Here, $P(t), Q(t)$ and $R(t)$ are Gaussian white noise functions of t and "$\lozenge$" denotes the Wick product as defined by Holden *et al.* [15].

2.2.4   *Wick-type stochastic Kersten–Krasil'shchik coupled KdV-mKdV equations*

Consider the Wick-type stochastic Kersten–Krasil'shchik coupled KdV-mKdV as follows:

$$U_t + H_1(t) \lozenge U_{xxx} - 6 H_2(t) \lozenge U \lozenge U_x + 3 H_3(t) \lozenge V \lozenge V_{xxx}$$

$$+ 3 H_4(t) \lozenge V_x \lozenge V_{xx} - 3 U_x \lozenge V^2 - 6 U \lozenge V \lozenge V_x = 0, \qquad (2.11)$$

$$V_t + V_{xxx} - 3 V^2 \lozenge V_x - 3 U \lozenge V_x - 3 U_x \lozenge V = 0,$$

which is the perturbation of the Kersten–Krasil'shchik coupled KdV-mKdV equations of the form [47]

$$u_t + u_{xxx} - 6 u u_x + 3 \nu \nu_{xxx} + 3 \nu_x \nu_{xx} - 3 u_x \nu^2 - 6 u \nu \nu_x = 0,$$
$$\nu_t + \nu_{xxx} - 3 \nu^2 \nu_x - 3 u \nu_x - 3 u_x \nu = 0. \qquad (2.12)$$

Here, $H_1(t)$, $H_2(t)$, $H_3(t)$ and $H_4(t)$ are Gaussian white noise functionals and "$\Diamond$" denotes the Wick product as defined by Holden *et al.* [15].

2.2.5 *Wick-type stochastic NLSE*

The Wick-type stochastic NLSE is of the form

$$iU_t + F(t)\Diamond U_{xx} + G(t)\Diamond U \Diamond |U|^{\Diamond 2} = 0, \tag{2.13}$$

which is the perturbation of the nonlinear Schrödinger equation as follows [61]:

$$iu_t + f(u)u_{xx} + g(t)u|u|^2 = 0. \tag{2.14}$$

Here, $F(t)$ and $G(t)$ are Gaussian white noise functionals and "$\Diamond$" denotes the Wick product as defined by Holden *et al.* [15].

2.3 Framework for SPDE Driven by White Noise

In this section, some important definitions and properties for SPDE driven by white noise have been discussed.

2.3.1 *Function and distribution space*

In this section, $(S(\mathbb{R}^d))$ and $(S(\mathbb{R}^d))^*$ are the Hida test function and distribution space on $\mathbb{R}^d$. Let the Hermite function

$$\xi_n(x) = e^{-(1/2)x^2} h_n(\sqrt{2}x)/(\pi(n-1)!)^{1/2}, \quad n \geq 1, \tag{2.15}$$

where $h_n(x)$ is the Hermite polynomial.

Then, the set $\{\xi_n(x)\}_{n\geq 1}$ represents an orthogonal basis for $L^2(\mathbb{R})$.

Let $\alpha = (\alpha_1, \ldots, \alpha_d)$ denote d-dimensional multi-indices with $\alpha_1, \ldots, \alpha_d \in \mathcal{N}$. It follows the family of tensor products $\xi_\alpha = \xi_{(\alpha_1,\ldots,\alpha_d)} = \xi_{\alpha_1} \otimes \cdots \otimes \xi_{\alpha_d} (\alpha \in \mathcal{N}^d)$ which form an orthogonal basis for $L^2(\mathbb{R}^d)$.

Let $\alpha^{(i)} = (\alpha_1^{(i)}, \ldots, \alpha_d^{(i)})$ be the ith multi-index number in some fixed ordering. Suppose, it possesses the property that $i < j$ implies $\alpha_1^{(i)} + \cdots + \alpha_d^{(i)} \leq (\alpha_1^{(j)} + \cdots + \alpha_d^{(j)})$.

Now, define

$$\eta_i = \xi_{\alpha_1^{(i)}} \otimes \cdots \otimes \xi_{\alpha_d^{(i)}}, \quad i \geq 1. \tag{2.16}$$

Consider the multi-indices (of arbitrary length) as elements of the space $(\mathcal{N}_0^{\mathcal{N}})_c$ for all sequences $\alpha = (\alpha_1, \alpha_2, \dots) \in \mathcal{J}$, where $\mathcal{J} = (\mathcal{N}_0^{\mathcal{N}})_c$ with elements $\alpha_i \in \mathcal{N}_0$. Then, for $\alpha \in \mathcal{J}$,

$$H_\alpha(\omega) = \prod_{i=1}^{\infty} h_{\alpha_i}(\langle \omega, \eta_i \rangle), \quad \omega \in (S(\mathbb{R}^d))^*. \tag{2.17}$$

2.3.2 *Kondratiev stochastic test function space and stochastic distribution space*

Consider the construction of the Kondratiev spaces of test functions with orthogonal bases given by a generating function $\gamma(\lambda)h(x; \alpha(\lambda))$, where h satisfies assumptions accepted in Ref. [9].

2.3.2.1 *Kondratiev spaces of test functions*

Consider a chain (a rigging of $\mathcal{H}_0$)

$$\mathcal{N}' \supset \cdots \supset \mathcal{H}_{-p} \supset \cdots \supset \mathcal{H}_0 \supset \cdots \supset \mathcal{H}_p \supset \cdots \supset \mathcal{N}, \tag{2.18}$$

where $\mathcal{N} = pr\lim_{p \in \mathbb{Z}_+} \mathrm{H}_p$ is the projective limit of the sequences of spaces $\{\mathcal{H}_p\}_{p \in \mathbb{Z}_+}$ (it means that $\mathcal{N} = \cap_{p \in \mathbb{Z}_+} \mathcal{H}_p$ with a topology of the projective limit — the weaker topology such that for each $p \in \mathbb{Z}_+$ the embedding $\mathcal{N}$ into $\mathcal{H}_p$ is continuous, see, e.g., Refs. [11,12] for details). Using properties of holomorphic functions [13] and the kernel theorem [11,12], it is shown in Ref. [10] that for each $x \in Q$ there exists an expansion

$$h(x; \lambda) = \sum_{n=0}^{\infty} \frac{1}{n!} \langle h_n(x), \lambda^{\otimes n} \rangle, \tag{2.19}$$

where $h_n(x) \in \mathcal{H}_{-2,\mathbb{C}}^{\hat{\otimes}n}$, $\lambda \in B_x := \{\lambda \in \mathcal{H}_{-2,\mathbb{C}} \cap U_0 : |\lambda|_2 < R_x, R_x > 0\}$, $\lambda^{\otimes 0} := 1$.

Definition 2.3.2.1. A Hilbert space of formal series

$$
\begin{aligned}
(\mathcal{H}_p)_q &:= \Bigg\{ f(x) = \sum_{n=0}^{\infty} \left\langle h_n(x), f^{(n)} \right\rangle, f^{(n)} \in \mathcal{H}_{p,\mathbb{C}}^{\hat{\otimes}n}, \ x \in Q : \|f\|_{(\mathcal{H}_p)_q}^2 \\
&:= \sum_{n=0}^{\infty} (n!)^2 K^{qn} |f^{(n)}|_p^2 < \infty \Bigg\},
\end{aligned}
\tag{2.20}
$$

with a corresponding scalar product $\|\cdot\|_{(\mathcal{H}_p)_q}$ is called the Kondratiev space of test functions.

Definition 2.3.2.2.

(a) Stochastic test function space
Let N be a natural number. For $0 \le \rho \le 1$, let

$$
(S)_\rho^N = (S)_\rho^{m;N}
$$

consists of those

$$
f = \sum_\alpha c_\alpha H_\alpha \in \mathbb{L}^2(\mu_m) = \bigoplus_{k=1}^{N} L^2(\mu_m) \quad \text{with } c_\alpha \in \mathbb{R}^N,
$$

such that

$$
\|f\|_{\rho,k}^2 = \sum_{\alpha \in J} c_\alpha^2 (\alpha!)^{1+\rho} (2\mathbb{N})^{k\alpha} < \infty \quad \text{for all } k \in \mathbb{N}_0,
$$

where $(2\mathbb{N})^{k\alpha} = (2{\cdot}1)^{k\alpha_1}(2{\cdot}2)^{k\alpha_2} \cdots (2{\cdot}m)^{k\alpha_m}$, if Index $(\alpha) = m$.
Set

$$
(S)_{\rho,k} := \{f : \|f\|_{\rho,k} < \infty\},
$$

and

$$
(S)_\rho := \bigcap_{k \in \mathbb{N}_0} (S)_{\rho,k},
$$

endowed with projective topology [10,14].

(b) Stochastic distribution space

Let N be a natural number. For $0 \leq \rho \leq 1$,

$$(S)^N_{-\rho} = (S)^{m;N}_{-\rho}$$

consists of those

$$F = \sum_{\alpha} b_{\alpha} H_{\alpha} \quad \text{with } b_{\alpha} \in \mathbb{R}^N,$$

therefore

$$\|F\|^2_{-\rho,-k} = \sum_{\alpha \in J} b^2_{\alpha}(\alpha!)^{1-\rho}(2\mathbb{N})^{-k\alpha} < \infty \quad \text{for all } k \in \mathbb{N}_0.$$

Set

$$(S)_{-\rho,-k} := \{F : \|F\|_{-\rho,-k} < \infty\},$$

and define

$$(S)_{-\rho} := \bigcup_{k \in \mathbb{N}_0} (S)_{-\rho,-k},$$

endowed with inductive topology [10,14].

The family of semi-norms $\|f\|_{\rho,k}$; $k \in \mathbb{N}$ advance to a topology on $(S)^N_{\rho}$, and it can represent $(S)^N_{-\rho}$ as the dual of $(S)^N_{\rho}$ by the following action:

$$\langle F, f \rangle = \sum_{\alpha} \langle b_{\alpha}, c_{\alpha} \rangle \alpha!,$$

if $F = \sum_{\alpha} b_{\alpha} H_{\alpha} \in (S)^N_{-\rho}$; $f = \sum_{\alpha} c_{\alpha} H_{\alpha} \in (S)^N_{\rho}$ and $\langle b_{\alpha}, c_{\alpha} \rangle$ is the usual inner product in $\mathbb{R}^N$.

2.3.3   *Hida stochastic test function space and distribution space*

Definition 2.3.3.1.

(a) The Hida test function space $(S)^N$ consists of

$$f = \sum_{\alpha} c_{\alpha} H_{\alpha} \in L^2(\mu_m) \quad \text{with } c_{\alpha} \in \mathbb{R}^N,$$

such that

$$\sup_{\alpha}\{c^2_{\alpha}\alpha!(2\mathbb{N})^{k\alpha}\} < \infty \quad \text{for all } k < \infty.$$

(b) The Hida distribution space $(S)^{*,N}$ consists of

$$F = \sum_\alpha b_\alpha H_\alpha \quad \text{with } b_\alpha \in \mathbb{R}^N,$$

such that

$$\sup_\alpha \{b_\alpha^2 \alpha! (2\mathbb{N})^{-q\alpha}\} < \infty \quad \text{for some } q < \infty.$$

2.3.4 *Wick product*

The Wick product $F \lozenge G$ is

$$F = \sum_\alpha a_\alpha H_\alpha,$$

and $G = \sum_\alpha b_\alpha H_\alpha \in (S)_{-1}^{m;N}$ with $a_\alpha, b_\alpha \in \mathbb{R}^N$ is defined as

$$F \lozenge G = \sum_{\alpha,\beta} (a_\alpha, b_\beta) H_{\alpha+\beta}.$$

2.3.5 *The Hermite transform*

The Hermite transform or $\mathcal{H}$-transform transforms Wick products into ordinary products and convergence in $(S)_{-1}$ into bounded, pointwise convergence in a certain neighbourhood of zero in $\mathbb{C}^N$.

Definition 2.3.5.1. Let $F = \sum_\alpha b_\alpha H_\alpha \in (S)_{-1}^N$ with $b_\alpha \in \mathbb{R}^N$. Then, the Hermite transform of F is defined as

$$\mathcal{H}F(z) = \tilde{F}(z) = \sum_\alpha b_\alpha z^\alpha \in \mathbb{C}^N, \quad \text{if convergent,}$$

where $z = (z_1, z_2, \ldots) \in \mathbb{C}^N$ and $z^\alpha = z_1^{\alpha_1}, z_2^{\alpha_2}, \ldots, z_n^{\alpha_n}, \ldots$, if $\alpha = (\alpha_1, \alpha_2, \ldots) \in \mathcal{J}$, where $z_j^0 = 1$.

From Definition 2.3.5.1, for $X, Y \in (S)^{N}_{-1}$,

$$\mathcal{H}(X \Diamond Y(z)) = \tilde{X}(z)\tilde{Y}(z), \tag{2.21}$$

and

$$(z^1_1, \ldots, z^1_n)(z^2_1, \ldots, z^2_n) = \sum_{k=1}^{n} z^1_k z^2_k, \quad \text{where } z^i_k \in \mathbb{C}.$$

Let $X = \sum_{\alpha} a_{\alpha} H_{\alpha} \in (S)^{N}_{-1}$ and $c_0 = X(0) \in \mathbb{R}^N$ be called the expectation of X, viz., $E(X)$. Assume $g : V \to \mathbb{C}^m$ is an analytic function, where V is a neighbour of $E(X)$. Assume that the Taylor series of g about $E(X)$ contains coefficients in $\mathbb{R}^m$. Then, the wick version is defined as

$$g^{\Diamond}(X) = \mathcal{H}^{-1}(g \circ \tilde{X}) \in (S)^{m}_{-1}.$$

If g has the Taylor series expansion

$$g(z) = \sum a_{\alpha}(z - E(X))^{\alpha}, \quad \text{where } a_{\alpha} \in \mathbb{R}^m,$$

then

$$g^{\Diamond}(z) = \sum a_{\alpha}(z - E(X))^{\Diamond \alpha} \in (S)^{m}_{-1}.$$

Consider an SPDE as

$$A(t, x, \partial_t, \nabla_x, U, \omega) = 0, \tag{2.22}$$

where A is some given function, $U = U(t, x, \omega)$ is an unknown (generalised) stochastic process and the operators are given as

$$\partial_t \equiv \frac{\partial}{\partial t}, \nabla_x \equiv \left(\frac{\partial}{\partial x_1}, \ldots, \frac{\partial}{\partial x_d} \right) \quad \text{when } x = (x_1, \ldots, x_d) \in \mathbb{R}^d.$$

The Wick-type SPDE is

$$A^{\Diamond}(t, x, \partial_t, \nabla_x, U, \omega) = 0. \tag{2.23}$$

Equation (2.23) undergoes Hermite transform so that the Wick products are converted into ordinary products and the equation can be written as

$$\tilde{A}(t, x, \partial_t, \nabla_x, \Phi, z_1, z_2, \ldots) = 0, \tag{2.24}$$

where $\Phi = \mathcal{H}(U)$ is the Hermite transform of U and $z_1, z_2, \ldots$ are complex numbers.

Suppose that a solution can be obtained for $u = u(t, x, z)$ of the equation $\tilde{A}(t, x, \partial_t, \nabla_x, \Phi, z_1, z_2, \ldots) = 0$ for each $z = (z_1, z_2, \ldots) \in \mathbb{K}_q(r)$ for some q, r, where $\mathbb{K}_q(r) = \{z = (z_1, z_2, \ldots) \in \mathbb{C}^N$ and $\sum_{\alpha \neq 0} |z^\alpha|^2 (2\mathbb{N})^{q\alpha} < r^2\}$.

2.3.6 *Inverse Hermite transform*

Theorem 2.3.6.1 ([15]). *Suppose $u(t, x, z)$ be a solution (in the usual strong, pointwise sense) of Eq. (2.24) for (t, x) in some bounded open set $G \subset \mathbb{R} \times \mathbb{R}^d$ and for all $z \in \mathbb{K}_q(r)$, for some q, r. Moreover, suppose that $u(t, x, z)$ and all its partial derivatives, which are involved in Eq. (2.24), are bounded for $(t, x, z) \in G \times \mathbb{K}_q(r)$, continuous with respect to $(t, x) \in G$ for all $z \in \mathbb{K}_q(r)$ and analytic with respect to $z \in \mathbb{K}_q(r)$, for all $(t, x) \in G$. Then, there exists $U(t, x) \in (S)_{-1}$ such that $u(t, x, z) = (\Phi(t, x)(z))$ for all $(t, x, z) \in G \times \mathbb{K}_q(r)$ and $U(t, x)$ solves (in the strong sense in $(S)_{-1}$) Eq. (2.23) in $(S)_{-1}$.*

2.4 Kudryashov Method for Solutions of Wick-Type Stochastic ZK Equation

2.4.1 *Kudryashov method*

The Kudryashov method allows to formulate solitary wave solutions for a wider class of nonlinear ODE. Exact solutions of high-order nonlinear advanced equations are determined by the proposed method as compared to other methods [76,77]. The Kudryashov method [78–81] has been presented step by step as follows:

Step 1. Consider a nonlinear PDE of the following form:

$$P(V, V_t, V_x, V_y, \ldots) = 0, \tag{2.25}$$

where $V(t, x, y)$ is the unknown function. P is the function in $V(t, x, y)$ along with their highest order partial derivatives and nonlinear terms of $V(t, x, y)$, respectively.

Step 2. The travelling wave transformation of Eq. (2.25) is given by

$$V(t, x, y) = \Phi(\xi), \quad \xi = k_1 t + k_2 x + k_3 y, \tag{2.26}$$

where k_1, k_2 and k_3 are constants, which are to be obtained later.

Using Eq. (2.26), the PDE (2.25) can be transformed into the following nonlinear ODE, which is given as

$$P(\Phi, k_1 \Phi'(\xi), k_2 \Phi'(\xi), k_3 \Phi'(\xi), \ldots) = 0. \tag{2.27}$$

Step 3. The exact solutions of Eq. (2.25) are assumed in the polynomial $\Phi(\xi)$ as follows:

$$\Phi(\xi) = a_0 + \sum_{i=1}^{N} a_i \varphi^i(\xi), \tag{2.28}$$

where $\varphi(\xi) = \frac{1}{1 \pm e^\xi}$ and $\varphi(\xi)$ satisfies the Riccati equation:

$$\varphi_\xi(\xi) = \varphi^2(\xi) - \varphi(\xi). \tag{2.29}$$

Step 4. According to the Kudryashov method, $\Phi = \xi^{-p}$ is substituted in Eq. (2.27) to obtain the highest order singularity. Then, the degrees of all terms of Eq. (2.27) are taken under consideration and the terms having the lower degree are selected. The maximum value of p is assumed to be the pole and is denoted as N. Kudryashov method only can be implemented for integer values of N. However, if N is not an integer, then Eq. (2.27) can be transformed and the above steps can be repeated.

Step 5. The derivatives of $\Phi(\xi)$ are presented in the following:

$$\Phi_\xi(\xi) = \sum_{i=1}^{N} a_i i (\varphi(\xi) - 1) \varphi^i(\xi),$$

$$\Phi_{\xi\xi}(\xi) = \sum_{i=1}^{N} a_i i [\varphi(\xi) + i(\varphi(\xi) - 1)](\varphi(\xi) - 1)\varphi^i(\xi), \tag{2.30}$$

$$\Phi_{\xi\xi\xi}(\xi) = \sum_{i=1}^{N} a_i i(i+1)(\varphi(\xi)-1)^2 \varphi^{i+1}(\xi)$$

$$+ \sum_{i=1}^{N} a_i i[\varphi(\xi) + i(\varphi(\xi)-1)]$$

$$(\varphi(\xi)-1)\varphi^{i+1}(\xi).$$

The derivatives of $\Phi(\xi)$ above are simplified using Eq. (2.29).

Step 6. Substituting Eq. (2.30) into Eq. (2.27) and equating the coefficients of φ^i $(i=0,1,2,\ldots)$ with zero, a set of algebraic equations is obtained. The obtained algebraic equations are solved, to obtain the unknowns a_i $(i=0,1,2,\ldots,N)$ and other constants. Then, all the obtained values of unknowns are substituted in Eq. (2.28) yielding the exact solution for Eq. (2.25) instantly.

2.4.2 *Solitary solutions of Wick-type stochastic ZK equation*

In this section, the Kudryashov method is employed here for finding the white noise solutions for the stochastic ZK equation (2.1).

Applying Hermite transform on Eq. (2.1),

$$V_t + (a(t) + \alpha \tilde{W}(t,z)VV_x + (b(t) + \beta \tilde{W}(t,z))V_{xxx}$$

$$+ (c(t) + \gamma \tilde{W}(t,z))V_{xyy} = 0, \tag{2.31}$$

where $V \equiv V(t,x,y,z) = \mathcal{H}(U(t,x,y))$, and the Hermite transform of $W(t)$ is given as

$$\tilde{W}(t,z) = \sum_{k=1}^{\infty} \eta_k(t) z_k,$$

where $z = (z_1, z_2, \ldots) \in \mathbb{C}^N$ is a parameter.

Let $A(t,z) = (a(t) + \alpha \tilde{W}(t,z)) = \mathcal{H}(P(t))$, $B(t,z) = (b(t) + \beta \tilde{W}(t,z)) = \mathcal{H}(Q(t))$ and $C(t,z) = (c(t) + \gamma \tilde{W}(t,z)) = \mathcal{H}(M(t))$. Then, Eq. (2.31) yields to

$$V_t + A(t,z)VV_x + B(t,z)V_{xxx} + C(t,z)V_{xyy} = 0. \tag{2.32}$$

Now, let $V(t, x, y, z) = \Phi(\xi, z)$, $\xi = k_1 x + k_2 y + k_3 t$, then Eq. (2.32) becomes

$$\Phi_\xi k_3 + A(t, z) k_1 \Phi \Phi_\xi + B(t, z) k_1^3 \Phi_{\xi\xi\xi} + C(t, z) k_1 k_2^2 \Phi_{\xi\xi\xi} = 0. \quad (2.33)$$

Consider the ansatz as follows:

$$\Phi(\xi, z) = a_0(t, z) + \sum_{i=1}^{N} a_i(t, z) \varphi^i(\xi), \quad (2.34)$$

where $\varphi(\xi)$ satisfies Eq. (2.29).

Now according to homogenous balancing principle, in Eq. (2.33), $A(t, z) k_1 \Phi \Phi_\xi$ and $(B(t, z) k_1^3 + C(t, z) k_1 k_2^2) \Phi_{\xi\xi\xi}$ are the dominant terms having the highest order singularity. The maximum value of the pole is 2 which implies here $N = 2$.

Consequently from Eq. (2.34), the ansatz becomes

$$\Phi(\xi) = a_0(t, z) + a_1(t, z) \varphi(\xi) + a_2(t, z) \varphi^2(\xi). \quad (2.35)$$

Substituting Eq. (2.35) along with Eq. (2.30) in Eq. (2.33) and equating each coefficient of φ^i $(i = 0, 1, 2, \ldots)$ with zero, a system of algebraic equations are obtained for k_1, k_2, k_3, a_0, a_1 and a_2 as follows:

$$\varphi(\xi) : -Aa_0 a_1 k_1 - a_1 B k_1^2 - a_1 C k_1 k_2^3 - a_1 k_3 = 0,$$

$$\varphi^2(\xi) : Aa_0 a_1 k_1 - Aa_1^2 k_1 - 2Aa_0 a_2 k_1 + 6a_1 B k_1^2 - 8a_2 B k_1^2$$
$$+ 6a_1 C k_1 k_2^3 - 8a_2 C k_1 k_2^3 + a_1(B k_1^2 + C k_1, k_2^3)$$
$$+ a_1 k_3 - 2a_3 k_3 = 0,$$

$$\varphi(\xi) : Aa_1^2 k_1 + 2Aa_0 a_2 k_1 - 3Aa_1 a_2 k_1 - 6a_1 B k_1^2 + 30a_2 B k_1^2$$
$$- 6a_1 C k_1 k_2^3 + 30a_2 C k_1 k_2^3 - 6a_1(B k_1^2 + C k_1 k_2^3)$$
$$+ 8a_2(B k_1^2 + C k_1 k_2^3) + 2a_2 k_3 = 0,$$

$$\varphi^4(\xi) : 3Aa_1 a_2 k_1 - 2Aa_2^2 k_1 - 24a_2 B k_1^2 - 24a_2 C k_1 k_2^3$$
$$+ 6a_1(B k_1^2 + C k k_2^3) - 30a_2(B k_1^2 + C k_1 k_2^3) = 0,$$

$$\varphi^5(\xi) : 2Aa_2^2 k_1 + 24a_2(B k_1^2 + C k_1 k_2^3) = 0. \quad (2.36)$$

Solving the above algebraic equation (2.36), a set of coefficients for the solution of Eq. (2.33) has been obtained,

$$a_0(t,z) = \frac{-B(t,z)k_1 - C(t,z)k_1 k_2^3 - k_3}{A(t,z)k_1},$$

$$a_1(t,z) = \frac{12(B(t,z)k_1 + C(t,z)k_2^3)}{A(t,z)},$$

$$a_2(t,z) = \frac{12(B(t,z)k_1 + C(t,z)k_2^3)}{A(t,z)}.$$

Thus, the solution has been obtained by substituting the values of a_0, a_1 and a_2 in Eq. (2.35) as

$$\Phi(\xi) = \frac{-B(t,z)k_1 - C(t,z)k_1 k_2^3 - k_3}{A(t,z)k_1} + \frac{12(B(t,z)k_1 + C(t,z)k_2^3)}{A(t,z)}\varphi\xi$$

$$- \frac{12(B(t,z)k_1 + C(t,z)k_2^3)}{A(t,z)}\varphi^2(\xi). \tag{2.37}$$

By Eq. (2.37) and the definition of $\tilde{W}$, from Theorem 2.3.6.1 and according to Xie [30], there exists a bounded open set $G \subset R_+ \times R, q > 0$ and $r > 0$ such that $\Phi(t,x,y,z)$, $\Phi_t(t,x,y,z)$ and $\Phi_{xxx}(t,x,y,z)$ are uniformly bounded for $(t,x,y,z) \in G \times \mathbb{K}_q(r)$, continuous with respect to $(t,x,y) \in G$ for all $z \in \mathbb{K}_q(r)$ and analytic with respect to $z \in \mathbb{K}_q(r)$ for all $(t,x,y) \in G$. Theorem 2.3.6.1 also suggests that there exists $U(t,x,y) \in (S)_{-1}$ such that $\Phi(t,x,y,z) = (\mathcal{H}U(t,x,y))(z)$ for all $(t,x,y,z) \in G \times \mathbb{K}_q(r)$ and that $U(t,x,y)$ solves Eq. (2.1). Therefore, $U = \mathcal{H}^{-1}\Phi \in (S)_{-1}$ and thereby obtain a solution U of the original Wick-type SPDE (2.1).

Hence, Eq. (2.37) implies that a stochastic solitary solution of Eq. (2.1) is

$$U(t,x,y) = \frac{-Q(t)k_1 - M(t)k_1 k_2^3 - k_3}{P(t)k_1}$$

$$+ \frac{12(Q(t)k_1 + M(t)k_2^3)}{P(t)}\Diamond\varphi^\Diamond(\Xi(t,x,y))$$

$$- \frac{12(Q(t)k_1 + M(t)k_2^3)}{P(t)}\Diamond\varphi^\Diamond(\Xi(t,x,y)), \tag{2.38}$$

where $\Xi(t,x,y) = \Psi(\xi(t,x,y))$.

2.5   Improved Sub-Equation Method for Solutions of the Wick-Type Stochastic KS Equation

2.5.1   *Improved sub-equation method*

In this section, the algorithm of the improved sub-equation method has been described [30,82–89]. The improved sub-equation method has been presented step by step as follows:

Step 1. Consider a nonlinear PDE of the following form:

$$P(V, V_t, V_x, \ldots) = 0, \tag{2.39}$$

where $V(t, x)$ is the unknown function. P is the function in $V(t, x)$ along with their highest order partial derivatives and nonlinear terms of $V(x)$, respectively.

Step 2. The travelling wave transformation of Eq. (2.39) is of the form

$$V(t, x) = \Phi(\xi), \quad \xi = k_1 x + k_2, \tag{2.40}$$

where k_1 and k_2 are constants, which are to be determined later.

Using Eq. (2.40), the PDE (2.39) can be transformed into the following nonlinear ODE, which is given as

$$P(\Phi, k_1\Phi'(\xi), k_2\Phi'(\xi), k_3\Phi'(\xi), \ldots) = 0. \tag{2.41}$$

Step 3. The exact solutions of Eq. (2.39) are assumed in the polynomial $\Phi(\xi)$ as follows:

$$\Phi(\xi) = \sum_{i=-N}^{-1} b_i\varphi^i(\xi) + a_0 + \sum_{i=1}^{N} a_i\varphi^i(\xi), \tag{2.42}$$

with $\varphi(\xi)$ satisfying the Riccati equation:

$$\varphi_\xi(\xi) = \varphi^2(\xi) + r, \tag{2.43}$$

where r is a constant.

According to Zhang *et al.* [89], the solutions of the Riccati equation (2.43) are as follows:

$$\varphi(\xi) = \begin{cases} -\sqrt{-r}\tanh(\sqrt{-r}\xi), & r < 0, \\ -\sqrt{-r}\coth(\sqrt{-r}\xi), & r < 0, \\ \sqrt{r}\tan(\sqrt{r}\xi), & r > 0, \\ -\sqrt{r}\cot(\sqrt{r}\xi), & r > 0, \\ -\dfrac{1}{\xi+\omega}, \omega \text{ is a constant}, & r = 0. \end{cases}$$

Step 4. Now, $\Phi = \xi^{-p}$ is substituted in Eq. (2.41) to obtain the highest order singularity. Then, the degrees of all terms of Eq. (2.41) are taken under consideration, and the terms having lower degree are selected. The maximum value of p is assumed to be the pole and is denoted as N. The improved sub-equation method can only be implemented for integer values of N. However, if N is not an integer, then Eq. (2.41) can be transformed and the above steps can be repeated.

Step 5. The derivatives of $\Phi(\xi)$ are presented in the following:

$$\Phi_\xi(\xi) = \sum_{i=-N}^{-1} b_i i(\varphi^2(\xi)+r)\varphi^{i-1}(\xi) + \sum_{i=1}^{N} a_i i(\varphi^2(\xi)+r)\varphi^{i-1}(\xi),$$

$$\Phi_{\xi\xi}(\xi) = \sum_{i=-N}^{-1} b_i i(\varphi^2(\xi)+r)\varphi^{i-2}(\xi)(\varphi^2(\xi)(1+i)+r(-1+i))$$

$$+ \sum_{i=1}^{N} a_i i(\varphi^2(\xi)+r)\varphi^{i-2}(\xi)(\varphi^2(\xi)(1+i)+r(-1+i)),$$

$$\Phi_{\xi\xi\xi}(\xi) = \sum_{i=-N}^{-1} b_i i(\varphi^2(\xi)+r)\varphi^{i-3}(\xi)(\varphi^2(\xi)(1+i)(2+i)$$

$$+ 2i^2 r\varphi^2(\xi)+r(-1+i)(-2+i))$$

$$+ \sum_{i=1}^{N} a_i i(\varphi^2(\xi) + r)\varphi^{i-3}(\xi)(\varphi^2(\xi)(1+i)(2+i)$$

$$+ 2i^2 r\varphi^2(\xi) + r(-1+i)(-2+i)), \tag{2.44}$$

and so on. The derivatives of $\Phi(\xi)$ above are simplified using Eq. (2.43).

Step 6. Substituting Eq. (2.44) into Eq. (2.41) and equating the coefficients of $\varphi^i (i = 0, 1, 2, \ldots)$ with zero, a set of algebraic equations is obtained. The obtained algebraic equations are solved, to obtain the unknowns $a_i(i = 0, 1, 2, \ldots, N)$, $b_i(i = -1, -2, \ldots, -N)$ and other constants. Then, all the obtained values of the unknowns are substituted in Eq. (2.42) yielding the exact solution for Eq. (2.39) instantly.

2.5.2 *Solutions of the Wick-type stochastic KS equation*

In this section, the improved sub-equation method is employed here for finding the white noise solutions for the stochastic Kudryashov–Sinelshchikov equation (2.7).

Applying Hermite transform on Eq. (2.7), it can be written as

$$V_t + (\gamma(t) + a\tilde{W}(t, z))VV_x + (1 + b\tilde{W}(t, z))V_{xxx}$$

$$- (\varepsilon(t) + \kappa(t) + c\tilde{W}(t, z))V_x V_{xx} - (\varepsilon(t) + d\tilde{W}(t, z))VV_{xxx}$$

$$- (\nu(t) + e\tilde{W}(t, z))V_{xx} - (\delta(t) + f\tilde{W}(t, z))V_x^2$$

$$- (\delta(t) + g\tilde{W}(t, z))VV_{xx} = 0, \tag{2.45}$$

where $V \equiv V(t, x, z) = \mathcal{H}(U(t, x))$ and the Hermite transform of $W(t)$ is defined by

$$\tilde{W}(t, z) = \sum_{k=1}^{\infty} \eta_k(t) z_k,$$

where $z = (z_1, z_2, \ldots) \in \mathcal{C}^N$ is a parameter.

Let

$$A(t, z) = (\gamma(t) + a\tilde{W}(t, z)) = \mathcal{H}(A(t)),$$
$$B(t, z) = 1 + b\tilde{W}(t, z) = \mathcal{H}(B(t)),$$
$$C(t, z) = \varepsilon(t) + \kappa(t) + c\tilde{W}(t, z) = \mathcal{H}(C(t)),$$
$$D(t, z) = \varepsilon(t) + d\tilde{W}(t, z) = \mathcal{H}(D(t)),$$
$$E(t, z) = \nu(t) + e\tilde{W}(t, z) = \mathcal{H}(E(t)),$$
$$F(t, z) = \delta(t) + f\tilde{W}(t, z) = \mathcal{H}(F(t)),$$
$$\text{and } G(t, z) = \delta(t) + g\tilde{W}(t, z) = \mathcal{H}(G(t)),$$

then Eq. (2.45) becomes

$$V_t + A(t, z)VV_x + B(t, z)V_{xxx} - C(t, z)V_x V_{xx} - D(t, z)VV_{xxx}$$
$$- E(t, z)V_{xx} - F(t, z)V_x^2 - G(t, z)VV_{xx} = 0. \tag{2.46}$$

Now, let $V(t, x, z) = \Phi(\xi, z)$, $\xi = k_1 x + k_2 t$, then Eq. (2.46) becomes

$$k_2\Phi_t + k_1 A(t, z)\Phi\Phi_x + k_1^3 B(t, z)\Phi_{xxx} - k_1^3 C(t, z)\Phi_x\Phi_{xx}$$
$$- k_1^3 D(t, z)\Phi\Phi_{xxx} - k_1^2 E(t, z)\Phi_{xx} - k_1^2 F(t, z)\Phi_x^2$$
$$- k_1^2 G(t, z)\Phi\Phi_{xx} = 0. \tag{2.47}$$

Consider the ansatz as

$$\Phi(\xi, z) = \sum_{i=-N}^{-1} b_i(t, z)\varphi^i(\xi) + a_0(t, z) + \sum_{i=1}^{N} a_i(t, z)\varphi^i(\xi), \tag{2.48}$$

where $\varphi(\xi)$ satisfies Eq. (2.43).

Now, according to the homogenous balancing principle, in Eq. (2.47), $k_1^3 B(t, z)\Phi_{xxx}$ and $k_1^2 G(t, z)\Phi\Phi_{xx}$ are the dominant terms having the highest order singularity. The maximum value of the pole is 1 which implies here $N = 1$.

Consequently, from Eq. (2.48), the ansatz becomes

$$\Phi(\xi, z) = b_1(t, z)\varphi^{-1}(\xi) + a_0(t, z) + a_1(t, z)\varphi(\xi). \qquad (2.49)$$

Substituting Eq. (2.49) along with Eq. (2.43) in Eq. (2.47) and equating each coefficient of φ^i $(i = 0, 1, 2, \ldots)$ with zero, a system of algebraic equations is obtained for $k_1, k_2, D(t, z), b_1(t, z), a_0(t, z)$ and $a_1(t, z)$ as follows:

$$\varphi^{-5}(\xi) : 6k_1^3 r^3 D b_1^2 + 2k_1^3 r^3 C b_1^2 = 0,$$

$$\varphi^{-4}(\xi) : -6k_1^3 r^3 B b_1 + 6k_1^3 r^3 D a_0 b_1 - k_1^2 r^2 F b_1^2 - 2k_1^2 r^2 G b_1^2 = 0,$$

$$\varphi^{-3}(\xi) : -2k_1^2 r^2 E b_1 - 2k_1^2 r^2 G a_0 b_1 + 6k_1^3 r^3 D a_1 b_1 - 2k_1^3 r^3 C a_1 b_1$$
$$- k_1 r A b_1^2 + 8k_1^3 r^2 D b_1^2 + 4k_1^3 r^2 C b_1^2 = 0,$$

$$\varphi^{-2}(\xi) : -k_2 r b_1 - 8k_1^3 r^2 B b_1 - k_1 r A a_0 b_1$$
$$+ 8k_1^3 r^2 D a_0 b_1 + 2k_1^2 r^2 F a_1 b_1$$
$$- 2k_1^2 r F b_1^2 - 2k_1^2 r^2 G a_1 b_1 - 2k_1^2 r F b_1^2 = 0,$$

$$\varphi^{-1}(\xi) : -2k_1^2 r E b_1 - 2k_1^2 r G a_0 b_1 + 6k_1^3 r^2 D a_1 b_1 - 2k_1^3 r^2 C a_1 b_1$$
$$- k_1 A b_1^2 + 2k_1^3 r D b_1^2 + 2k_1^3 r C b_1^2 = 0,$$

$$\varphi^0(\xi) : k_2 r a_1 + 2k_1^3 r^2 B a_1 + k_1 r A a_0 a_1 - 2k_1^3 r^2 D a_0 a_1 - k_1^2 r^2 F a_1^2$$
$$- k_2 b_1 - 2k_1^3 r B b_1 - k_1 A a_0 b_1 + 2k_1^3 r D a_0 b_1$$
$$- 4k_1^2 r G a_1 b_1 - k_1^2 F b_1^2 + 4k_1^2 r F a_1 b_1 = 0,$$

$$\varphi(\xi) : -2k_1^2 r E a_1 - 2k_1^2 r G a_0 a_1 + k_1 r A a_1^2 - 2k_1^3 r^2 D a_1^2$$
$$- 2k_1^3 r^2 C a_1^2 - 6k_1^3 r D a_1 b_1 + 2k_1^3 r C a_1 b_1 = 0,$$

$$\varphi^2(\xi) : k_2 a_1 + 8k_1^3 r B a_1 + k_1 A a_0 a_1 - 8k_1^3 r D a_0 a_1 - 2k_1^2 r F a_1^2$$
$$- 2k_1^2 r G a_1^2 + 2k_1^2 F a_1 b_1 - 2k_1^2 G a_1 b_1 = 0,$$

$$\varphi^3(\xi) : -2k_1^2 E a_1 - 2k_1^2 G a_0 a_1 + k_1 A a_1^2 - 8k_1^3 r D a_1^2$$
$$- 4k_1^3 r C a_1^2 - 6k_1^3 D a_1 b_1 + 2k_1^3 r C a_1 b_1 = 0,$$

$$\varphi^4(\xi) : 6k_1^3 a_1 B - 6k_1^3 D a_0 a_1 - k_1^2 F a_1^2 - 2k_1^2 G a_1^2 = 0,$$

$$\varphi^5(\xi) : -6k_1^3 D a_1^2 - 2k_1^3 C a_1^2 = 0. \qquad (2.50)$$

2.5.3 *Results and discussion*

Solving the above algebraic equation (2.50), a set of coefficients have been obtained for the solution of Eq. (2.47).

Case 1.

$$b_1(t,z) = \frac{6k_1 r(C(t,z)E(t,z) - 3B(t,z)G(t,z))}{-3C(t,z)A(t,z) + 4k_1^2 rC^2(t,z) + 3G(t,z)(F(t,z) + 2G(t,z))},$$

$$a_0(t,z) = \frac{3(B(t,z)(-3A(t,z) + 4k_1^2 rC(t,z) + E(t,z)(F(t,z) + 2G(t,z))}{3C(t,z)A(t,z) + 4k_1^2 rC^2(t,z) + 3G(t,z)(F(t,z) + 2G(t,z))},$$

$$a_1(t,z) = 0.$$

Case 2.

$$b_1(t,z) = 0,$$

$$a_0(t,z) = \frac{3(B(t,z)(-3A(t,z) + 4k_1^2 rC(t,z) + E(t,z)(F(t,z) + 2G(t,z))}{3C(t,z)A(t,z) + 4k_1^2 rC^2(t,z) + 3G(t,z)(F(t,z) + 2G(t,z))},$$

$$a_1(t,z) = \frac{6k_1(C(t,z)E(t,z) - 3B(t,z)G(t,z))}{-3C(t,z)A(t,z) + 4k_1^2 rC^2(t,z) + 3G(t,z)(F(t,z) + 2G(t,z))}.$$

Case 3.

$$b_1(t,z) = \frac{6k_1 r(C(t,z)E(t,z) - 3B(t,z)G(t,z))}{-3C(t,z)A(t,z) + 16k_1^2 rC^2(t,z) + 3G(t,z)(F(t,z) + 2G(t,z))},$$

$$a_0(t,z) = -\frac{3(B(t,z)(-3A(t,z) + 16k_1^2 rC(t,z)) + E(t,z)(F(t,z) + 2G(t,z)))}{3C(t,z)A(t,z) + 16k_1^2 rC^2(t,z) - 3G(t,z)(F(t,z) + 2G(t,z))},$$

$$a_1(t,z) = \frac{6k_1 r(C(t,z)E(t,z) - 3B(t,z)G(t,z))}{-3C(t,z)A(t,z) + 16k_1^2 rC^2(t,z) + 3G(t,z)(F(t,z) + 2G(t,z))}.$$

Thus, the solution has been obtained by substituting the values of a_0, b_1 and a_1 in Eq. (2.49) as follows:

Set 1 solutions for Case 1.

For $\varphi(\xi) = -\sqrt{-r}\tanh(\sqrt{-r}\xi), r < 0,$

$$\Phi_{1,1}(\xi,z) = -\frac{6k_1(C(t,z)E(t,z) - 3B(t,z)G(t,z))}{-3C(t,z)A(t,z) + 4k_1^2 rC^2(t,z) + 3G(t,z)(F(t,z) + 2G(t,z))}$$

$$\times \sqrt{-r}\coth(\sqrt{-r}\xi)$$

$$+ \frac{3(B(t,z)(-3A(t,z) + 4k_1^2 rC(t,z)) + E(t,z)(F(t,z) + 2G(t,z)))}{3C(t,z)A(t,z) - 4k_1^2 rC^2(t,z) - 3G(t,z)(F(t,z) + 2G(t,z))}.$$

For $\varphi(\xi) = -\sqrt{-r}\coth(\sqrt{-r}\xi), r < 0$,

$$\Phi_{1,2}(\xi,z) = -\frac{6k_1(C(t,z)E(t,z) - 3B(t,z)G(t,z))}{-3C(t,z)A(t,z) + 4k_1^2 rC^2(t,z) + 3G(t,z)(F(t,z) + 2G(t,z))}$$
$$\times \sqrt{-r}\tanh(\sqrt{-r}\xi)$$
$$+ \frac{3(B(t,z)(-3A(t,z) + 4k_1^2 rC(t,z)) + E(t,z)(F(t,z) + 2G(t,z)))}{3C(t,z)A(t,z) - 4k_1^2 rC^2(t,z) - 3G(t,z)(F(t,z) + 2G(t,z))}.$$

For $\varphi(\xi) = -\sqrt{-r}\tan(\sqrt{-r}\xi), r < 0$,

$$\Phi_{1,3}(\xi,z) = \frac{6k_1(C(t,z)E(t,z) - 3B(t,z)G(t,z))}{-3C(t,z)A(t,z) + 4k_1^2 rC^2(t,z) + 3G(t,z)(F(t,z) + 2G(t,z))}$$
$$\times \sqrt{r}\cot(\sqrt{r}\xi)$$
$$+ \frac{3(B(t,z)(-3A(t,z) + 4k_1^2 rC(t,z)) + E(t,z)(F(t,z) + 2G(t,z)))}{3C(t,z)A(t,z) - 4k_1^2 rC^2(t,z) - 3G(t,z)(F(t,z) + 2G(t,z))}.$$

For $\varphi(\xi) = -\sqrt{-r}\cot(\sqrt{-r}\xi), r < 0$,

$$\Phi_{1,4}(\xi,z) = -\frac{6k_1(C(t,z)E(t,z) - 3B(t,z)G(t,z))}{-3C(t,z)A(t,z) + 4k_1^2 rC^2(t,z) + 3G(t,z)(F(t,z) + 2G(t,z))}$$
$$\times \sqrt{r}\tan(\sqrt{r}\xi)$$
$$+ \frac{3(B(t,z)(-3A(t,z) + 4k_1^2 rC(t,z)) + E(t,z)(F(t,z) + 2G(t,z)))}{3C(t,z)A(t,z) - 4k_1^2 rC^2(t,z) - 3G(t,z)(F(t,z) + 2G(t,z))}.$$

For $\varphi(\xi) = -\frac{1}{\xi+\omega}, \omega$ is a constant, $r = 0$,

$$\Phi_{1,5}(\xi,z) = -\frac{6k_1(C(t,z)E(t,z) - 3B(t,z)G(t,z))}{-3C(t,z)A(t,z) + 4k_1^2 rC^2(t,z) + 3G(t,z)(F(t,z) + 2G(t,z))}(\xi + \omega)$$
$$+ \frac{3(B(t,z)(-3A(t,z) + 4k_1^2 rC(t,z)) + E(t,z)(F(t,z) + 2G(t,z)))}{3C(t,z)A(t,z) - 4k_1^2 rC^2(t,z) - 3G(t,z)(F(t,z) + 2G(t,z))}.$$

Set 2 solutions for Case 2.

For $\varphi(\xi) = -\sqrt{-r}\tanh(\sqrt{-r}\xi), r < 0$,

$$\Phi_{2,1}(\xi,z) = \frac{3(B(t,z)(-3A(t,z) + 4k_1^2 rC(t,z)) + E(t,z)(F(t,z) + 2G(t,z)))}{3C(t,z)A(t,z) - 4k_1^2 rC^2(t,z) - 3G(t,z)(F(t,z) + 2G(t,z))}$$

$$+ \frac{6k_1(C(t,z)E(t,z) - 3B(t,z)G(t,z))}{-3C(t,z)A(t,z) + 4k_1^2 rC^2(t,z) + 3G(t,z)(F(t,z) + 2G(t,z))}$$

$$\times \sqrt{-r}\tanh(\sqrt{-r}\xi).$$

For $\varphi(\xi) = -\sqrt{-r}\cot(\sqrt{-r}\xi), r < 0,$

$$\Phi_{2,2}(\xi,z) = \frac{3(B(t,z)(-3A(t,z) + 4k_1^2 rC(t,z)) + E(t,z)(F(t,z) + 2G(t,z)))}{3C(t,z)A(t,z) - 4k_1^2 rC^2(t,z) - 3G(t,z)(F(t,z) + 2G(t,z))}$$

$$+ \frac{6k_1(C(t,z)E(t,z) - 3B(t,z)G(t,z))}{-3C(t,z)A(t,z) + 4k_1^2 rC^2(t,z) + 3G(t,z)(F(t,z) + 2G(t,z))}$$

$$\times \sqrt{-r}\coth(\sqrt{-r}\xi).$$

For $\varphi(\xi) = \sqrt{r}\tan(\sqrt{r}\xi), r > 0,$

$$\Phi_{2,3}(\xi,z) = \frac{3(B(t,z)(-3A(t,z) + 4k_1^2 rC(t,z)) + E(t,z)(F(t,z) + 2G(t,z)))}{3C(t,z)A(t,z) - 4k_1^2 rC^2(t,z) - 3G(t,z)(F(t,z) + 2G(t,z))}$$

$$- \frac{6k_1(C(t,z)E(t,z) - 3B(t,z)G(t,z))}{-3C(t,z)A(t,z) + 4k_1^2 rC^2(t,z) + 3G(t,z)(F(t,z) + 2G(t,z))}$$

$$\times \sqrt{r}\tan(\sqrt{r}\xi).$$

For $\varphi(\xi) = -\sqrt{r}\cot(\sqrt{r}\xi), r > 0,$

$$\Phi_{2,4}(\xi,z) = \frac{3(B(t,z)(-3A(t,z) + 4k_1^2 rC(t,z)) + E(t,z)(F(t,z) + 2G(t,z)))}{3C(t,z)A(t,z) - 4k_1^2 rC^2(t,z) - 3G(t,z)(F(t,z) + 2G(t,z))}$$

$$+ \frac{6k_1(C(t,z)E(t,z) - 3B(t,z)G(t,z))}{-3C(t,z)A(t,z) + 4k_1^2 rC^2(t,z) + 3G(t,z)(F(t,z) + 2G(t,z))}$$

$$\times \sqrt{r}\cot(\sqrt{r}\xi).$$

For $\varphi(\xi) = -\frac{1}{\xi+\omega}, \omega$ is a constant, $r = 0,$

$$\Phi_{2,5}(\xi,z) = \frac{3(B(t,z)(-3A(t,z) + 4k_1^2 rC(t,z)) + E(t,z)(F(t,z) + 2G(t,z)))}{3C(t,z)A(t,z) - 4k_1^2 rC^2(t,z) - 3G(t,z)(F(t,z) + 2G(t,z))}$$

$$+ \frac{6k_1(C(t,z)E(t,z) - 3B(t,z)G(t,z))}{-3C(t,z)A(t,z) + 4k_1^2 rC^2(t,z) + 3G(t,z)(F(t,z) + 2G(t,z))}$$

$$\times \frac{1}{(\xi + \omega)}.$$

Set 3 solutions for Case 3.

For $\varphi(\xi) = -\sqrt{-r}\tanh(\sqrt{-r}\xi), r < 0,$

$$\Phi_{3,1}(\xi, z) = -\frac{6k_1(C(t,z)E(t,z) - 3B(t,z)G(t,z))}{-3C(t,z)A(t,z) + 16k_1^2 rC^2(t,z) + 3G(t,z)(F(t,z) + 2G(t,z))}$$

$$\times \sqrt{-r}\coth(\sqrt{-r}\xi)$$

$$- \frac{3(B(t,z)(-3A(t,z) + 16k_1^2 rC(t,z)) + E(t,z)(F(t,z) + 2G(t,z)))}{-3C(t,z)A(t,z) + 16k_1^2 rC^2(t,z) + 3G(t,z)(F(t,z) + 2G(t,z))}$$

$$+ \frac{6k_1(C(t,z)E(t,z) - 3B(t,z)G(t,z))}{-3C(t,z)A(t,z) + 16k_1^2 rC^2(t,z) + 3G(t,z)(F(t,z) + 2G(t,z))}$$

$$\times \sqrt{-r}\tanh(\sqrt{-r}\xi).$$

For $\varphi(\xi) = -\sqrt{-r}\coth(\sqrt{-r}\xi), r < 0,$

$$\Phi_{3,2}(\xi, z) = -\frac{6k_1(C(t,z)E(t,z) - 3B(t,z)G(t,z))}{-3C(t,z)A(t,z) + 16k_1^2 rC^2(t,z) + 3G(t,z)(F(t,z) + 2G(t,z))}$$

$$\times \sqrt{-r}\tanh(\sqrt{-r}\xi)$$

$$- \frac{3(B(t,z)(-3A(t,z) + 16k_1^2 rC(t,z)) + E(t,z)(F(t,z) + 2G(t,z)))}{-3C(t,z)A(t,z) + 16k_1^2 rC^2(t,z) + 3G(t,z)(F(t,z) + 2G(t,z))}$$

$$+ \frac{6k_1(C(t,z)E(t,z) - 3B(t,z)G(t,z))}{-3C(t,z)A(t,z) + 16k_1^2 rC^2(t,z) + 3G(t,z)(F(t,z) + 2G(t,z))}$$

$$\times \sqrt{-r}\coth(\sqrt{-r}\xi).$$

For $\varphi(\xi) = \sqrt{r}\tan(\sqrt{r}\xi), r > 0,$

$$\Phi_{3,3}(\xi, z) = \frac{6k_1(C(t,z)E(t,z) - 3B(t,z)G(t,z))}{-3C(t,z)A(t,z) + 16k_1^2 rC^2(t,z) + 3G(t,z)(F(t,z) + 2G(t,z))}$$

$$\times \sqrt{r}\cot(\sqrt{r}\xi)$$

$$- \frac{3(B(t,z)(-3A(t,z) + 16k_1^2 rC(t,z)) + E(t,z)(F(t,z) + 2G(t,z)))}{-3C(t,z)A(t,z) + 16k_1^2 rC^2(t,z) + 3G(t,z)(F(t,z) + 2G(t,z))}$$

$$- \frac{6k_1(C(t,z)E(t,z) - 3B(t,z)G(t,z))}{-3C(t,z)A(t,z) + 16k_1^2 rC^2(t,z) + 3G(t,z)(F(t,z) + 2G(t,z))}$$

$$\times \sqrt{r}\tan(\sqrt{r}\xi).$$

For $\varphi(\xi) = \sqrt{r}\cot(\sqrt{r}\xi), r > 0,$

$$\Phi_{3,4}(\xi, z) = -\frac{6k_1(C(t,z)E(t,z) - 3B(t,z)G(t,z))}{-3C(t,z)A(t,z) + 16k_1^2 rC^2(t,z) + 3G(t,z)(F(t,z) + 2G(t,z))}$$
$$\times \sqrt{r}\tan(\sqrt{r}\zeta)$$
$$-\frac{3(B(t,z)(-3A(t,z) + 16k_1^2 rC(t,z)) + E(t,z)(F(t,z) + 2G(t,z)))}{-3C(t,z)A(t,z) + 16k_1^2 rC^2(t,z) + 3G(t,z)(F(t,z) + 2G(t,z))}$$
$$+\frac{6k_1(C(t,z)E(t,z) - 3B(t,z)G(t,z))}{-3C(t,z)A(t,z) + 16k_1^2 rC^2(t,z) + 3G(t,z)(F(t,z) + 2G(t,z))}$$
$$\times \sqrt{r}\cot(\sqrt{r}\zeta).$$

For $\varphi(\xi) = -\frac{1}{\xi + \omega}, \omega$ is a constant, $r = 0,$

$$\Phi_{3,5}(\xi, z) = -\frac{6k_1(C(t,z)E(t,z) - 3B(t,z)G(t,z))}{-3C(t,z)A(t,z) + 16k_1^2 rC^2(t,z) + 3G(t,z)(F(t,z) + 2G(t,z))}(\xi + \omega)$$
$$-\frac{3(B(t,z)(-3A(t,z) + 16k_1^2 rC(t,z)) + E(t,z)(F(t,z) + 2G(t,z)))}{-3C(t,z)A(t,z) + 16k_1^2 rC^2(t,z) + 3G(t,z)(F(t,z) + 2G(t,z))}$$
$$+\frac{6k_1(C(t,z)E(t,z) - 3B(t,z)G(t,z))}{-3C(t,z)A(t,z) + 16k_1^2 rC^2(t,z) + 3G(t,z)(F(t,z) + 2G(t,z))}\frac{1}{(\xi + \omega)}.$$

By Set 1, Set 2 and Set 3, from Theorem 2.3.6.1 and according to Xie [30], there exists a bounded open set $G \subset \mathbb{R}_+ \times \mathbb{R}, q > 0$ and $r > 0$ such that $\Phi(t, x, z)$, $\Phi_t(t, x, z)$, $\Phi_x(t, x, z)$ and $\Phi_{xxx}(t, x, z)$ are uniformly bounded for $(t, x, z) \in G \times \mathbb{K}_q(r)$, continuous with respect to $(t, x) \in G$ for all $z \in \mathbb{K}_q(r)$ and analytic with respect to $z \in \mathbb{K}_q(r)$ for all $(t, x) \in G$. Theorem 2.3.6.1 also suggests that there exists $U(t, x) \in (S)_{-1}$ such that $\Phi(t, x, z) = (\mathcal{H}U(t, x))(z)$ for all $(t, x, z) \in G \times \mathbb{K}_q(r)$ and that $U(t, x)$ solves Eq. (2.7). Therefore, $U = \mathcal{H}^{-1}\Phi \in (S)_{-1}$ and thereby obtain a solution U of the original Wick-type SPDE (2.7).

The above three sets of solutions for the different cases contain some arbitrary functions and constants. The arbitrary functions and constants provide enough freedom to construct solutions that may be related to the real physical phenomena. If $r = -1$, $A = 0.25$, $B = -2.5$, $C = -0.65$, $E = 0.5$, $F = 2.0$ and $G = 1.0$. Then, the

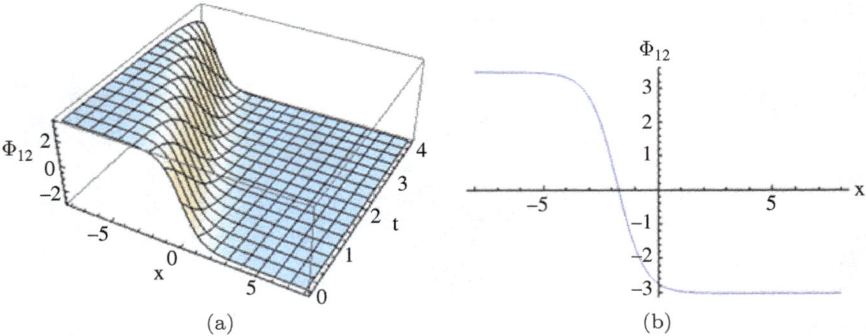

(a)   (b)

Fig. 2.1.   (a) The three-dimensional solitary wave solution for $\Phi_{1,2}$ and (b) the corresponding two-dimensional solution for $\Phi_{1,2}$ at $t = 1$.

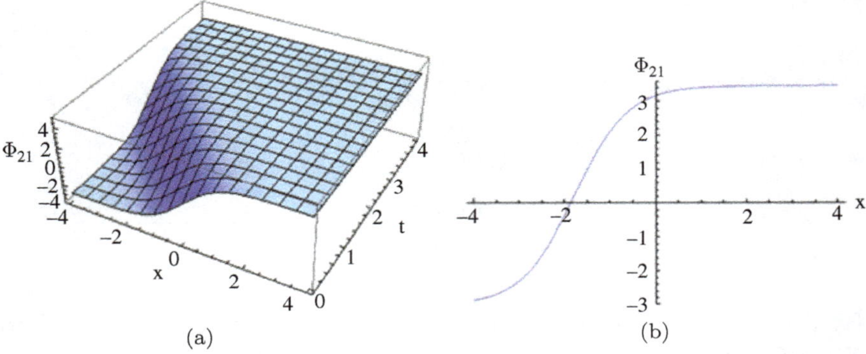

(a)   (b)

Fig. 2.2.   (a) The three-dimensional solitary wave solution for $\Phi_{2,1}$ and (b) the corresponding two-dimensional solution for $\Phi_{2,1}$ at $t = 1$.

solutions are given by

$$\Phi_{1,2} = 0.2168 - 3.24791\tanh(1.51794t + 0.85x), \tag{2.51}$$

$$\Phi_{2,1} = 0.2168 + 3.24791\tanh(1.51794t + 0.85x). \tag{2.52}$$

The behaviours of the solutions (2.51) and (2.52) are shown graphically in Figs. 2.1 and 2.2, respectively.

From the graph of Fig. 2.1(a), it can be observed that the solution surface for $\Phi_{1,2}$ represents antikink-type solitary wave solution. The corresponding two-dimensional graph of Fig. 2.1(b) is symmetric about the point of intersection of the solution curve for $\Phi_{1,2}$ with the x-axis. Also, solitary wave solution is bounded and asymptotically parallel to the x-axis as $x \to +\infty$ when $t = 1$. On the other hand, the solution surface for $\Phi_{2,1}$ shown in Fig. 2.2(a) represents kink-type solitary wave solution. Also, the corresponding two-dimensional graph in Fig. 2.2(b) for the solitary wave solution curve of $\Phi_{2,1}$ is bounded and asymptotically parallel to the x-axis as $x \to +\infty$ when $t = 1$.

2.5.3.1 *Stochastic solutions for Set 1*

Using Set 1, Set 2 and Set 3, the stochastic solutions of Eq. (2.7) are as follows:

$$
\begin{aligned}
U_{1,1}(t,x) = {}&-\sqrt{-r}\frac{6k_1(C(t)\Diamond E(t) - 3B(t)\Diamond G(t))}{-3C(t)\Diamond A(t) + 4k_1^2 rC^{\Diamond 2}(t) + 3G(t)\Diamond(F(t) + 2G(t))} \\
&\times \Diamond \coth^{\Diamond}(\sqrt{-r}\Xi_{1,1}(t,x)) \\
&+ \frac{3(B(t)\Diamond(-3A(t) + 4k_1^2 rC(t)) + E(t)\Diamond(F(t) + 2G(t)))}{3C(t)\Diamond A(t) - 4k_1^2 rC^{\Diamond 2}(t) - 3G(t)\Diamond(F(t) + 2G(t))},
\end{aligned}
$$

where $\Xi_{1,1}(t,x) = \Psi_{1,1}(\xi(t,x))$.

$$
\begin{aligned}
U_{1,2}(t,x) = {}&-\sqrt{-r}\frac{6k_1(C(t)\Diamond E(t) - 3B(t)\Diamond G(t))}{-3C(t)\Diamond A(t) + 4k_1^2 rC^{\Diamond 2}(t) + 3G(t)\Diamond(F(t) + 2G(t))} \\
&\times \Diamond \tanh^{\Diamond}(\sqrt{-r}\Xi_{1,2}(t,x)) \\
&+ \frac{3(B(t)\Diamond(-3A(t) + 4k_1^2 rC(t)) + E(t)\Diamond(F(t) + 2G(t)))}{3C(t)\Diamond A(t) - 4k_1^2 rC^{\Diamond 2}(t) - 3G(t)\Diamond(F(t) + 2G(t))},
\end{aligned}
$$

where $\Xi_{1,2}(t,x) = \Psi_{1,2}(\xi(t,x))$.

$$
\begin{aligned}
U_{1,3}(t,x) = {}&\sqrt{r}\frac{6k_1(C(t)\Diamond E(t) - 3B(t)\Diamond G(t))}{-3C(t)\Diamond A(t) + 4k_1^2 rC^{\Diamond 2}(t) + 3G(t)\Diamond(F(t) + 2G(t))} \\
&\times \Diamond \cot^{\Diamond}(\sqrt{-r}\Xi_{1,1}(t,)) \\
&+ \frac{3(B(t)\Diamond(-3A(t) + 4k_1^2 rC(t)) + E(t)\Diamond(F(t) + 2G(t)))}{3C(t)\Diamond A(t) - 4k_1^2 rC^{\Diamond 2}(t) - 3G(t)\Diamond(F(t) + 2G(t))},
\end{aligned}
$$

where $\Xi_{1,3}(t,x) = \Psi_{1,3}(\xi(t,x))$.

$$
\begin{aligned}
U_{1,4}(t,x) = {}&-\sqrt{r}\frac{6k_1(C(t)\Diamond E(t) - 3B(t)\Diamond G(t))}{-3C(t)\Diamond A(t) + 4k_1^2 rC^{\Diamond 2}(t) + 3G(t)\Diamond(F(t) + 2G(t))} \\
&\times \Diamond \tan^{\Diamond}(\sqrt{r}\Xi_{1,4}(t,x)) \\
&+ \frac{3(B(t)\Diamond(-3A(t) + 4k_1^2 rC(t)) + E(t)\Diamond(F(t) + 2G(t)))}{3C(t)\Diamond A(t) - 4k_1^2 rC^{\Diamond 2}(t) - 3G(t)\Diamond(F(t) + 2G(t))},
\end{aligned}
$$

where $\Xi_{1,4}(t, x) = \Psi_{1,4}(\xi(t, x))$.

$$U_{1,5}(t, x) = -\frac{6k_1(C(t)\Diamond E(t) - 3B(t)\Diamond G(t))}{-3C(t)\Diamond A(t) + 4k_1^2 rC^{\Diamond 2}(t) + 3G(t)\Diamond(F(t) + 2G(t))}$$
$$\times (\Xi_{1,5}(t, x) + \omega)$$
$$+ \frac{3(B(t)\Diamond(-3A(t) + 4k_1^2 rC(t)) + E(t)\Diamond(F(t) + 2G(t)))}{3C(t)\Diamond A(t) - 4k_1^2 rC^{\Diamond 2}(t) - 3G(t)\Diamond(F(t) + 2G(t))},$$

where $\Xi_{1,5}(t, x) = \Psi_{1,5}(\xi(t, x))$.

2.5.3.2   *Stochastic solutions for Set 2*

$$U_{2,1}(t, x) = \frac{3(B(t)\Diamond(-3A(t) + 4k_1^2 rC(t)) + E(t)\Diamond(F(t) + 2G(t)))}{3C(t)\Diamond A(t) - 4k_1^2 rC^{\Diamond 2}(t) - 3G(t)\Diamond(F(t) + 2G(t))}$$
$$+ \sqrt{-r}\frac{6k_1(C(t)\Diamond E(t) - 3B(t)\Diamond G(t))}{-3C(t)\Diamond A(t) + 4k_1^2 rC^{\Diamond 2}(t) + 3G(t)\Diamond(F(t) + 2G(t))}$$
$$\times \Diamond \tanh^{\Diamond}(\sqrt{-r}\Xi_{2,1}(t, x),$$

where $\Xi_{2,1}(t, x) = \Psi_{2,1}(\xi(t, x))$.

$$U_{2,2}(t, x) = \frac{3(B(t)\Diamond(-3A(t) + 4k_1^2 rC(t)) + E(t)\Diamond(F(t) + 2G(t)))}{3C(t)\Diamond A(t) - 4k_1^2 rC^{\Diamond 2}(t) - 3G(t)\Diamond(F(t) + 2G(t))}$$
$$+ \sqrt{-r}\frac{6k_1(C(t)\Diamond E(t) - 3B(t)\Diamond G(t))}{-3C(t)\Diamond A(t) + 4k_1^2 rC^{\Diamond 2}(t) + 3G(t)\Diamond(F(t) + 2G(t))}$$
$$\times \Diamond \coth^{\Diamond}(\sqrt{-r}\Xi_{2,2}(t, x)),$$

where $\Xi_{2,2}(t, x) = \Psi_{2,2}(\xi(t, x))$.

$$U_{2,3}(t, x) = \frac{3(B(t)\Diamond(-3A(t) + 4k_1^2 rC(t)) + E(t)\Diamond(F(t) + 2G(t)))}{3C(t)\Diamond A(t) - 4k_1^2 rC^{\Diamond 2}(t) - 3G(t)\Diamond(F(t) + 2G(t))}$$
$$- \sqrt{r}\frac{6k_1(C(t)\Diamond E(t) - 3B(t)\Diamond G(t))}{-3C(t)\Diamond A(t) + 4k_1^2 rC^{\Diamond 2}(t) + 3G(t)\Diamond(F(t) + 2G(t))}$$
$$\times \Diamond \tan^{\Diamond}(\sqrt{r}\Xi_{2,3}(t, x)),$$

where $\Xi_{2,3}(t,x) = \Psi_{2,3}(\xi(t,x))$.

$$U_{2,4}(t,x) = \frac{3(B(t)\Diamond(-3A(t) + 4k_1^2 rC(t)) + E(t)\Diamond(F(t) + 2G(t)))}{3C(t)\Diamond A(t) - 4k_1^2 rC^{\Diamond 2}(t) - 3G(t)\Diamond(F(t) + 2G(t))}$$
$$+ \sqrt{r}\frac{6k_1(C(t)\Diamond E(t) - 3B(t)\Diamond G(t))}{-3C(t)\Diamond A(t) + 4k_1^2 rC^{\Diamond 2}(t) + 3G(t)\Diamond(F(t) + 2G(t))}$$
$$\times \Diamond\cot^\Diamond(\sqrt{r}\Xi_{2,4}(t,x)),$$

where $\Xi_{2,4}(t,x) = \Psi_{2,4}(\xi(t,x))$.

$$U_{2,5}(t,x) = \frac{3(B(t)\Diamond(-3A(t) + 4k_1^2 rC(t)) + E(t)\Diamond(F(t) + 2G(t)))}{3C(t)\Diamond A(t) - 4k_1^2 rC^{\Diamond 2}(t) - 3G(t)\Diamond(F(t) + 2G(t))}$$
$$+ \frac{6k_1(C(t)\Diamond E(t) - 3B(t)\Diamond G(t))}{-3C(t)\Diamond A(t) + 4k_1^2 rC^{\Diamond 2}(t) + 3G(t)\Diamond(F(t) + 2G(t))}$$
$$\times \frac{1}{(\Xi_{2,5}(t,x) + \omega)},$$

where $\Xi_{2,5}(t,x) = \Psi_{2,5}(\xi(t,x))$.

2.5.3.3 *Stochastic solutions for Set 3*

$$U_{3,1}(t,x) = -\sqrt{-r}\frac{6k_1(C(t)\Diamond E(t) - 3B(t)\Diamond G(t))}{-3C(t)\Diamond A(t) + 16k_1^2 rC^{\Diamond 2}(t) + 3G(t)\Diamond(F(t) + 2G(t))}$$
$$\times \Diamond\coth^\Diamond(\sqrt{-r}\Xi_{3,1}(t,x))$$
$$- \frac{3(B(t)\Diamond(-3A(t) + 16k_1^2 rC(t)) + E(t)\Diamond(F(t) + 2G(t)))}{-3C(t)\Diamond A(t) + 16k_1^2 rC^{\Diamond 2}(t) + 3G(t)\Diamond(F(t) + 2G(t))}$$
$$+ \sqrt{-r}\frac{6k_1(C(t)\Diamond E(t) - 3B(t)\Diamond G(t))}{-3C(t)\Diamond A(t) + 16k_1^2 rC^{\Diamond 2}(t) + 3G(t)\Diamond(F(t) + 2G(t))}$$
$$\times \Diamond\tanh^\Diamond(\sqrt{-r}\Xi_{3,1}(t,x)),$$

where $\Xi_{3,1}(t,x) = \Psi_{3,1}(\xi(t,x))$.

$$U_{3,2}(t,x) = -\sqrt{-r}\frac{6k_1(C(t)\Diamond E(t) - 3B(t)\Diamond G(t))}{-3C(t)\Diamond A(t) + 16k_1^2 rC^{\Diamond 2}(t) + 3G(t)\Diamond(F(t) + 2G(t))}$$
$$\times \Diamond\tanh^\Diamond(\sqrt{-r}\Xi_{3,2}(t,x))$$
$$- \frac{3(B(t)\Diamond(-3A(t) + 16k_1^2 rC(t)) + E(t)\Diamond(F(t) + 2G(t)))}{-3C(t)\Diamond A(t) + 16k_1^2 rC^{\Diamond 2}(t) + 3G(t)\Diamond(F(t) + 2G(t))}$$

$$+ \sqrt{-r}\frac{6k_1(C(t)\Diamond E(t) - 3B(t)\Diamond G(t))}{-3C(t)\Diamond A(t) + 16k_1^2 rC^{\Diamond 2}(t) + 3G(t)\Diamond(F(t) + 2G(t))}$$

$$\times\ \Diamond\coth^{\Diamond}(\sqrt{-r}\Xi_{3,2}(t,x)),$$

where $\Xi_{3,2}(t,x) = \Psi_{3,2}(\xi(t,x))$.

$$U_{3,3}(t,x) = \sqrt{r}\frac{6k_1(C(t)\Diamond E(t) - 3B(t)\Diamond G(t))}{-3C(t)\Diamond A(t) + 16k_1^2 rC^{\Diamond 2}(t) + 3G(t)\Diamond(F(t) + 2G(t))}$$

$$\times\ \Diamond\cot^{\Diamond}(\sqrt{-r}\Xi_{3,3}(t,x))$$

$$-\ \frac{3(B(t)\Diamond(-3A(t) + 16k_1^2 rC(t)) + E(t)\Diamond(F(t) + 2G(t)))}{-3C(t)\Diamond A(t) + 16k_1^2 rC^{\Diamond 2}(t) + 3G(t)\Diamond(F(t) + 2G(t))}$$

$$-\ \sqrt{r}\frac{6k_1(C(t)\Diamond E(t) - 3B(t)\Diamond G(t))}{-3C(t)\Diamond A(t) + 4k_1^2 rC^{\Diamond 2}(t) + 3G(t)\Diamond(F(t) + 2G(t))}$$

$$\times\ \Diamond\tan^{\Diamond}(\sqrt{-r}\Xi_{3,3}(t,x)),$$

where $\Xi_{3,3}(t,x) = \Psi_{3,3}(\xi(t,x))$.

$$U_{3,4}(t,x) = -\sqrt{r}\frac{6k_1(C(t)\Diamond E(t) - 3B(t)\Diamond G(t))}{-3C(t)\Diamond A(t) + 16k_1^2 rC^{\Diamond 2}(t) + 3G(t)\Diamond(F(t) + 2G(t))}$$

$$\times\ \Diamond\tan^{\Diamond}(\sqrt{r}\Xi_{3,4}(t,x))$$

$$-\ \frac{3(B(t)\Diamond(-3A(t) + 16k_1^2 rC(t)) + E(t)\Diamond(F(t) + 2G(t)))}{-3C(t)\Diamond A(t) + 16k_1^2 rC^{\Diamond 2}(t) + 3G(t)\Diamond(F(t) + 2G(t))}$$

$$+\ \sqrt{r}\frac{6k_1(C(t)\Diamond E(t) - 3B(t)\Diamond G(t))}{-3C(t)\Diamond A(t) + 16k_1^2 rC^{\Diamond 2}(t) + 3G(t)\Diamond(F(t) + 2G(t))}$$

$$\times\ \Diamond\cot^{\Diamond}(\sqrt{r}\Xi_{3,4}(t,x)),$$

where $\Xi_{3,4}(t,x) = \Psi_{3,4}(\xi(t,x))$.

$$U_{3,5}(t,x) = -\frac{6k_1(C(t)\Diamond E(t) - 3B(t)\Diamond G(t))}{-3C(t)\Diamond A(t) + 16k_1^2 rC^{\Diamond 2}(t) + 3G(t)\Diamond(F(t) + 2G(t))}$$

$$\times\ (\Xi_{3,5}(t,x) + \omega)$$

$$-\ \frac{3(B(t)\Diamond(-3A(t) + 16k_1^2 rC(t)) + E(t)\Diamond(F(t) + 2G(t)))}{-3C(t)\Diamond A(t) + 16k_1^2 rC^{\Diamond 2}(t) + 3G(t)\Diamond(F(t) + 2G(t))}$$

$$+\ \frac{6k_1(C(t)\Diamond E(t) - 3B(t)\Diamond G(t))}{-3C(t)\Diamond A(t) + 16k_1^2 rC^{\Diamond 2}(t) + 3G(t)\Diamond(F(t) + 2G(t))}$$

$$\times\ \frac{1}{(\Xi_{3,5}(t,x) + \omega)},$$

where $\Xi_{3,5}(t,x) = \Psi_{3,5}(\xi(t,x))$.

2.6 Improved Sub-Equation Method for the Solutions of the Wick-Type Stochastic Modified Boussinesq Equation

In this section, the improved sub-equation method is employed here for finding the white noise solutions for the stochastic MBE (2.9).

Applying Hermite transform on Eq. (2.9), it can be obtained as

$$V_{tt} - V_{xx} + P(t,z)VV_x^2 + Q(t,z)V^2V_{xx} + R(t,z)V_{xxxx} = 0, \quad (2.53)$$

where $V \equiv V(t,x,z) = \mathcal{H}(U(t,x))$ and where $z = (z_1, z_2, \ldots) \in \mathbb{C}^N$ is a parameter.

Now, let $V(t,x,z) = \Phi(\xi,z)$, $\xi = k_1x + k_2t$, then Eq. (2.53) becomes

$$k_2^2\Phi_{tt} - k_1^2\Phi_{xx} + k_1^2\Phi_x^2\Phi P(t,z) + k_1^2\Phi_{xx}\Phi^2 Q(t,z)$$
$$+ k_1^4\Phi_{xxxx}R(t,z) = 0. \quad (2.54)$$

Consider the ansatz as follows:

$$\Phi(\xi,z) = \sum_{i=-N}^{-1} b_i(t,z)\varphi^i(\xi) + a_0(t,z) + \sum_{i=1}^{N} a_i(t,z)\varphi^i(\xi), \quad (2.55)$$

where $\varphi(\xi)$ satisfies Eq. (2.43).

Now, according to the homogenous balancing principle, in Eq. (2.55), $k_1^2\Phi_{xx}\Phi^2 Q(t,z)$ and $k_1^4\Phi_{xxxx}R(t,z)$ are the dominant terms having the highest order singularity. The maximum value of the pole is 1 which implies here $N = 1$.

Consequently, from Eq. (2.55), the ansatz becomes

$$\Phi(\xi,z) = b_1(t,z)\varphi^{-1}(\xi) + a_0(t,z) + a_1(t,z)\varphi(\xi). \quad (2.56)$$

Substituting Eq. (2.56) along with Eq. (2.44) in Eq. (2.54) and equating each coefficient of φ^i ($i = 0, 1, 2, \ldots$) with zero, a system of algebraic equations are obtained for $k_1, k_2, P(t,z), Q(t,z), R(t,z),$

$b_1(t, z), a_0(t, z)$ and $a_1(t, z)$ as follows:

$$\varphi^{-5}(\xi) : k_1^2 r^2 b_1^3 P + 2k_1^2 r^2 b_1^3 Q + 24k_1^4 r^4 b_1 R = 0,$$

$$\varphi^{-4}(\xi) : k_1^2 r^2 a_0 b_1^2 P + 4k_1^2 r^2 a_0 b_1^2 Q = 0,$$

$$\varphi^{-3}(\xi) : -2k_1^2 r^2 b_1 + 2k_2 r^2 b_1 - k_1^2 r^2 a_1 b_1^2 P + 2k_1^2 r b_1^3 P + 2k_1^2 r^2 a_0^2 b_1 Q$$
$$+ 4k_1^2 r^2 a_1 b_1^2 Q + 2k_1^2 r b_1^3 Q + 40k_1^4 r^3 b_1 R = 0,$$

$$\varphi^{-2}(\xi) : -2k_1^2 r^2 a_0 a_1 b_1 P + 2k_1^2 r a_0 b_1^2 P$$
$$+ 4k_1^2 r^2 a_0 a_1 b_1 Q + 4k_1^2 r a_0 b_1^2 Q = 0,$$

$$\varphi^{-1}(\xi) : -2k_1^2 r b_1 + 2k_2^2 r b_1 - k_1^2 r^2 a_1^2 b_1 P - 2k_1^2 r a_1 b_1^2 P + k_1^2 b_1^3 P$$
$$+ 2k_1^2 r a_0^2 b_1 Q + 2k_1^2 r^2 a_1^2 b_1 Q + 6k_1^2 r a_1 b_1^2 Q + 16k_1^4 r^2 b_1 R = 0,$$

$$\varphi^{0}(\xi) : k_1^2 r^2 a_0 a_1^2 P - 4k_1^2 r a_0 a_1 b_1 P + k_1^2 a_0 b_1^2 P + 8k_1^2 r a_0 a_1 b_1 Q = 0,$$

$$\varphi(\xi) : -2k_1^2 r a_1 + 2k_2^2 r a_1 + k_1^2 r^2 a_1^3 P - 2k_1^2 r a_1^2 b_1 P - k_1^2 a_1 b_1^2 P$$
$$+ 2k_1^2 r a_0^2 a_1 Q + 6k_1^2 r a_1^2 b_1 Q + 2k_1^2 a_1 b_1^2 Q + 16k_1^4 r^2 a_1 R = 0,$$

$$\varphi^{2}(\xi) : 2k_1^2 r a_0 a_1^2 P - 2k_1^2 a_0 a_1 b_1 P + 4k_1^2 r a_0 a_1^2 Q + k_1^2 a_0 a_1 b_1 Q = 0,$$

$$\varphi^{3}(\xi) : -2k_1^2 a_1 + 2k_2^2 a_1 + 2k_1^2 r a_1^3 P - k_1^2 a_1^2 b_1 P + 2k_1^2 a_0^2 a_1 Q$$
$$+ 2k_1^2 r a_1^3 Q + 4k_1^2 a_1^2 b Q + 40k_1^4 r a_1 R = 0,$$

$$\varphi^{4}(\xi) : k_1^2 a_0 a_1^2 P + 4k_1^2 a_0 a_1^2 Q = 0,$$

$$\varphi^{5}(\xi) : k_1^2 a_1^3 P + 2k_1^2 a_1^3 Q + 24k_1^4 a_1 R = 0. \tag{2.57}$$

2.6.1 *Results and discussion*

Solving the above algebraic equation (2.57), a set of coefficients for the solution of Eq. (2.54) have been obtained.

Case 1.

$$b_1(t, z) = \frac{i\sqrt{6}k_1\sqrt{R(t, z)}}{\sqrt{Q(t, z)}},$$

$$a_0(t, z) = 0,$$

$$a_1(t, z) = -\frac{i\sqrt{6}k_1\sqrt{R(t, z)}}{\sqrt{Q(t, z)}}.$$

Case 2.

$$b_1(t, z) = \frac{i\sqrt{6}k_1\sqrt{R(t, z)}}{\sqrt{Q(t, z)}},$$

$$a_0(t, z) = 0,$$

$$a_1(t, z) = -\frac{i\sqrt{6}k_1\sqrt{R(t, z)}}{\sqrt{Q(t, z)}}.$$

Case 3.

$$b_1(t, z) = \frac{i\sqrt{6}k_1\sqrt{R(t, z)}}{\sqrt{Q(t, z)}},$$

$$a_0(t, z) = 0,$$

$$a_1(t, z) = -\frac{i\sqrt{6}k_1\sqrt{R(t, z)}}{\sqrt{Q(t, z)}}.$$

Case 4.

$$b_1(t, z) = \frac{i\sqrt{6}k_1\sqrt{R(t, z)}}{\sqrt{Q(t, z)}},$$

$$a_0(t, z) = 0,$$

$$a_1(t, z) = -\frac{i\sqrt{6}k_1\sqrt{R(t, z)}}{\sqrt{Q(t, z)}}.$$

Thus, the solution has been obtained by substituting the values of a_1, b_1 and a_1 in Eq. (2.56) as follows:

Set 1 solutions for Case 1.

For $\varphi(\xi) = -\sqrt{t}\tanh(\sqrt{-r}\xi), r < 0,$

$$\Phi_{1,1} = \frac{i\sqrt{6}k_1\sqrt{R(t, z)}}{\sqrt{Q(t, z)}}\sqrt{-r}\coth(\sqrt{-r}\xi) + \frac{i\sqrt{6}k_1\sqrt{R(t, z)}}{\sqrt{Q(t, z)}}$$
$$\times \sqrt{-r}\tanh(\sqrt{-r}\xi).$$

For $\varphi(\xi) = -\sqrt{-r}\coth(\sqrt{-r}\xi), r < 0,$

$$\Phi_{1,2} = \frac{i\sqrt{6}k_1\sqrt{R(t,z)}}{\sqrt{Q(t,z)}}\sqrt{-r}\tanh(\sqrt{-r}\xi) + \frac{i\sqrt{6}k_1\sqrt{R(t,z)}}{\sqrt{Q(t,z)}}$$
$$\times \sqrt{-r}\coth(\sqrt{-r}\xi).$$

For $\varphi(\xi) = -\sqrt{r}\tan(\sqrt{r}\xi), r > 0,$

$$\Phi_{1,3} = \frac{i\sqrt{6}k_1\sqrt{R(t,z)}}{\sqrt{Q(t,z)}}\sqrt{r}\cot(\sqrt{r}\xi) + \frac{i\sqrt{6}k_1\sqrt{R(t,z)}}{\sqrt{Q(t,z)}}$$
$$\times \sqrt{r}\tan(\sqrt{r}\xi).$$

For $\varphi(\xi) = -\sqrt{r}\cot(\sqrt{r}\xi), r > 0,$

$$\Phi_{1,4} = \frac{i\sqrt{6}k_1\sqrt{R(t,z)}}{\sqrt{Q(t,z)}}\sqrt{r}\tan(\sqrt{r}\xi) + \frac{i\sqrt{6}k_1\sqrt{R(t,z)}}{\sqrt{Q(t,z)}}$$
$$\times \sqrt{r}\coth(\sqrt{-r}\xi).$$

For $\varphi(\xi) = -\frac{1}{\xi+\omega}, \omega,$ is a constant, $r = 0,$

$$\Phi_{1,5} = \frac{i\sqrt{6}k_1\sqrt{R(t,z)}}{\sqrt{Q(t,z)}}(\xi + \omega) + \frac{i\sqrt{6}k_1\sqrt{R(t,z)}}{\sqrt{Q(t,z)}}\frac{1}{(\xi + \omega)}.$$

Set 2 solutions for Case 2.

For $\varphi(\xi) = -\sqrt{-r}\tanh(\sqrt{-r}\xi), r < 0,$

$$\Phi_{2,1} = \frac{i\sqrt{6}k_1\sqrt{R(t,z)}}{\sqrt{Q(t,z)}}\sqrt{-r}\coth(\sqrt{-r}\xi) - \frac{i\sqrt{6}k_1\sqrt{R(t,z)}}{\sqrt{Q(t,z)}}$$
$$\times \sqrt{-r}\tanh(\sqrt{-r}\xi).$$

For $\varphi(\xi) = -\sqrt{-r}\coth(\sqrt{-r}\xi), r < 0,$

$$\Phi_{2,2} = \frac{i\sqrt{6}k_1\sqrt{R(t,z)}}{\sqrt{Q(t,z)}}\sqrt{-r}\tanh(\sqrt{-r}\xi) - \frac{i\sqrt{6}k_1\sqrt{R(t,z)}}{\sqrt{Q(t,z)}}$$
$$\times \sqrt{-r}\coth(\sqrt{-r}\xi).$$

For $\varphi(\xi) = \sqrt{r}\tan(\sqrt{r}\xi), r > 0$,

$$\Phi_{2,3} = \frac{i\sqrt{6}k_1\sqrt{R(t,z)}}{\sqrt{Q(t,z)}}\sqrt{r}\cot(\sqrt{r}\xi) + \frac{i\sqrt{6}k_1\sqrt{R(t,z)}}{\sqrt{Q(t,z)}}$$
$$\times \sqrt{r}\tan(\sqrt{r}\xi).$$

For $\varphi(\xi) = \sqrt{r}\cot(\sqrt{r}\xi), r > 0$,

$$\Phi_{2,4} = \frac{i\sqrt{6}k_1\sqrt{R(t,z)}}{\sqrt{Q(t,z)}}\sqrt{r}\tan(\sqrt{r}\xi) + \frac{i\sqrt{6}k_1\sqrt{R(t,z)}}{\sqrt{Q(t,z)}}$$
$$\times \sqrt{r}\coth(\sqrt{r}\xi).$$

For $\varphi(\xi) = \frac{1}{\xi+\omega}, \omega$ is a constant, $r = 0$,

$$\Phi_{2,5} = \frac{i\sqrt{6}k_1\sqrt{R(t,z)}}{\sqrt{Q(t,z)}}(\xi + \omega) - \frac{i\sqrt{6}k_1\sqrt{R(t,z)}}{\sqrt{Q(t,z)}}\frac{1}{(\xi+\omega)}.$$

Set 3 solutions for Case 3.

For $\varphi(\xi) = -\sqrt{-r}\tanh(\sqrt{-r}\xi), r < 0$,

$$\Phi_{3,1} = \frac{i\sqrt{6}k_1\sqrt{R(t,z)}}{\sqrt{Q(t,z)}}\sqrt{-r}\coth(\sqrt{-r}\xi) + \frac{i\sqrt{6}k_1\sqrt{R(t,z)}}{\sqrt{Q(t,z)}}$$
$$\times \sqrt{-r}\tanh(\sqrt{-r}\xi).$$

For $\varphi(\xi) = -\sqrt{-r}\cot(\sqrt{-r}\xi), r < 0$,

$$\Phi_{3,2} = \frac{i\sqrt{6}k_1\sqrt{R(t,z)}}{\sqrt{Q(t,z)}}\sqrt{-r}\tanh(\sqrt{-r}\xi) + \frac{i\sqrt{6}k_1\sqrt{R(t,z)}}{\sqrt{Q(t,z)}}$$
$$\times \sqrt{-r}\coth(\sqrt{-r}\xi).$$

For $\varphi(\xi) = -\sqrt{r}\tan(\sqrt{r}\xi), r > 0$,

$$\Phi_{3,3} = \frac{i\sqrt{6}k_1\sqrt{R(t,z)}}{\sqrt{Q(t,z)}}\sqrt{r}\cot(\sqrt{r}\xi) - \frac{i\sqrt{6}k_1\sqrt{R(t,z)}}{\sqrt{Q(t,z)}}$$
$$\times \sqrt{r}\tan(\sqrt{r}\xi).$$

For $\varphi(\xi) = -\sqrt{r}\cot(\sqrt{r}\xi), r > 0,$

$$\Phi_{3,4} = \frac{i\sqrt{6}k_1\sqrt{R(t,z)}}{\sqrt{Q(t,z)}}\sqrt{r}\tan(\sqrt{r}\xi) - \frac{i\sqrt{6}k_1\sqrt{R(t,z)}}{\sqrt{Q(t,z)}}$$
$$\times \ \sqrt{r}\coth(\sqrt{r}\xi).$$

For $\varphi(\xi) = -\frac{1}{\xi+\omega}, \omega$ is a constant, $r = 0,$

$$\Phi_{3,5} = \frac{i\sqrt{6}k_1\sqrt{R(t,z)}}{\sqrt{Q(t,z)}}(\xi+\omega) + \frac{i\sqrt{6}k_1\sqrt{R(t,z)}}{\sqrt{Q(t,z)}}\frac{1}{\xi+\omega}.$$

Set 4 solutions for Case 4.

For $\varphi(\xi) = -\sqrt{-r}\tanh(\sqrt{-r}\xi), r < 0,$

$$\Phi_{4,1} = \frac{i\sqrt{6}k_1\sqrt{R(t,z)}}{\sqrt{Q(t,z)}}\sqrt{-r}\coth(\sqrt{-r}\xi) - \frac{i\sqrt{6}k_1\sqrt{R(t,z)}}{\sqrt{Q(t,z)}}$$
$$\times \ \sqrt{-r}\tanh(\sqrt{-r}\xi).$$

For $\varphi(\xi) = -\sqrt{-r}\cot(\sqrt{-r}\xi), r < 0,$

$$\Phi_{4,2} = \frac{i\sqrt{6}k_1\sqrt{R(t,z)}}{\sqrt{Q(t,z)}}\sqrt{-r}\tanh(\sqrt{-r}\xi) - \frac{i\sqrt{6}k_1\sqrt{R(t,z)}}{\sqrt{Q(t,z)}}$$
$$\times \ \sqrt{-r}\coth(\sqrt{-r}\xi).$$

For $\varphi(\xi) = -\sqrt{r}\tan(\sqrt{-r}\xi), r > 0,$

$$\Phi_{4,3} = \frac{i\sqrt{6}k_1\sqrt{R(t,z)}}{\sqrt{Q(t,z)}}\sqrt{-r}\cot(\sqrt{r}\xi) + \frac{i\sqrt{6}k_1\sqrt{R(t,z)}}{\sqrt{Q(t,z)}}$$
$$\times \ \sqrt{r}\tan(\sqrt{-r}\xi).$$

For $\varphi(\xi) = -\sqrt{r}\cot(\sqrt{r}\xi), r > 0,$

$$\Phi_{4,4} = \frac{i\sqrt{6}k_1\sqrt{R(t,z)}}{\sqrt{Q(t,z)}}\sqrt{r}\tan(\sqrt{r}\xi) - \frac{i\sqrt{6}k_1\sqrt{R(t,z)}}{\sqrt{Q(t,z)}}$$
$$\times \ \sqrt{r}\coth(\sqrt{r}\xi).$$

For $\varphi(\xi) = -\frac{1}{\xi+\omega}, \omega$ is a constant, $r = 0,$

$$\Phi_{4,5} = \frac{i\sqrt{6}k_1\sqrt{R(t,z)}}{\sqrt{Q(t,z)}}(\xi+\omega) - \frac{i\sqrt{6}k_1\sqrt{R(t,z)}}{\sqrt{Q(t,z)}}\frac{1}{(\xi+\omega)}.$$

2.6.1.1 *Stochastic set of solutions*

Now, from Set 1, Set 2, Set 3 and Set 4 from Theorem 2.3.6.1 and according to Xie [30], there exists a bounded open set $G \subset \mathbb{R}_+ \times \mathbb{R}, q > 0$ and $r > 0$ such that $\Phi(t, x, z)$, $\Phi_t(t, x, z)$ and $\Phi_{xxx}(t, x, z)$ are uniformly bounded for $(t, x, z) \in G \times \mathrm{K}_q(r)$, continuous with respect to $(t, x) \in G$ for all $z \in \mathbb{K}_q(r)$ and analytic with respect to $z \in \mathbb{K}_q(r)$ for all $(t, x) \in G$. Theorem 2.3.6.1 also suggests that there exists $U(t, x) \in (S)_{-1}$ such that $\Phi(t, x, z) = (\mathcal{H}U(t, x))(z)$ for all $(t, x, z) \in G \times \mathrm{K}_q(r)$ and that $U(t, x)$ solves Eq. (2.9). Therefore, $U = \mathcal{H}^{-1}\Phi \in (S)_{-1}$ and thereby obtain a solution U of the original Wick-type SPDE (2.9).

Hence, Set 1, Set 2, Set 3 and Set 4 provide the stochastic solutions of Eq. (2.9) as follows:

Stochastic set of solutions for Set 1.

$$U_{1,1}(t, x) = \sqrt{-r}\frac{i\sqrt{6}k_1\sqrt{R(t)}}{\sqrt{Q(t)}} \diamond \coth^{\diamond}(\sqrt{-r}\Xi_{1,1}(t, x))$$

$$+ \sqrt{-r}\frac{i\sqrt{6}k_1\sqrt{R(t)}}{\sqrt{Q(t)}} \diamond \tanh^{\diamond}(\sqrt{-r}\Xi_{1,1}(t, x)),$$

where $\Xi_{1,1}(t, x) = \Psi_{1,1}(\xi(t, x))$.

$$U_{1,2}(t, x) = \sqrt{-r}\frac{i\sqrt{6}k_1\sqrt{R(t)}}{\sqrt{Q(t)}} \diamond \tanh^{\diamond}(\sqrt{-r}\Xi_{1,2}(t, x))$$

$$+ \sqrt{-r}\frac{i\sqrt{6}k_1\sqrt{R(t)}}{\sqrt{Q(t)}} \diamond \coth^{\diamond}(\sqrt{-r}\Xi_{1,2}(t, x)),$$

where $\Xi_{1,2}(t, x) = \Psi_{1,2}(\xi(t, x))$.

$$U_{1,3}(t, x) = -\sqrt{r}\frac{i\sqrt{6}k_1\sqrt{R(t)}}{\sqrt{Q(t)}} \diamond \cot^{\diamond}(\sqrt{-r}\Xi_{1,3}(t, x))$$

$$- \sqrt{r}\frac{i\sqrt{6}k_1\sqrt{R(t)}}{\sqrt{Q(t)}} \diamond \tan^{\diamond}(\sqrt{-r}\Xi_{1,3}(t, x)),$$

where $\Xi_{1,3}(t,x) = \Psi_{1,3}(\xi(t,x))$.

$$U_{1,4}(t,x) = \sqrt{r}\,\frac{i\sqrt{6}k_1\sqrt{R(t)}}{\sqrt{Q(t)}}\Diamond \tan^{\Diamond}(\sqrt{r}\,\Xi_{1,4}(t,x))$$

$$+ \frac{i\sqrt{6}k_1\sqrt{R(t)}}{\sqrt{Q(t)}}\Diamond \cot^{\Diamond}(\sqrt{r}\,\Xi_{1,4}(t,x)),$$

where $\Xi_{1,4}(t,x) = \Psi_{1,4}(\xi(t,x))$.

$$U_{1,5}(t,x) = \frac{i\sqrt{6}k_1\sqrt{R(t)}}{\sqrt{Q(t)}}(\Xi_{1,5}(t,x) + \omega)$$

$$+ \frac{i\sqrt{6}k_1\sqrt{R(t)}}{\sqrt{Q(t)}}\frac{1}{(\Xi_{1,5}(t,x) + \omega)},$$

where $\Xi_{1,5}(t,x) = \Psi_{1,5}(\xi(t,x))$.

Stochastic set of solutions for Set 2.

$$U_{2,1}(t,x) = \sqrt{-r}\,\frac{i\sqrt{6}k_1\sqrt{R(t)}}{\sqrt{Q(t)}}\Diamond \coth^{\Diamond}(\sqrt{-r}\,\Xi_{2,1}(t,x))$$

$$- \sqrt{-r}\,\frac{i\sqrt{6}k_1\sqrt{R(t)}}{\sqrt{Q(t)}}\Diamond \tanh^{\Diamond}(\sqrt{-r}\,\Xi_{2,1}(t,x)),$$

where $\Xi_{2,1}(t,x) = \Psi_{2,1}(\xi(t,x))$.

$$U_{2,2}(t,x) = \sqrt{-r}\,\frac{i\sqrt{6}k_1\sqrt{R(t)}}{\sqrt{Q(t)}}\Diamond \tanh^{\Diamond}(\sqrt{-r}\,\Xi_{2,2}(t,x))$$

$$- \sqrt{-r}\,\frac{i\sqrt{6}k_1\sqrt{R(t)}}{\sqrt{Q(t)}}\Diamond \coth^{\Diamond}(\sqrt{-r}\,\Xi_{2,2}(t,x)),$$

where $\Xi_{2,2}(t,x) = \Psi_{2,2}(\xi(t,x))$.

$$U_{2,3}(t,x) = \sqrt{-r}\,\frac{i\sqrt{6}k_1\sqrt{R(t)}}{\sqrt{Q(t)}}\Diamond \cot^{\Diamond}(\sqrt{r}\,\Xi_{2,3}(t,x))$$

$$+ \sqrt{r}\,\frac{i\sqrt{6}k_1\sqrt{R(t)}}{\sqrt{Q(t)}}\Diamond \tan^{\Diamond}(\sqrt{r}\,\Xi_{2,3}(t,x)),$$

where $\Xi_{2,3}(t,x) = \Psi_{2,3}(\xi(t,x))$.

$$U_{2,4}(t,x) = \sqrt{r}\frac{i\sqrt{6}k_1\sqrt{R(t)}}{\sqrt{Q(t)}} \Diamond \tan^\Diamond(\sqrt{r}\Xi_{2,4}(t,x))$$
$$- \sqrt{r}\frac{i\sqrt{6}k_1\sqrt{R(t)}}{\sqrt{Q(t)}} \Diamond \cot^\Diamond(\sqrt{r}\Xi_{2,4}(t,x)),$$

where $\Xi_{2,4}(t,x) = \Psi_{2,4}(\xi(t,x))$.

$$U_{2,5}(t,x) = \frac{i\sqrt{6}k_1\sqrt{R(t)}}{\sqrt{Q(t)}}(\Xi_{2,5}(t,x) + \omega)$$
$$- \frac{i\sqrt{6}k_1\sqrt{R(t)}}{\sqrt{Q(t)}}\frac{1}{(\Xi_{2,5}(t,x) + \omega)},$$

where $\Xi_{2,5}(t,x) = \Psi_{2,5}(\xi(t,x))$.

Stochastic set of solutions for Set 3.

$$U_{3,1}(t,x) = \sqrt{-r}\frac{i\sqrt{6}k_1\sqrt{R(t)}}{\sqrt{Q(t)}} \Diamond \coth^\Diamond(\sqrt{-r}\Xi_{1,1}(t,x))$$
$$+ \sqrt{-r}\frac{i\sqrt{6}k_1\sqrt{R(t)}}{\sqrt{Q(t)}} \Diamond \tanh^\Diamond(\sqrt{-r}\Xi_{1,2}(t,x)),$$

where $\Xi_{3,1}(t,x) = \Psi_{3,1}(\xi(t,x))$.

$$U_{3,2}(t,x) = \sqrt{-r}\frac{i\sqrt{6}k_1\sqrt{R(t)}}{\sqrt{Q(t)}} \Diamond \tanh^\Diamond(\sqrt{-r}\Xi_{3,2}(t,x))$$
$$+ \sqrt{-r}\frac{i\sqrt{6}k_1\sqrt{R(t)}}{\sqrt{Q(t)}} \Diamond \coth^\Diamond(\sqrt{-r}\Xi_{3,2}(t,x)),$$

where $\Xi_{3,2}(t,x) = \Psi_{3,2}(\xi(t,x))$.

$$U_{3,3}(t,x) = \sqrt{r}\frac{i\sqrt{6}k_1\sqrt{R(t)}}{\sqrt{Q(t)}} \Diamond \cot^\Diamond(\sqrt{r}\Xi_{3,3}(t,x))$$
$$- \sqrt{r}\frac{i\sqrt{6}k_1\sqrt{R(t)}}{\sqrt{Q(t)}} \Diamond \tan^\Diamond(\sqrt{-r}\Xi_{3,3}(t,x)),$$

where $\Xi_{3,3}(t,x) = \Psi_{3,3}(\xi(t,x))$.

$$U_{3,4}(t,x) = -\sqrt{r}\,\frac{i\sqrt{6}k_1\sqrt{R(t)}}{\sqrt{Q(t)}}\Diamond\tan^{\Diamond}(\sqrt{-r}\Xi_{3,4}(t,x))$$
$$+ \sqrt{r}\,\frac{i\sqrt{6}k_1\sqrt{R(t)}}{\sqrt{Q(t)}}\Diamond\cot^{\Diamond}(\sqrt{-r}\Xi_{3,4}(t,x)),$$

where $\Xi_{3,4}(t,x) = \Psi_{3,4}(\xi(t,x))$.

$$U_{3,5}(t,x) = -\frac{i\sqrt{6}k_1\sqrt{R(t)}}{\sqrt{Q(t)}}(\Xi_{3,5}(t,x)+\omega)$$
$$+ \frac{i\sqrt{6}k_1\sqrt{R(t)}}{\sqrt{Q(t)}}\frac{1}{(\Xi_{3,5}(t,x)+\omega)},$$

where $\Xi_{3,5}(t,x) = \Psi_{3,5}(\xi(t,x))$.

Stochastic set of solutions for Set 4.

$$U_{4,1}(t,x) = -\sqrt{-r}\,\frac{i\sqrt{6}k_1\sqrt{R(t)}}{\sqrt{Q(t)}}\Diamond\coth^{\Diamond}(\sqrt{-r}\Xi_{4,1}(t,x))$$
$$- \sqrt{-r}\,\frac{i\sqrt{6}k_1\sqrt{R(t)}}{\sqrt{Q(t)}}\Diamond\tanh^{\Diamond}(\sqrt{-r}\Xi_{4,1}(t,x)),$$

where $\Xi_{4,1}(t,x) = \Psi_{4,1}(\xi(t,x))$.

$$U_{4,2}(t,x) = -\sqrt{-r}\,\frac{i\sqrt{6}k_1\sqrt{R(t)}}{\sqrt{Q(t)}}\Diamond\tanh^{\Diamond}(\sqrt{-r}\Xi_{4,2}(t,x))$$
$$- \sqrt{-r}\,\frac{i\sqrt{6}k_1\sqrt{R(t)}}{\sqrt{Q(t)}}\Diamond\coth^{\Diamond}(\sqrt{-r}\Xi_{4,2}(t,x)),$$

where $\Xi_{4,2}(t,x) = \Psi_{4,2}(\xi(t,x))$.

$$U_{4,3}(t,x) = \sqrt{r}\,\frac{i\sqrt{6}k_1\sqrt{R(t)}}{\sqrt{Q(t)}}\Diamond\cot^{\Diamond}(\sqrt{r}\Xi_{4,3}(t,x))$$
$$+ \sqrt{r}\,\frac{i\sqrt{6}k_1\sqrt{R(t)}}{\sqrt{Q(t)}}\Diamond\tan^{\Diamond}(\sqrt{-r}\Xi_{4,3}(t,x)),$$

where $\Xi_{4,3}(t,x) = \Psi_{4,3}(\xi(t,x))$.

$$U_{4,4}(t,x) = -\sqrt{r}\frac{i\sqrt{6}k_1\sqrt{R(t)}}{\sqrt{Q(t)}}\lozenge \tan^{\lozenge}(\sqrt{-r}\Xi_{4,4}(t,x))$$
$$-\sqrt{r}\frac{i\sqrt{6}k_1\sqrt{R(t)}}{\sqrt{Q(t)}}\lozenge \cot^{\lozenge}(\sqrt{-r}\Xi_{4,4}(t,x)),$$

where $\Xi_{4,4}(t,x) = \Psi_{4,4}(\xi(t,x))$.

$$U_{4,5}(t,x) = -\frac{i\sqrt{6}k_1\sqrt{R(t)}}{\sqrt{Q(t)}}(\Xi_{4,5}(t,x) + \omega)$$
$$+\frac{i\sqrt{6}k_1\sqrt{R(t)}}{\sqrt{Q(t)}}\frac{1}{(\Xi_{4,5}(t,x) + \omega)},$$

where $\Xi_{4,5}(t,x) = \Psi_{4,5}(\xi(t,x))$.

2.7 JEF Expansion Method for Solutions of Wick-Type Kersten–Krasil'shchik Coupled KdV-mKdV Equations

2.7.1 *JEF expansion method*

The JEF expansion method [90–94] has been presented step by step as follows:

Step 1. Consider a nonlinear PDE of the following form:

$$P(U, U_t, U_x, \ldots) = 0, \tag{2.58}$$

where $U(t,x)$ is the unknown function. P is the function in $U(t,x)$ along with their highest order partial derivatives and nonlinear terms of $U(t,x)$, respectively.

Step 2. The travelling wave transformation of Eq. (2.58) is

$$U(t,x) = \Phi(\xi), \xi(x,t,z) = kx + c\int_0^t l(\tau,z)d\tau + \xi_0, \tag{2.59}$$

where k and c are constants, which are to be determined later.

Using Eq. (2.59), the PDE (2.58) can be transformed into the following nonlinear ODE, which is given as

$$P(\Phi, \Phi'(\xi), \Phi'(\xi), \Phi'(\xi), \ldots) = 0. \tag{2.60}$$

Step 3. The exact solutions of Eq. (2.58) are assumed in the polynomial $\Phi(\xi)$ as follows:

$$\Phi(\xi) = \sum_{i=0}^{N} a_i(t, z) sn^i \xi, \tag{2.61}$$

where $a_i(t, z)(i = 0, 1, 2, \ldots, n)$ are functions to be determined later and $sn(\xi)$ is a Jacobian elliptic sine function.

Step 4. Now, $\Phi = \xi^{-p}$ is substituted in Eq. (2.60) to obtain the highest order singularity. Then, the degrees of all terms of Eq. (2.60) are taken under consideration and the terms having lower degree are selected. The maximum value of p is assumed to be the pole and is denoted as N. The Jacobi elliptic function expansion method can only be implemented for integer values of N. However, if N is not an integer, then Eq. (2.60) can be transformed and the above steps can be repeated.

Step 5. The derivatives of $\Phi(\xi)$ are presented in the following:

$$\Phi_\xi(\xi) = \sum_{i=0}^{N} i a_i(t, z) sn^{i-1}\xi cn\xi dn\xi,$$

$$\Phi_{\xi\xi}(\xi) = \sum_{i=0}^{N} i a_i(t, z)(-sn^{i-2}\xi(dn^2\xi sn^2\xi$$

$$+ \, cn^2\xi(-(i-1)dn^2\xi + m^2 sn^2\xi))), \tag{2.62}$$

$$\Phi_{\xi\xi\xi}(\xi) = \sum_{i=0}^{N} i a_i(t, z)(cn\,\xi dn\,\xi sn^{i-3}\xi(sn^2\xi((2 - 3i)dn^2\xi + 4m^2 sn^2\xi)$$

$$+ \, cn^2\xi((2 - 3i + i^2)dn^2\xi + (2 - 3i)m^2 sn^2\xi))).$$

Step 6. Substituting Eq. (2.62) into Eq. (2.60) and equating the coefficients of $\varphi^i (i = 0, 1, 2, \dots)$ with zero, a set of algebraic equations is obtained. The obtained algebraic equations are solved, to obtain the unknowns $a_i(i = 0, 1, 2, \dots, N)$ and other constants. Then, all the obtained values of the unknowns are substituted in Eq. (2.61) yielding the exact solution for Eq. (2.58) instantly.

2.7.2 *Solutions of the stochastic Kersten–Krasil'shchik coupled KdV-mKdV equations*

In this section, the JEF expansion method is employed here for finding the white noise solutions for the stochastic Kersten–Krasil'shchik coupled KdV-mKdV equations (2.11).

Applying Hermite transform on Eq. (2.11), it can be obtained as

$$\tilde{U}_t + H_1(t, z)\tilde{U}_{xxx} - 6H_2(t, z)\tilde{U}\tilde{U}_x + 3H_3(t, z)\tilde{V}\tilde{V}_{xxx}$$

$$+ \, 3H_4(t, z)\tilde{V}_x\tilde{V}_{xx} - 3\tilde{U}_x\tilde{V}^2 - 6\tilde{U}\tilde{V}\tilde{V}_x = 0, \tag{2.63}$$

$$\tilde{V}_t + \tilde{V}_{xxx} - 3\tilde{V}^2\tilde{V}_x - 3\tilde{U}\tilde{V}_x - 3\tilde{U}_x\tilde{V} = 0,$$

where $\tilde{U} \equiv \tilde{U}(t, x, z) = \mathcal{H}(U(t, x))$, $\tilde{V} \equiv \tilde{V}(t, x, z) = \mathcal{H}(V, t, x))$, $H_1(t, z) = \mathcal{H}(H_1(t))$, $H_2(t, z) = \mathcal{H}(H_2(t))$, $H_3(t, z) = \mathcal{H}(H_3(t))$ and $H_4(t, z) = \mathcal{H}(H_4(t))$, where $z = (z_1, z_2, \dots) \in \mathbb{C}^N$ is a parameter.

Now, let $\tilde{U}(t, x, z) = \Phi(\xi, z)$, $\tilde{V}(t, x, z) = \Psi(\xi, z)$, $\xi(x, t, z) = kx + c\int_0^t l(\tau, z)d\tau + \xi_0$, then Eq. (2.63) becomes

$$cl(\tau, z)\Phi_\xi + k^3 H_1\Phi_{\xi\xi\xi} - 6kH_2\Phi\Phi_\xi + 3k^3 H_3\Psi\Psi_{\xi\xi\xi}$$

$$+ \, 3k^3 H_4\Psi_\xi\Psi_{\xi\xi} - 3k\Phi_\xi\Psi^2 - 6k\Phi\Psi\Psi_\xi = 0, \tag{2.64}$$

$$cl(\tau, z)\Psi_\xi + k^3\Psi_{\xi\xi} - 3k\Psi^2\Psi_\xi - 3k\Phi\Psi_\xi - 3k\Phi\Psi_\xi - 3k\Phi_\xi\Psi = 0.$$

Consider the ansatz in the following form:

$$\Phi(\xi, z) = \sum_{i=0}^{N} a_i(t, z)sn^i\xi,$$

$$\Psi(\xi, z) = \sum_{i=0}^{n} b_i(t, z)sn^i\xi. \tag{2.65}$$

Now, according to the homogenous balancing principle, in Eq. (2.64), $k^3 H_1 \Phi_{\xi\xi\xi}$ and $6k H_2 \Phi\Phi_\xi$ are the dominant terms having the highest order singularity. Similarly, $k^3 \Psi_{\xi\xi\xi}$ and $3k\Psi^2\Psi_\xi$ are the dominant terms having the highest order singularity. Thus, the maximum values of the poles are 2 and 1 respectively which imply here $N = 2$ and $n = 1$.

From Eq. (2.65), the ansatz becomes

$$\Phi(\xi, z) = a_0(t, z) + a_1(t, z)sn\xi + a_2(t, z)sn^2\xi,$$

$$\Psi(\xi, z) = b_0(t, z) + b_1(t, z)sn\xi. \tag{2.66}$$

Substituting Eq. (2.66) in Eq. (2.64) and equating each coefficient of $sn^i(\xi)$, $sn^i\xi cn\xi dn\xi (i = 0, 1, 2, \ldots)$ with zero, a system of algebraic equations are obtained for $k, c, a_0(t, z), a_1(t, z), a_2(t, z), b_0(t, z)$ and $b_1(t, z)$.

$$cn\xi dn\xi : -H_1(t, z)k^3 a_1(t, z) - H_1(t, z)k^3 m^2 a_1(t, z) + cl(t, z)a_1$$
$$- 6H_2(t, z)ka_0(t, z)a_1(t, z) - 3ka_1(t, z)b_0^2(t, z)$$
$$- 3H_3(t, z)k^3 b_0(t, z)b_1(t, z) - 3H_3(t, z)k^3 m^2 b_0(t, z)b_1(t, z)$$
$$- 6ka_0(t, z)b_0(t, z)b_1(t, z) = 0,$$

$$sn\xi cn\xi dn\xi : -6H_2(t, z)ka_1^2(t, z) - 8H_1(t, z)k^3 a_2(t, z)$$
$$- 8H_1(t, z)k^3 m^2 a_2(t, z) + 2cl(t, z)a_2(t, z)$$
$$-12H_2(t, z)ka_0(t, z)a_2(t, z) - 6ka_2(t, z)b_0^2(t, z)$$
$$- 12ka_1(t, z)b_0(t, z)b_1(t, z) - 3H_3(t, z)k^3 b_1^2(t, z)$$
$$-3H_4(t, z)k^3 b_1^2(t, z) - 3H_3(t, z)k^3 m^2 b_1^2(t, z)$$
$$- 3H_4(t, z)k^3 m^2 b_1^2(t, z) - 6ka_0(t, z)b_1^2(t, z) = 0,$$

$$sn^2\xi cn\xi dn\xi : 6H_1(t, z)k^3 m^2 a_1(t, z) - 18H_2(t, z)ka_1(t, z)a_2(t, z)$$
$$+18H_3 k^3 m^2 b_0(t, z)b_1(t, z) - 18ka_2(t, z)b_0(t, z)b_1(t, z)$$
$$-9ka_1(t, z)b_1^2(t, z) = 0,$$

$$sn^3\xi cn\xi dn\xi : 24H_1(t, z)k^3 m^2 a_2(t, z) - 12H_2(t, z)ka_2^2(t, z)$$
$$+18H_3(t, z)k^3 m^2 b_1^2(t, z) - 6H_4(t, z)k^3 m^2 b_1^2(t, z)$$
$$-12ka_2(t, z)b_1^2(t, z) = 0,$$

$$cn\xi dn\xi : -3ka_1(t,z)b_0(t,z) - k^3 b_1(t,z) - k^3 m^2 b_1(t,z)$$
$$+ cl(t,z)b_1(t,z) - 3ka_0(t,z)b_1(t,z) - 3kb_0^2(t,z)b_1(t,z) = 0,$$
$$sn\xi cn\xi dn\xi : -6ka_2(t,z)b_0(t,z) - 6ka_1(t,z)b_1(t,z)$$
$$- 6kb_0(t,z)b_1^2(t,z) = 0,$$
$$sn^2\xi cn\xi dn\xi : 6k^3 m^2 b_1(t,z) - 9ka_2(t,z)b_1(t,z) - 3kb_1^3(t,z) = 0.$$
$$(2.67)$$

2.7.3 *Results and discussion*

Solving the above algebraic equation (2.67), a set of coefficients for the solution of Eq. (2.64) have been obtained.

Let $N(t,z) = -2 + H_2(t,z)$,

$$F(t,z) = 16 + 16H_1^2(t,z) + 81H_3^3(t,z) - 24H_4(t,z)$$
$$+ 16H_2(t,z)H_4(t,z) + 9H_4^2(t,z)$$
$$- 8H_1(t,z)(4 + 9H_3(t,z) + 3H_4(t,z))$$
$$+ 6H_3(t,z)(-12 + 8H_2(t,z)$$
$$+ 9H_4(t,z)),$$

$$M(t,z) = \sqrt{F(t,z)},$$
$$K(t,z) = (-4 + 4H_1(t,z)N(t,z) - 9N(t,z)H_3(t,z)$$
$$+ 6H_4(t,z) - 3H_2(t,z)H_4(t,z)),$$
$$R(t,z) = (-2 + 8H_1 - 9H_3(t,z) - 9H_4(t,z)),$$
$$G(t,z) = (6 + 8H_1(t,z) - 4H_2(t,z) - 9H_3(t,z) - 9H_4(t,z)),$$
$$H(t,z) = 8 + 27H_3^2(t,z) + 6(-3 + 2H_2(t,z))H_4(t,z) + 9H_4^2(t,z)$$
$$+ H_3(t,z)(-30 + 20H_2(t,z) + 36H_4(t,z)),$$

$$P(t,z) = -k^2(1 + m^2)$$
$$\times \left[\frac{\begin{array}{c}(32H_1^2(t,z) - 4H_1(t,z)(2 + 4H_2(t,z) + 27H_3(t,z) + 15H_4(t,z)) \\ + 3H(t,z) + G(t,z)M(t,z)\end{array}}{4(K(t,z) + N(t,z)M(t,z))}\right],$$

$$Q(t,z) = k^3(1+m^2)$$

$$\times \left[\frac{(32H_1^2(t,z) - 4H_1(t,z)(2 + 4H_2(t,z) + 27H_3(t,z) + 15H_4(t,z))}{4(-K(t,z) + N(t,z)M(t,z))} \right].$$

Set 1. For $l(t,z) = \frac{1}{c}P(t,z)$ and $\xi = kx + \int_0^t P(\tau,z)d\tau\xi_0$, the following results have been obtained:

Case 1.

$$a_0(t,z) = -k^2(1+m^2)\left[\frac{\begin{array}{l}8 + 32H_1^2(t,z) + 81H_3^2(t,z) - 30H_4(t,z)\\ +24H_2(t,z)H_4(t,z) + 27H_4^2(t,z)\end{array}}{12(K(t,z) + N(t,z)M(t,z))}\right.$$

$$\left. + \frac{\begin{array}{l}-4H_1(t,z)(10 + 27H_3(t,z) + 15H_4(t,z)) + 6H_3(t,z)(-3 + 4H_2(t,z)\\ +18H_4(t,z)) + R(t,z)M(t,z)\end{array}}{12(K(t,z) + N(t,z)M(t,z))}\right],$$

$$a_1(t,z) = 0, \quad a_2(t,z) = \frac{(-4 + 4H_1(t,z) - 9H_3(t,z) - 3H_4(t,z)) + M(t,z)k^2m^2}{4(-3 + H_2(t,z))},$$

$$b_0(t,z) = 0, \quad b_1(t,z) = -\frac{1}{2}\sqrt{\frac{\begin{array}{l}((-12 - 12H_1(t,z) + 8H_2(t,z) + 27H_3(t,z)\\ +9H_4(t,z)) - 3M(t,z))k^3m^2\end{array}}{(-3 + H_2(t,z))}}.$$

Thus, the following double periodic wave solutions of Eq. (2.63) have been obtained in terms of Jacobi elliptic function:

$$U(\xi,z) = -k^2(1+m^2)\left[\frac{\begin{array}{l}8 + 32H_1^2(t,z) + 81H_3^2(t,z) - 30H_4(t,z)\\ +24H_2(t,z)H_4(t,z) + 27H_4^2(t,z)\end{array}}{12(K(t,z) + N(t,z)M(t,z))}\right.$$

$$\left. + \frac{\begin{array}{l}-4H_1(t,z)(10 + 27H_3(t,z) + 15H_4(t,z))\\ +6H_3(t,z)(-3 + 4H_2(t,z) + 18H_4(t,z)) + R(t,z)M(t,z)\end{array}}{12(K(t,z) + N(t,z)M(t,z))}\right],$$

$$+ \frac{(-4 + 4H_1(t, z) - 9H_3(t, z) - 3H_4(t, z)) + M(t, z)k^2 m^2}{4(-3 + H_2(t, z))} sn^2 \xi,$$

$$\text{and} \quad V(\xi, z) = -\frac{1}{2} \sqrt{\frac{((-12 - 12H_1(t, z) + 8H_2(t, z) + 27H_3(t, z) + 9H_4(t, z)) - 3M(t, z))k^3 m^2}{(-3 + H_2(t, z))}} sn\xi.$$

Case 2.

$$a_0(t, z) = -k^2(1 + m^2) \left[\frac{8 + 32H_1^2(t, z) + 81H_3^2(t, z) - 30H_4(t, z) + 24H_2(t, z)H_4(t, z) + 27H_4^2(t, z)}{12(K(t, z) + N(t, z)M(t, z))} \right.$$

$$\left. + \frac{-4H_1(t, z)(10 + 27H_3(t, z) + 15H_4(t, z)) + 6H_3(t, z)(-3 + 4H_2(t, z) + 18H_4(t, z)) + R(t, z)M(t, z)}{12(K(t, z) + N(t, z)M(t, z))} \right],$$

$$a_1(t, z) = 0, \quad a_2(t, z) = \frac{(-4 + 4H_1(t, z) - 9H_3(t, z) - 3H_4(t, z)) + M(t, z)k^2 m^2}{4(-3 + H_2(t, z))},$$

$$b_0(t, z) = 0, \quad b_1(t, z) = -\frac{1}{2} \sqrt{\frac{((-12 - 12H_1(t, z) + 8H_2(t, z) + 27H_3(t, z) + 9H_4(t, z)) - 3M(t, z))k^3 m^2}{(-3 + H_2(t, z))}}.$$

Here, the following double periodic wave solutions of Eq. (2.63) have been obtained in terms of Jacobi elliptic function:

$$U(\xi, z) = -k^2(1 + m^2) \left[\frac{8 + 32H_1^2(t, z) + 81H_3^2(t, z) - 30H_4(t, z) + 24H_2(t, z)H_4(t, z) + 27H_4^2(t, z)}{12(K(t, z) + N(t, z)M(t, z))} \right.$$

$$\left. + \frac{-4H_1(t, z)(10 + 27H_3(t, z) + 15H_4(t, z)) + 6H_3(t, z)(-3 + 4H_2(t, z) + 18H_4(t, z)) + R(t, z)M(t, z)}{12(K(t, z) + N(t, z)M(t, z))} \right],$$

$$+ \frac{(-4 + 4H_1(t, z) - 9H_3(t, z) - 3H_4(t, z)) + M(t, z)k^2 m^2}{4(-3 + H_2(t, z))} sn^2 \xi,$$

$$\text{and} \quad V(\xi, z) = -\frac{1}{2} \sqrt{\frac{((-12 - 12H_1(t, z) + 8H_2(t, z) + 27H_3(t, z) + 9H_4(t, z)) - 3M(t, z))k^3 m^2}{(-3 + H_2(t, z))}} sn\xi.$$

Set 2. For $l(t, z) = \frac{1}{c}Q(t, z)$ and $\xi = kx + \int_0^t Q(\tau, z)d\tau + \xi_0$, the following results have been obtained.

Case 1.

$$a_0(t, z) = -k^2(1 + m^2)\left[\frac{\begin{array}{c}8 + 32H_1^2(t, z) + 81H_3^2(t, z) - 30H_4(t, z) \\ + 24H_2(t, z)H_4(t, z) + 27H_4^2(t, z)\end{array}}{12(-K(t, z) + N(t, z)M(t, z))}\right.$$

$$\left.+ \frac{\begin{array}{c}-4H_1(t, z)(10 + 27H_3(t, z) + 15H_4(t, z)) \\ + 6H_3(t, z)(-3 + 4H_2(t, z) + 18H_4(t, z)) + R(t, z)M(t, z)\end{array}}{12(-K(t, z) + N(t, z)M(t, z))}\right],$$

$$a_1(t, z) = 0, \quad a_2(t, z) = \frac{(-4 + 4H_1(t, z) - 9H_3(t, z) - 3H_4(t, z)) + M(t, z)k^2m^2}{4(-3 + H_2(t, z))},$$

$$b_0(t, z) = 0, \quad b_1(t, z) = -\frac{1}{2}\sqrt{\frac{\begin{array}{c}((-12 - 12H_1(t, z) + 8H_2(t, z) + 27H_3(t, z) \\ + 9H_4(t, z)) + 3M(t, z))k^3m^2\end{array}}{(-3 + H_2(t, z))}}.$$

Here, the following double periodic wave solutions of Eq. (2.63) have been obtained in terms of Jacobi elliptic function:

$$U(\xi, z) = -k^2(1 + m^2)\left[\frac{\begin{array}{c}8 + 32H_1^2(t, z) + 81H_3^2(t, z) - 30H_4(t, z) \\ + 24H_2(t, z)H_4(t, z) + 27H_4^2(t, z)\end{array}}{12(K(t, z) + N(t, z)M(t, z))}\right.$$

$$\left.+ \frac{\begin{array}{c}-4H_1(t, z)(10 + 27H_3(t, z) + 15H_4(t, z)) + 6H_3(t, z)(-3 \\ + 4H_2(t, z) + 18H_4(t, z)) - R(t, z)M(t, z)\end{array}}{12(-K(t, z) + N(t, z)M(t, z))}\right],$$

$$+ \frac{(-4 + 4H_1(t, z) - 9H_3(t, z) - 3H_4(t, z)) + M(t, z)k^2m^2}{4(-3 + H_2(t, z))}sn^2\xi,$$

$$\text{and} \quad V(\xi, z) = -\frac{1}{2}\sqrt{\frac{\begin{array}{c}((-12 - 12H_1(t, z) + 8H_2(t, z) + 27H_3(t, z) \\ + 9H_4(t, z)) + 3M(t, z))k^3m^2\end{array}}{(-3 + H_2(t, z))}}\,sn\xi.$$

Case 2.

$$a_0(t,z) = k^2(1+m^2)\left[\frac{\begin{array}{c}8+32H_1^2(t,z)+81H_3^2(t,z)-30H_4(t,z)\\+24H_2(t,z)H_4(t,z)+27H_4^2(t,z)\end{array}}{12(-K(t,z)+N(t,z)M(t,z))}\right.$$

$$\left.+\frac{\begin{array}{c}-4H_1(t,z)(10+27H_3(t,z)+15H_4(t,z))\\+6H_3(t,z)(-3+4H_2(t,z)+18H_4(t,z))+R(t,z)M(t,z)\end{array}}{12(-K(t,z)+N(t,z)M(t,z))}\right],$$

$$a_1(t,z)=0,\quad a_2(t,z)=\frac{(-4+4H_1(t,z)-9H_3(t,z)-3H_4(t,z))+M(t,z)k^2m^2}{4(-3+H_2(t,z))},$$

$$b_0(t,z)=0,\quad b_1(t,z)=-\frac{1}{2}\sqrt{\frac{\begin{array}{c}((-12-12H_1(t,z)+8H_2(t,z)+27H_3(t,z)\\+9H_4(t,z))+3M(t,z))k^3m^2\end{array}}{(-3+H_2(t,z))}}.$$

Thus, the following double periodic wave solutions of Eq. (2.63) have been obtained:

$$U(\xi,z)=-k^2(1+m^2)\left[\frac{\begin{array}{c}8+32H_1^2(t,z)+81H_3^2(t,z)-30H_4(t,z)\\+24H_2(t,z)H_4(t,z)+27H_4^2(t,z)\end{array}}{12(K(t,z)+N(t,z)M(t,z))}\right.$$

$$\left.+\frac{\begin{array}{c}-4H_1(t,z)(10+27H_3(t,z)+15H_4(t,z))+6H_3(t,z)(-3\\+4H_2(t,z)+18H_4(t,z))-R(t,z)M(t,z)\end{array}}{12(-K(t,z)+N(t,z)M(t,z))}\right],$$

$$+\frac{(-4+4H_1(t,z)-9H_3(t,z)-3H_4(t,z))+M(t,z)k^2m^2}{4(-3+H_2(t,z))}sn^2\xi,$$

$$\text{and}\quad V(\xi,z)=-\frac{1}{2}\sqrt{\frac{\begin{array}{c}((-12-12H_1(t,z)+8H_2(t,z)+27H_3(t,z)\\+9H_4(t,z))+3M(t,z))k^3m^2\end{array}}{(-3+H_2(t,z))}}sn\xi.$$

2.7.3.1 *Stochastic set of solutions*

Using the definition of inverse Hermite transform [15] and according to Xie [30] from Set 1 and Set 2, the following stochastic solutions of Eq. (2.11) have been obtained:

Set 1. For $l(t) = \frac{1}{c}P(t)$ and $\Xi(t,x) = kx + \int_0^t P(\tau)d\tau + \Xi_0$, where $\Xi(t,x) = \mathcal{H}^{-1}(\xi(t,x,z))$, the following results have been obtained:

Case 1.

$$U(x,t) = -k^2(1+m^2)\left[\frac{8 + 32H_1^{\Diamond 2}(t) + 81H_3^{\Diamond 2}(t) - 30H_4(t) + 24H_2(t)\Diamond H_4(t) + 27H_4^{\Diamond 2}(t)}{12(K(t) + N(t)\Diamond M(t))}\right.$$

$$\left. + \frac{-4H_1(t)\Diamond(10 + 27H_3(t) + 15H_4(t)) + 6H_3(t)\Diamond(-3 + 4H_2(t) + 18H_4(t)) + R(t)\Diamond M(t)}{12(K(t) + N(t)\Diamond M(t))}\right],$$

$$+ \frac{(-4 + 4H_1(t) - 9H_3(t) - 3H_4(t)) + M(t)k^2m^2}{4(-3 + H_2(t))}sn^{\Diamond 2}\xi,$$

$$\text{and} \quad V(x,t) = -\frac{1}{2}\sqrt{\frac{((-12 - 12H_1(t) + 8H_2(t) + 27H_3(t) + 9H_4(t)) - 3M(t))k^3m^2}{(-3 + H_2(t))}}sn^{\Diamond}\Xi.$$

Case 2.

$$U(x,t) = k^2(1+m^2)\left[\frac{8 + 32H_1^{\Diamond 2}(t) + 81H_3^{\Diamond 2}(t) - 30H_4(t) + 24H_2(t)\Diamond H_4(t) + 27H_4^{\Diamond 2}(t)}{12(K(t) + N(t)\Diamond M(t))}\right.$$

$$\left. + \frac{-4H_1(t)\Diamond(10 + 27H_3(t) + 15H_4(t)) + 6H_3(t)\Diamond(-3 + 4H_2(t) + 18H_4(t)) + R(t)\Diamond M(t)}{12(K(t) + N(t)\Diamond M(t))}\right],$$

$$+ \frac{(-4 + 4H_1(t) - 9H_3(t) - 3H_4(t)) + M(t,z)k^2m^2}{4(-3 + H_2(t))}sn^{\Diamond 2}\Xi,$$

$$\text{and} \quad V(x,t) = -\frac{1}{2}\sqrt{\frac{((-12 - 12H_1(t) + 8H_2(t) + 27H_3(t) + 9H_4(t)) - 3M(t))k^3m^2}{(-3 + H_2(t))}}sn^{\Diamond}\Xi.$$

Set 2. For $l(t) = \frac{1}{c}Q(t)$ and $\Xi(t,x) = kx + \int_0^t Q(\tau)d\tau + \Xi_0$, where $\Xi(t,x) = \mathcal{H}^{-1}(\xi(t,x,z))$, the following results have been obtained:

Case 1.

$$U(x,t) = -k^2(1+m^2)\left[\frac{8 + 32H_1^{\Diamond 2}(t) + 81H_3^{\Diamond 2}(t) - 30H_4(t) + 24H_2(t)\Diamond H_4(t) + 27H_4^{\Diamond 2}(t)}{12(K(t) + N(t)\Diamond M(t))}\right.$$

$$\left.+ \frac{-4H_1(t)\Diamond(10 + 27H_3(t) + 15H_4(t)) + 6H_3(t)\Diamond(-3 + 4H_2(t) + 18H_4(t)) - R(t)\Diamond M(t)}{12(K(t) + N(t)\Diamond M(t))}\right],$$

$$+ \frac{(-4 + 4H_1(t) - 9H_3(t) - 3H_4(t)) + M(t)k^2m^2}{4(-3 + H_2(t))}sn^{\Diamond 2}\Xi,$$

$$\text{and} \quad V(x,t) = -\frac{1}{2}\sqrt{\frac{((-12 - 12H_1(t) + 8H_2(t) + 27H_3(t) + 9H_4(t)) + 3M(t))k^2m^2}{(-3 + H_2(t))}}sn^{\Diamond}\Xi.$$

Case 2.

$$U(x,t) = -k^2(1+m^2)\left[\frac{8 + 32H_1^{\Diamond 2}(t) + 81H_3^{\Diamond 2}(t) - 30H_4(t) + 24H_2(t)\Diamond H_4(t) + 27H_4^{\Diamond 2}(t)}{12(-K(t) + N(t)\Diamond M(t))}\right.$$

$$\left.+ \frac{-4H_1(t)\Diamond(10 + 27H_3(t) + 15H_4(t)) + 6H_3(t)\Diamond(-3 + 4H_2(t) + 18H_4(t)) + R(t)\Diamond M(t)}{12(-K(t) + N(t)\Diamond M(t))}\right],$$

$$+ \frac{(-4 + 4H_1(t) - 9H_3(t) - 3H_4(t)) + M(t)k^2m^2}{4(-3 + H_2(t))}sn^{\Diamond 2}\Xi,$$

$$\text{and} \quad V(x,t) = -\frac{1}{2}\sqrt{\frac{((-12 - 12H_1(t) + 8H_2(t) + 27H_3(t) + 9H_4(t)) + 3M(t))k^3m^2}{(-3 + H_2(t))}}sn^{\Diamond}\Xi.$$

Taking the module of the Jacobian elliptic function $m \to 1$, $sn\xi \to \tanh\xi$, then the stochastic soliton-like solutions of Eq. (2.11) have been obtained.

Set 1. For $l(t) = \frac{1}{c}P(t)$ and $\Xi(t, x) = kx + \int_0^t P(\tau)d\tau + \Xi_0$, the following results have been obtained:

Case 1.

$$U(x,t) = -k^2(1+m^2)\left[\frac{8 + 32H_1^{\Diamond 2}(t) + 81H_3^{\Diamond 2}(t) - 30H_4(t) + 24H_2(t)\Diamond H_4(t) + 27H_4^{\Diamond 2}(t)}{12(-K(t) + N(t)\Diamond M(t))}\right.$$

$$\left. + \frac{-4H_1(t)\Diamond(10 + 27H_3(t) + 15H_4(t)) + 6H_3(t)\Diamond(-3 + 4H_2(t) + 18H_4(t)) + R(t)\Diamond M(t)}{12(-K(t) + N(t)\Diamond M(t))}\right],$$

$$+ \frac{(-4 + 4H_1(t) - 9H_3(t) - 3H_4(t)) + M(t)k^2m^2}{4(-3 + H_2(t))}\tanh^{\Diamond 2}\Xi,$$

$$\text{and} \quad V(x,t) = -\frac{1}{2}\sqrt{\frac{((-12 - 12H_1(t) + 8H_2(t) + 27H_3(t) + 9H_4(t)) - 3M(t))k^3m^2}{(-3 + H_2(t))}}\tanh^{\Diamond}\Xi.$$

Case 2.

$$U(x,t) = k^2(1+m^2)\left[\frac{8 + 32H_1^{\Diamond 2}(t) + 81H_3^{\Diamond 2}(t) - 30H_4(t) + 24H_2(t)\Diamond H_4(t) + 27H_4^{\Diamond 2}(t)}{12(K(t) + N(t)\Diamond M(t))}\right.$$

$$\left. + \frac{-4H_1(t)\Diamond(10 + 27H_3(t) + 15H_4(t)) + 6H_3(t)\Diamond(-3 + 4H_2(t) + 18H_4(t)) + R(t)\Diamond M(t)}{12(-K(t) + N(t)\Diamond M(t))}\right],$$

$$+ \frac{(-4 + 4H_1(t) - 9H_3(t) - 3H_4(t)) + M(t)k^2m^2}{4(-3 + H_2(t))}\tanh^{\Diamond 2}\Xi,$$

$$\text{and} \quad V(x,t) = -\frac{1}{2}\sqrt{\frac{((-12 - 12H_1(t) + 8H_2(t) + 27H_3(t) + 9H_4(t)) - 3M(t))k^3m^2}{(-3 + H_2(t))}}\tanh^{\Diamond}\Xi.$$

Set 2. For $l(t) = \frac{1}{c}P(t)$ and $\Xi(t, x) = kx + \int_0^t Q(\tau)d\tau + \Xi_0$, the following results have been obtained:

Case 1.

$$U(x,t) = -k^2(1+m^2)\left[\frac{\begin{array}{c}8 + 32H_1^{\diamond 2}(t) + 81H_3^{\diamond 2}(t) - 30H_4(t)\\ +24H_2(t)\diamond H_4(t) + 27H_4^{\diamond 2}(t)\end{array}}{12(-K(t) + N(t)\diamond M(t))}\right.$$

$$\left.+\frac{\begin{array}{c}-4H_1(t)\diamond(10 + 27H_3(t) + 15H_4(t)) + 6H_3(t)\diamond(-3\\ +4H_2(t) + 18H_4(t)) - R(t)\diamond M(t)\end{array}}{12(-K(t) + N(t)\diamond M(t))}\right],$$

$$+\frac{(-4 + 4H_1(t) - 9H_3(t) - 3H_4(t)) + M(t)k^2m^2}{4(-3 + H_2(t))}\tanh^{\diamond 2}\Xi,$$

$$\text{and}\quad V(x,t) = -\frac{1}{2}\sqrt{\frac{\begin{array}{c}((-12 - 12H_1(t) + 8H_2(t) + 27H_3(t)\\ +9H_4(t)) + 3M(t))k^3m^2\end{array}}{(-3 + H_2(t))}}\tanh^{\diamond}\Xi.$$

Case 2.

$$U(x,t) = -k^2(1+m^2)\left[\frac{\begin{array}{c}8 + 32H_1^{\diamond 2}(t) + 81H_3^{\diamond 2}(t) - 30H_4(t)\\ +24H_2(t)\diamond H_4(t) + 27H_4^{\diamond 2}(t)\end{array}}{12(-K(t) + N(t)\diamond M(t))}\right.$$

$$\left.+\frac{\begin{array}{c}-4H_1(t)\diamond(10 + 27H_3(t) + 15H_4(t)) + 6H_3(t)\diamond(-3\\ +4H_2(t) + 18H_4(t)) + R(t)\diamond M(t)\end{array}}{12(-K(t) + N(t)\diamond M(t))}\right],$$

$$+\frac{(-4 + 4H_1(t) - 9H_3(t) - 3H_4(t)) + M(t)k^2m^2}{4(-3 + H_2(t))}\tanh^{\diamond 2}\Xi,$$

$$\text{and}\quad V(x,t) = -\frac{1}{2}\sqrt{\frac{\begin{array}{c}((-12 - 12H_1(t) + 8H_2(t) + 27H_3(t)\\ +9H_4(t)) + 3M(t))k^3m^2\end{array}}{(-3 + H_2(t))}}\tanh^{\diamond}\Xi.$$

2.8 Extended Auxiliary Equation Method for Solutions of the Wick-Type Stochastic NLSE

2.8.1 *Extended auxiliary equation method*

In this section, the algorithm of the extended auxiliary equation method has been described [62,63]. The extended auxiliary equation method has been presented step by step as follows:

Step 1. Consider a nonlinear PDE of the following form:

$$P(U, U_t, \ldots, U_x, \ldots) = 0, \tag{2.68}$$

where $U(t, x)$ is the unknown function. P is the function in $U(t, x)$ along with their highest order partial derivatives and nonlinear terms of $U(t, x)$, respectively.

Step 2. The travelling wave transformation of Eq. (2.68) can be written as

$$U(t, x) = \Psi(\xi)e^{i(kx - ct)}, \quad \xi(x, t, z) = \lambda x + \mu \int_0^t l(\tau, z)d\tau + \xi_0, \tag{2.69}$$

where k and c are constants, which are to be determined later.

Using Eq. (2.69), the PDE (2.68) can be transformed into the following nonlinear ODE, which is given as

$$P(\Psi, \Psi'(\xi), \Psi'(\xi), \Psi'(\xi), \ldots) = 0. \tag{2.70}$$

Step 3. The exact solutions of Eq. (2.68) are assumed in the polynomial $\Phi(\xi)$ as follows:

$$\Psi(\xi) = \sum_{i=0}^{2N} \alpha_i \Phi^i(\xi), \tag{2.71}$$

where $\Phi(\xi)$ satisfies the first-order ordinary differential equation

$$\Phi'(\xi) = \sqrt{c_0 + c_2\Phi^2(\xi) + c_4\Phi^4(\xi) + c_6\Phi^6(\xi)}, \tag{2.72}$$

where $c_j(j = 0, 2, 4, 6)$ and $\alpha_i(i = 0, \ldots, 2N)$ are arbitrary constants to be determined later.

Step 4. Now, $\Psi = \xi^{-p}$ is substituted in Eq. (2.70) to obtain the highest order singularity. Then, the degrees of all terms of Eq. (2.70) are taken under consideration and the terms having lower degree are selected. The maximum value of p is assumed to be the pole and is denoted as N. The extended auxiliary equation method can only be implemented for integer values of N. However, if N is not an integer, then Eq. (2.70) can be transformed and the above steps can be repeated.

Step 5. Substituting Eq. (2.71) along with (2.72) into Eq. (2.70) and equating the coefficients of $\Phi^j(\Phi')^l$ $(j = 0, 1, 2, \ldots)$ and $(l = 0, 1)$ with zero, a set of algebraic equations are obtained. The obtained algebraic equations are solved, to obtain the unknowns c_j $(j = 0, 2, 4, 6)$ and α_i $(i = 0, \ldots, 2N)$, λ and μ. Then, all the obtained values of the unknowns are substituted in Eq. (2.71) yielding the exact solution for Eq. (2.68) instantly.

Step 6. Equation (2.72) has the following solutions [62,63]:

$$\Phi(\xi) = \frac{1}{2}\left[-\frac{c_4}{c_6}(1 \pm f(\xi))\right]^{\frac{1}{2}}, \tag{2.73}$$

where the function $f(\xi)$ could be expressed through the Jacobi elliptic function $sn(\xi, m)$, $cn(\xi, m)$, $dn(\xi, m)$ and so on, where $0 < m < 1$ is the modulus of the JEF. When m approaches 1 or 0, the Jacobi elliptic functions degenerate to hyperbolic functions and trigonometric functions, respectively [95].

The function $f(\xi)$ given by Eq. (2.73) has 12 forms as follows:

(1) If $c_0 = \frac{c_4^3(m^2-1)}{32c_6^2m^2}$, $c_2 = \frac{c_4^2(5m^2-1)}{16c_6m^2}$, $c_6 > 0$, then

$$f(\xi) = sn(\rho\xi) \quad \text{or} \quad f(\xi) = \frac{1}{msn(\rho\xi)}, \tag{2.74}$$

where $\rho = \frac{c_4}{2m}\sqrt{\frac{1}{c_6}}$.

(2) If $c_0 = \frac{c_4^3(1-m^2)}{32c_6^2}$, $c_2 = \frac{c_4^2(5-m^2)}{16c_6}$, $c_6 > 0$, then

$$f(\xi) = msn(\rho\xi) \quad \text{or} \quad f(\xi) = \frac{1}{sn(\rho\xi)}, \tag{2.75}$$

where $\rho = \frac{c_4}{2}\sqrt{\frac{1}{c_6}}$.

(3) If $c_0 = \frac{c_4^3}{32m^2 c_6^2}$, $c_2 = \frac{c_4^2(4m^2+1)}{16c_6 m^2}$, $c_6 < 0$, then

$$f(\xi) = cn(\rho\xi) \quad \text{or} \quad f(\xi) = \frac{\sqrt{1 - m^2}sn(\rho\xi)}{dn(\rho\xi)}, \qquad (2.76)$$

where $\rho = \frac{c_4\sqrt{-c_6}}{2mc_6}$.

(4) If $c_0 = \frac{c_4^3 m^2}{32c_6^2(m^2-1)}$, $c_2 = \frac{c_4^2(5m^2-4)}{16c_6(m^2-1)}$, $c_6 < 0$, then

$$f(\xi) = \frac{dn(\rho\xi)}{\sqrt{1 - m^2}} \quad \text{or} \quad f(\xi) = \frac{1}{dn(\rho\xi)}, \qquad (2.77)$$

where $\rho = \frac{c_4\sqrt{c_6(m^2-1)}}{2c_6(m^2-1)}$.

(5) If $c_0 = \frac{c_4^3}{32c_6^2(1-m^2)}$, $c_2 = \frac{c_4^2(4m^2-5)}{16c_6(m^2-1)}$, $c_6 < 0$, then

$$f(\xi) = \frac{1}{cn(\rho\xi)} \quad \text{or} \quad f(\xi) = \frac{dn(\rho\xi)}{\sqrt{1 - m^2}sn(\rho\xi)}, \qquad (2.78)$$

where $\rho = \frac{c_4\sqrt{c_6(m^2-1)}}{2c_6(1-m^2)}$.

(6) If $c_0 = \frac{m^2 c_4^3}{32c_6^2}$, $c_2 = \frac{c_4^2(m^2+4)}{16c_6}$, $c_6 < 0$, then

$$f(\xi) = dn(\rho\xi) \quad \text{or} \quad f(\xi) = \frac{\sqrt{1 - m^2}}{dn(\rho\xi)}, \qquad (2.79)$$

where $\rho = \frac{c_4\sqrt{-c_6}}{2c_6}$.

Step 7. Substituting the values of $c_j (j = 0, 2, 4, 6)$ and $\alpha_i (i = 0, \ldots, 2N)$, λ and μ as well as the solutions (2.74)–(2.79) into Eq. (2.71) yields the Jacobi elliptic function solutions of Eq. (2.68).

2.8.2 *Solutions of the stochastic NLSE*

In this section, EAEM is employed here for finding the white noise solutions for the stochastic NLS equation (2.13).

Applying Hermite transform on Eq. (2.13), it can be obtained as

$$i\tilde{U}_t + \tilde{F}(t,z)\tilde{U}_{xx} + \tilde{G}(t,z)\tilde{U}|\tilde{U}|^2 = 0, \tag{2.80}$$

where $\tilde{U} \equiv \tilde{U}(t,x,z) = \mathcal{H}(U(t,x))$, $\tilde{F}(t,z) = \mathcal{H}(F(t))$, $\tilde{G}(t,z) = \mathcal{H}(G(t))$ where $z = (z_1, z_2, \ldots) \in C^N$ is a parameter.

Let $u(t,x,z) \equiv \tilde{U}(t,x,z)$, $f(t,z) = \tilde{F}(t,z)$ and $g(t,z) = \tilde{G}(t,z)$. Now, let

$$u(t,x,z) = \Psi(\xi)e^{i(kx-ct)}, \tag{2.81}$$

and $\xi(x,t,z) = \lambda x + \mu \int_0^t l(\tau,z)d\tau + \xi_0$, where $\lim_{\tau \to 0+} l(\tau,z) = 0$.

Then, Eq. (2.80) becomes

$$i\mu l(t,z)\Psi_\xi + f(t,z)(-k^2\Psi + 2ik\lambda\Psi_\xi + \lambda^2\Psi_{\xi\xi}) + g(t,z)\Psi^3 = 0. \tag{2.82}$$

Now, the real part and imaginary part of Eq. (2.82) can be written as follows:

$$-k^2 f(t,z)\Psi + g(t,z)\Psi^3 + \lambda^2 f(t,z)\Psi'' = 0, \tag{2.83}$$

$$(2k\lambda f(t,z) + \mu l(t,z))\Psi' = 0. \tag{2.84}$$

From Eq. (2.84), it can be derived as $l(t,z) = \frac{-2k\lambda f(t,z)}{\mu}$.

Consider the ansatz as follows:

$$\Psi(\xi) = \sum_{i=0}^{2N} \alpha_i \Phi^i(\xi). \tag{2.85}$$

Now, according to the homogenous balancing principle, in Eq. (2.83), $g(t,z)\Psi^3$ and $\lambda^2 f(t,z)\Psi''$ are the dominant terms having the highest order singularity. Thus, the maximum value of the poles is 1 which implies here $N = 1$.

Consequently, from Eq. (2.85), the ansatz becomes

$$\Psi(\xi) = \alpha_0 + \alpha_1 \Phi(\xi) + \alpha_2 \Phi^2(\xi), \tag{2.86}$$

where $\Phi(\xi) = \frac{1}{2}\left[-\frac{c_4}{c_6}(1 \pm f(\xi))\right]^{\frac{1}{2}}$.

Substituting Eq. (2.86) in Eq. (2.83) and equating each coefficient of Φ^i, $(i = 0, 1, \ldots, 6)$ with zero, a system of algebraic equations are obtained for $c_j (j = 0, 2, 4, 6)$ and $\alpha_i (i = 0, \ldots, 2N)$, λ and μ.

$$\Phi^0(\xi) : c\alpha_0 - k^2 \alpha_0 f(t, z) + 2c_0 \alpha_2 \lambda^2 f(t, z) + \alpha_0^3 g(t, z) = 0,$$

$$\Phi^1(\xi) : c\alpha_1 - k^2 \alpha_1 f(t, z) + c_2 \alpha_1 \lambda^2 f(t, z) + 3\alpha_0 \alpha_1 g(t, z) = 0,$$

$$\Phi^2(\xi) : c\alpha_2 - k^2 \alpha_2 f(t, z) + 4c_2 \alpha_2 \lambda^2 f(t, z)$$
$$+ 3\alpha_0 \alpha_1^2 g(t, z) + 3\alpha_0^2 \alpha_2 g(t, z) = 0,$$

$$\Phi^3(\xi) : 2c_4 \alpha_1 \lambda^2 f(t, z) + \alpha_1^3 g(t, z) + 6\alpha_0 \alpha_1 \alpha_2 g(t, z) = 0, \qquad (2.87)$$

$$\Phi^4(\xi) : 6c_4 \alpha_2 \lambda^2 f(t, z) + 3\alpha_1^2 \alpha_2 g(t, z) + 3\alpha_0 \alpha_2^2 g(t, z) = 0,$$

$$\Phi^5(\xi) : 3c_3 \alpha_1 \lambda^2 f(t, z) + 3\alpha_1 \alpha_2^2 g(t, z) = 0,$$

$$\Phi^6(\xi) : 8c_6 \alpha_2 \lambda^2 f(t, z) + \alpha_2^3 g(t, z) = 0.$$

2.8.3   *Results and discussion*

Solving the above algebraic equations (2.87), the following set of coefficients for the solution of Eq. (2.83) have been obtained:

$$c_0 = -\frac{\alpha_0(-c - k^2 f(t, z) + \alpha_0^2 g(t, z))}{2\alpha_2 \lambda^2 f(t, z)},$$

$$c_2 = \frac{-c + k^2 f(t, z) - 3\alpha_0^2 g(t, z)}{4\lambda^2 f(t, z)}, \qquad (2.88)$$

$$c_4 = -\frac{\alpha_0 \alpha_2 g(t, z)}{2\lambda^2 f(t, z)}, \quad c_6 = \frac{\alpha_2^2 g(t, z)}{8\lambda^2 f(t, z)},$$

$$\alpha_0 = \alpha_0, \quad \alpha_1 = 0, \quad \alpha_2 = \alpha_2.$$

Substituting Eq. (2.88) into Eq. (2.86), the following Jacobi elliptic function solutions of Eq. (2.83) have been obtained:

$$\Phi(\xi) = \left[-\frac{\alpha_0}{\alpha_2}(1 \pm f(\xi)) \right]^{\frac{1}{2}}, \qquad (2.89)$$

where $c_6 < 0$. Thus, $f(\xi)$ satisfies only the functions (2.76), (2.77) and (2.79).

For $l(t,z) = \frac{-2k\lambda f(t,z)}{\mu}$ and $\xi = \lambda x + \mu \int_0^t l(\tau, z)d\tau + \xi_0$, the following results have been obtained.

Set 1. Using Eq. (2.76) and Eq. (2.89), from Eq. (2.81), the following Jacobi elliptic function solutions of Eq. (2.80) have been obtained:

$$u_{11}(t,x,z) = \left[\alpha_0 - \alpha_0 \left[1 \pm cn\left(\frac{\alpha_0}{\sqrt{2}m\alpha_2}\sqrt{\frac{\alpha_2^2 g(t,z)}{\lambda^2 f(t,z)}}\xi\right)\right]\right] e^{i(kx-ct)},$$

$$u_{12}(t,x,z) = \left[\alpha_0 - \alpha_0 \left[1 \pm \sqrt{1-m^2}\frac{sn\left(\frac{\alpha_0}{\sqrt{2}m\alpha_2}\sqrt{\frac{\alpha_2^2 g(t,z)}{\lambda^2 f(t,z)}}\xi\right)}{dn\left(\frac{\alpha_0}{\sqrt{2}m\alpha_2}\sqrt{\frac{\alpha_2^2 g(t,z)}{\lambda^2 f(t,z)}}\xi\right)}\right]\right]$$
$$\times e^{i(kx-ct)}.$$

If $m \to 1$, it can be written as $cn\xi \to \sec h\xi$, then the hyperbolic function solutions of Eq. (2.80) have been obtained.

$$u_{13}(t,x,z) = \left[\alpha_0 - \alpha_0 \left[1 \pm \sec h\left(\frac{\alpha_0}{\sqrt{2}m\alpha_2}\sqrt{\frac{\alpha_2^2 g(t,z)}{\lambda^2 f(t,z)}}\xi\right)\right]\right]$$
$$\times e^{i(kx-ct)}.$$

Set 2. Using Eq. (2.77) and Eq. (2.89), from Eq. (2.81), the following Jacobi elliptic function solutions of Eq. (2.80) have been obtained:

$$u_{21}(t,x,z) = \left[\alpha_0 - \alpha_0 \left[1 \pm \frac{dn\left(\frac{\alpha_0}{\sqrt{2}\alpha_2(m^2-1)}\sqrt{-\frac{(m^2-1)\alpha_2^2 g(t,z)}{\lambda^2 f(t,z)}}\xi\right)}{\sqrt{1-m^2}}\right]\right]$$
$$\times e^{i(kx-ct)},$$

$$u_{22}(t,x,z) = \left[\alpha_0 - \alpha_0 \left[1 \pm \frac{1}{dn\left(\frac{\alpha_0}{\sqrt{2}\alpha_2(m^2-1)}\sqrt{-\frac{(m^2-1)\alpha_2^2 g(t,z)}{\lambda^2 f(t,z)}}\xi\right)}\right]\right]$$
$$\times e^{i(kx-ct)}.$$

Set 3. Using Eq. (2.79) and Eq. (2.89), from Eq. (2.81), the following Jacobi elliptic function solutions of Eq. (2.80) have been obtained:

$$u_{31}(t, x, z) = \left[\alpha_0 - \alpha_0 \left[1 \pm dn\left(\frac{\alpha_0}{\sqrt{2}\alpha_2}\sqrt{\frac{\alpha_2^2 g(t, z)}{\lambda^2 f(t, z)}}\xi\right)\right]\right] e^{i(kx-ct)},$$

$$u_{32}(t, x, z) = \left[\alpha_0 - \alpha_0 \left[1 \pm \frac{\sqrt{1 - m^2}}{dn\left(\left(\frac{\alpha_0}{\sqrt{2}\alpha_2}\sqrt{\frac{\alpha_2^2 g(t,z)}{\lambda^2 f(t,z)}}\xi\right)\right)}\right]\right] e^{i(kx-ct)}.$$

2.8.3.1 *Stochastic set of solutions*

Using the definition of inverse Hermite transform [15] and according to Xie [30], from Set 1, Set 2 and Set 3, the following stochastic solutions of Eq. (2.13) can be obtained.

For $l(t) = \frac{-2k\lambda F(t)}{\mu}$ and $\Xi(t, x) = \lambda x + \mu \int_0^t \frac{-2k\lambda F(\tau)}{\mu} d\tau + \Xi_0$, where $\Xi(t, x) = \mathcal{H}^{-1}(\xi(t, x, z))$, the following results have been obtained:

Case 1. Using Set 1,

$$U_{11}(t, x) = \left[\alpha_0 - \alpha_0 \left[1 \pm cn^{\lozenge}\left(\frac{\alpha_0}{\sqrt{2m}\alpha_2}\sqrt{\frac{\alpha_2^2 G(t)}{\lambda^2 F(t)}}\xi\right)\right]\right] e^{i(kx-ct)},$$

$$U_{12}(t, x) = \left[\alpha_0 - \alpha_0 \left[1 \pm \sqrt{1 - m^2}\frac{sn^{\lozenge}\left(\frac{\alpha_0}{\sqrt{2m}\alpha_2}\sqrt{\frac{\alpha_2^2 G(t)}{\lambda^2 F(t)}}\xi\right)}{dn^{\lozenge}\left(\frac{\alpha_0}{\sqrt{2m}\alpha_2}\sqrt{\frac{\alpha_2^2 G(t)}{\lambda^2 F(t)}}\xi\right)}\right]\right]$$
$$\times \ e^{i(kx-ct)}.$$

If $m \to 1$, it can be written as $cn\xi \to \sec h\xi$, then the stochastic soliton-like solutions of Eq. (2.13) have been obtained.

$$U_{13}(t, x) = \left[\alpha_0 - \alpha_0 \left[1 \pm \sec h^{\lozenge}\left(\frac{\alpha_0}{\sqrt{2m}\alpha_2}\sqrt{\frac{\alpha_2^2 G(t)}{\lambda^2 F(t)}}\xi\right)\right]\right] e^{i(kx-ct)}.$$

Case 2. Using Set 2,

$$
U_{21}(t,x) = \left[\alpha_0 - \alpha_0 \left[1 \pm \frac{dn^{\diamond}\left(\frac{\alpha_0}{\sqrt{2}\alpha_2(m^2-1)} \sqrt{-\frac{(m^2-1)\alpha_2^2 g G(t)}{\lambda^2 F(t)}}\xi \right)}{\sqrt{1-m^2}} \right] \right]
$$
$$
\times\; e^{i(kx-ct)},
$$

$$
U_{22}(t,x) = \left[\alpha_0 - \alpha_0 \left[1 \pm \frac{1}{dn^{\diamond}\left(\frac{\alpha_0}{\sqrt{2}\alpha_2(m^2-1)} \sqrt{-\frac{(m^2-1)\alpha_2^2 g G(t)}{\lambda^2 F(t)}}\xi \right)} \right] \right]
$$
$$
\times\; e^{i(kx-ct)}.
$$

Case 3. Using Set 3,

$$
U_{31}(t,x) = \left[\alpha_0 - \alpha_0 \left[1 \pm dn^{\diamond}\left(\frac{\alpha_0}{\sqrt{2}\alpha_2} \sqrt{\frac{\alpha_2^2 G(t)}{\lambda^2 F(t)}}\xi \right) \right] \right] e^{i(kx-ct)},
$$

$$
U_{32}(t,x) = \left[\alpha_0 - \alpha_0 \left[1 \pm \frac{\sqrt{1-m^2}}{dn^{\diamond}\left(\frac{\alpha_0}{\sqrt{2}\alpha_2} \sqrt{\frac{\alpha_2^2 G(t)}{\lambda^2 F(t)}}\xi \right)} \right] \right] e^{i(kx-ct)}.
$$

2.9 Summary

The general idea in this chapter is that the stochastic Wick-type non-linear evolution equations are being reduced to ODEs by travelling wave variable transformation. Second, it is supposed the solution can be expressed in a polynomial in a variable. The degree of the polynomial can be determined by the homogeneous balance method, and the coefficients can be obtained by solving a set of algebraic equations. Then, the obtained deterministic solution undergoes inverse Hermite transform to obtain the stochastic solutions of Wick-type stochastic partial differential equations.

Kudryashov method has been applied here for finding the solution for the Wick-type stochastic ZK equation. The Hermite transform has been applied which transforms Wick products into ordinary products. Then, the proposed method has been implemented for determining the exact solution for deterministic ZK equation. Finally, inverse Hermite transform is used for getting the exact solution in Wick type.

Improved sub-equation method has been applied here for finding the solution of the Wick-type stochastic Kudryashov–Sinelshchikov equation and Wick-type stochastic MBE. The Hermite transform transforms Wick products into ordinary products. Then, the proposed method has been applied for determining the exact solutions for deterministic Kudryashov–Sinelshchikov equation and deterministic MBE. Finally, inverse Hermite transform is applied for getting the exact solutions in Wick type for the equations in discussion. The solutions obtained are new and graphical representation of solutions of Wick-type stochastic Kudryashov–Sinelshchikov equation has also been shown which further shows the behaviour of the obtained solutions.

The Jacobi elliptic function expansion method has been successfully applied here for finding the solutions for the Wick-type stochastic Kersten–Krasil'shchik coupled KdV-mKdV equations. The Hermite transform transforms Wick products into ordinary products. Then, the proposed method has been implemented for determining the exact solutions for deterministic modified Kersten–Krasil'shchik coupled KdV-mKdV equations. Finally, inverse Hermite transform is applied for getting the exact solutions in Wick type.

The extended auxiliary equation method has been successfully applied here for finding the solutions for the Wick-type stochastic NLSE. The Hermite transform transforms Wick products into ordinary products. Finally, inverse Hermite transform has been implemented for getting the exact solutions in Wick type.

However, obtaining analytical solutions is sometimes a tedious and difficult process, so in the upcoming chapters, various numerical techniques and their applicability in solving various stochastic models arising in modern-day mathematics have been discussed.

Chapter 3

Numerical Solutions of Stochastic Integral Equation

3.1 Introduction

Integral equations arise in various fields of science and numerous applications in elasticity, plasticity, heat and mass transfer, fluid dynamics, filtration theory, electrostatics, electrodynamics, biomechanics, game theory, control, queuing theory, electrical engineering, economics, etc. Many researchers have been investigating neutron diffusion and biological species coexisting together with increasing and decreasing rates of generating. Furthermore, integral equations arise as representation forms of differential equations. Fredholm and Volterra integral equations arise from different origins and applications, such as boundary value problems as in Fredholm equations, and from initial value problems as in Volterra equations. Based on the fact that integral equations arise from distinct origins, different techniques and approaches will be used to determine the solution of each type of integral equation. In recent years, different basic functions such as wavelets have been implemented to estimate the solution of various integral equations [96,97].

Due to nondeterministic behaviour of problems occurring in the general areas of the biological, engineering, oceanographic and physical sciences, the mathematical descriptions of their physical phenomena result in random or stochastic integral equations. The stochastic integral equation has numerous applications to the problems in chemotherapy, chemical kinetics, physiological systems, population growth, telephone engineering, turbulence and systems theory [20]. In many cases, the behaviour of a dynamical system is dependent on Gaussian white noise; modelling such phenomena mathematically requires the use of various stochastic integral equations or stochastic integro-differential equations of Itô type [16,98–107]. Thus, it is more realistic in this situation to construct a stochastic model for the system rather than a deterministic model.

The Volterra–Fredholm integral equation arises from parabolic boundary value problems, the mathematical modelling of the spatio-temporal development of an epidemic and various physical models. The essential features of these models are of wide applicability. Wavelet theory is a new and emerging area in mathematics [108]. It is applied in a wide range of engineering disciplines [109,110] particularly; wavelets have been successfully used in signal analysis for waveform representation and segmentation, time-frequency analysis and construction of fast but easily implemented algorithms.

Recently, various wavelets, such as CAS wavelet [111], Legendre wavelet [112,113], Legendre multiwavelet [114], Haar wavelet [115], Chebyshev wavelet [116] and so on, were implemented to generate solutions of integral equations, integro-differential equations and so on. Shekarbi and Damercheli [117] obtained solutions by Block-Pulse function matrix of integration. Maleknejad *et al.* [118] obtained approximate solutions of nonlinear Volterra–Fredholm–Hammerstein integral equations by applying Bernstein operational matrix of integration. Mirzaee and Samadyar [119] obtained approximate results of nonlinear stochastic Itô–Volterra integral equations having fractional Brownian motion. Sahu and Saha Ray [120] have applied Bernstein collocation method to solve nonlinear Fredholm integral equations. Vanani *et al.* [121] solved nonlinear delay Volterra integral equations by homotopy perturbation method.

3.2 Outline of Present Study

In this chapter, the numerical solutions of stochastic Volterra–Fredholm integral equations (SVFIEs) have been obtained by hybrid Legendre Block-Pulse functions (HLBPFs) and stochastic operational matrix (SOM). The HLBPFs are orthonormal and have compact support on [0, 1). The numerical results obtained by the above functions have been compared with those obtained by second-kind Chebyshev wavelets (SKCWs). Furthermore, results of the proposed computational method establish its accuracy and efficiency.

Also, space-time Brownian motion (BM) and its applications to mixed-type stochastic integral equations have been studied. Approximate solutions of stochastic mixed Volterra–Fredholm integral equation (SMVFIE) have been obtained by using two-dimensional second-kind Chebyshev wavelets (two-dimensional CWs). Furthermore, some examples have been presented to justify the efficiency of two-dimensional second-kind Chebyshev wavelets.

3.2.1 *Stochastic Volterra–Fredholm integral equations*

Consider the SVFIE

$$X(t) = f(t) + \int_\alpha^\beta k_1(s,t)X(s)ds + \int_0^t k_2(s,t)X(s)ds$$

$$+ \int_0^t k_3(s,t)X(s)dB(s), t \in [0,T), \tag{3.1}$$

where $X(t), f(t), k_1(s,t), k_2(s,t)$ and $k_3(s,t)$, for $s,t \in [0,T)$, are the stochastic processes defined on $(\Omega, \mathcal{F}, \mathbb{P})$ and $X(t)$ is unknown. Also, $B(t)$ is a BM process and $\int_0^t k_3(s,t)X(s)dB(s)$ is the Itô integral.

3.2.2 *Stochastic mixed Volterra–Fredholm integral equation*

Let the SMVFIE [117] driven by space-time BM is of the following form:

$$u(t,x) = \int_0^x G(t,x,y)u_0(y)dy + \int_0^t \int_0^1 G(t-s,x,y)f(s,y,u(s,y))dyds$$

$$+ \int_0^t \int_0^1 G(t-s,x,y)\sigma(s,y,u(s,y))dB(s,y), t, x \in [0,1],$$

$$\tag{3.2}$$

where $G(t,x,y)$ is Green's function and unknown function $u(t,x)$ is to be obtained. $B = \{|B(t,x)|t \geq 0, x \in [0,1]\}$ is the Brownian sheet on $\mathbb{R}_+ \times [0,1]$. The nonlinear terms $f(t, x, u(t, x))$ and $\sigma(t, x, u(t, x))$ are locally bounded Borel functions mapping $\mathbb{R}_+ \times [0,1] \times \mathbb{R}$ into $\mathbb{R}$. Let $(\Omega, \mathcal{F}, \mathbb{P})$ be a stochastic basis carrying an $\mathcal{F}_t$-adapted Brownian sheet B. Recall that B is a zero mean Gaussian random field with covariance function

$$E(B(t,s)B(s,y)) = \min(t,s)\min(x,y).$$

3.3 HLBPFs for SVFIE

3.3.1 *BPFs*

A set of BPFs [99] (BPFs) $\phi_n(x), n = 1, 2, \ldots, N$ on $[0,1)$ are defined as follows:

$$\phi_n(t) = \begin{cases} 1, & \frac{n-1}{N} \leq t \leq \frac{n}{N}, \\ 0, & \text{otherwise}, \end{cases}$$

with $t \in [0,1), n = 1, 2, \ldots, N$ and $h = \frac{1}{N}$.

The properties of BPFs are as follows:

(i) The BPFs on the interval $[0,1)$ are disjoint

$$\phi_n(t)\phi_m(t) = \delta_{nm}\phi_n(t),$$

where $n, m = 1, 2, \ldots, N$ and δ_{nm} is Kronecker delta.

(ii) The BPFs are orthogonal on the interval $[0,1)$.

$$\int_0^1 \phi_n(t)\phi_m(t)dt = h\delta_{nm}, \quad n, m = 1, 2, \ldots, N.$$

(iii) If $N \to \infty$, then the BPFs set is complete; for every $f \in L^2([0,1))$, Parseval's identity holds

$$\int_0^1 f^2(t)dt = \sum_{i-1}^{\infty} f_n^2 \|\phi_n(t)\|^2,$$

where

$$f_n = \frac{1}{h}\int_0^1 f(t)\phi_n(t)dt.$$

Consider the first N terms of BPFs and those have been written concisely as N-vector

$$\Phi(t) = (\phi_1(t), \phi_2(t), \dots, \phi_N(t))^T, \quad t \in [0, 1). \tag{3.3}$$

The above representation and disjointness property follows

$$\Phi(t)\Phi^T(t) = \begin{pmatrix} \phi_1(t) & 0 & \cdots & 0 \\ 0 & \phi_2(t) & \cdots & 0 \\ \vdots & \vdots & \ddots & \vdots \\ 0 & 0 & \cdots & \phi_N(t) \end{pmatrix}_{N \times N}.$$

Furthermore, $\Phi^T(t)\Phi(t) = 1$ and $\Phi(t)\Phi^T(t)F^T = D_F\Phi(t)$, where D_F usually denotes a diagonal matrix whose diagonal entries are related to a constant vector $F = (f_1, f_2, \dots, f_N)^T$.

3.3.2   HLBPFs

Consider the Legendre polynomials $L_m(x)$ on the interval $[-1, 1]$

$$L_0(x) = 1,$$
$$L_1(x) = x,$$
$$L_m(x) = \tfrac{2m-1}{m}xL_{m-1}(x) - \tfrac{m-1}{m}L_{m-2}(x), m = 2, 3, \dots$$

The set $\{L_m(x) : m = 0, 1, \dots\}$ in Hilbert space $L^2[-1, 1]$ is a complete orthogonal set.

For $m = 0, 1, \dots, M-1$ and $n = 1, 2, \dots, N$, the HLBPFs on $[0,1)$ are defined in [100,101] as follows:

$$\psi(n, m, t) = \begin{cases} \sqrt{N(2m + 1)}L_m(2Nt - 2n + 1), & \frac{n-1}{N} \leq t < \frac{n}{N}, \\ 0, & \text{otherwise}, \end{cases}$$

where m and n are the order of the Legendre polynomials and BPFs, respectively. $\sqrt{N(2m + 1)}$ is for the orthonormality.

3.3.2.1   *Function approximation*

Any function $f(t)$ over the $[0,1)$ which is square integrable, i.e., $f(t) \in L^2[0,1)$, can be expanded in terms of HLBPFs as

$$f(t) = \sum_{n=1}^{\infty} \sum_{m=0}^{\infty} X(n \cdot m)\psi(n,m,t), \qquad (3.4)$$

where $X(n,m) = \langle f(t), \psi(n,m,t) \rangle$. If Eq. (3.4) is truncated at some values of N and M, then it yields

$$f(t) \approx \sum_{n=1}^{N} \sum_{m=0}^{m-1} X(n,m)\psi(n,m,t) = X^T \Psi(t), \qquad (3.5)$$

where X and $\Psi(t)$ are $\hat{m} = NM$ column vectors.

$$X = [X(1,0), X(1,1), \ldots, X(1,M-1), \ldots, X(N,0),$$
$$X(N,1), \ldots, X(N,M-1)]^T, \qquad (3.6)$$

and

$$\Psi(t) = [\psi(1,0,t), \psi(1,1,t), \ldots, \psi(1,M-1,t), \ldots, \psi(N,0,t), \ldots,$$
$$\psi(N,m-1,t)]^T. \qquad (3.7)$$

Since $\int_0^1 \Psi(t)\Psi^T(t)dt = 1$, where the I is an identity matrix of dimension NM, then

$$X(n,m) = \int_0^1 f(t)\psi(n,m,t)dt, \quad n = 1,2,\ldots,N \text{ and}$$
$$m = 0,1,\ldots,M-1.$$

Similarly, let $K(s,t) \in L^2([0,1) \times [0,1))$, where

$$K(s,t) \approx \Psi^T(s)K\Psi(t) = \Psi^T(t)K^T\Phi(s), \qquad (3.8)$$

where $\Psi(s)$ and $\Psi(t)$ are $\hat{m}$-dimensional HLBP vectors, respectively, and $K = [k_{ij}]_{\hat{m} \times \hat{m}}$ is the $\hat{m} \times \hat{m}$ HLBP coefficient matrix with

$$k_{ij} = \int_0^1 \int_0^1 K(s,t)\Psi_i(s)\Psi_j(t)dtds = \langle \Psi_i(s), \langle K(s,t), \Psi_j(t) \rangle \rangle.$$

3.3.2.2 *Integration OM*

In this section, a stochastic OM for the HLBPFs has been derived.

Lemma 3.3.2.1. *Let $\Phi(t)$ be the m-dimensional BPFs vector defined in Eq. (3.3). The integral of this vector is*

$$\int_0^t \Phi(s)ds \cong P\Phi(t),\tag{3.9}$$

where P is called the OM of integration for BPFs and

$$P = \frac{h}{2}\begin{bmatrix} 1 & 2 & 2 & \cdots & 2 \\ 0 & 1 & 2 & \cdots & 2 \\ 0 & 0 & 1 & \vdots & \vdots \\ \vdots & \vdots & \vdots & \ddots & 2 \\ 0 & 0 & 0 & \cdots & 1 \end{bmatrix}_{\hat{m}\times\hat{m}}.$$

Proof. It may be referred to Ref. [99]. $\square$

Lemma 3.3.2.2. *Let $\Phi(t)$ be the $\hat{m}$-dimensional BPFs vector. The Itô integral of this vector is*

$$\int_0^t \Phi(s)dB(s) \cong P_s\Phi(t),\tag{3.10}$$

where P_s is called the stochastic OM of integration for BPFs and

$$P_s = \begin{bmatrix} B\left(\frac{h}{2}\right) & B(h) & B(h) & \cdots & B(h) \\ 0 & B\left(\frac{3h}{2}\right)-B(h) & B(2h)-B(h) & \cdots & B(2h)-B(h) \\ 0 & 0 & B\left(\frac{5h}{2}\right)-B(2h) & \cdots & B(3h)-B(2h) \\ \vdots & \vdots & \vdots & \ddots & \vdots \\ 0 & 0 & 0 & \cdots & \begin{array}{c} B\left(\frac{(2\hat{m}-1)h}{2}\right) \\ -B((\hat{m}-1)h) \end{array} \end{bmatrix}_{\hat{m}\times\hat{m}}.$$

Proof. It may be further referred to Ref. [99]. $\square$

3.3.2.3   *Hybrid Legendre Block-Pulse functions and BPFs*

In the following section, the relation between the HLBPFs and BPFs has been reviewed.

Theorem 3.3.2.3. *Let $\Psi(t)$ and $\Phi(t)$ be the $\hat{m}$-dimensional HLBPFs and BPFs vector. The vector $\Psi(t)$ can be expanded by BPFs vector $\Phi(t)$ as*

$$\Psi(t) = Q\Phi(t), \tag{3.11}$$

where Q is an $\hat{m} \times \hat{m}$ block matrix and

$$Q_{ij} = \Psi_1\left(\frac{2j-1}{2\hat{m}}\right), \quad i,j = 1,2,\ldots,\hat{m}.$$

Proof.   Let $\psi_i(t), i = 1,2,\ldots,\hat{m}$, be the ith element of HLBPFs vector. Then,

$$\psi_i(t) = \sum_{j=1}^{\hat{m}} Q_{ij}\phi_j(t), \quad i = 1,2,\ldots,\hat{m},$$

where Q_{ij} is the (i,j)th element of matrix Q and

$$Q_{ij} = \frac{1}{h}\int_0^1 \psi_i(t)\phi_j(t)dt = \frac{1}{h}\int_{\frac{j-1}{m}}^{\frac{j}{\hat{m}}} \psi_i(t)dt$$

$$= \hat{m}\int_{\frac{j-1}{m}}^{\frac{j}{\hat{m}}} \psi_i(t)dt, \quad \text{where } \hat{m} = \frac{1}{h}.$$

Now, using the mean value theorem, it yields

$$Q_{ij} = \hat{m}\left(\frac{j}{\hat{m}} - \frac{j-1}{\hat{m}}\right)\psi_i(\eta_j) = \psi_i(\eta_j), \quad n_j \in \left(\frac{j-1}{\hat{m}}, \frac{j}{\hat{m}}\right).$$

By choosing $\eta_j = \frac{2j-1}{2\hat{m}}$, so

$$Q_{ij} = \psi_i\left(\frac{2j-1}{2\hat{m}}\right), \quad i,j = 1,2,\ldots,\hat{m}. \qquad \square$$

Remark 3.3.2.4. For an $\hat{m}$-vector F, it yields

$$\Psi(t)\Psi^T(t)F = \tilde{F}\Psi(t),$$

in which $\tilde{F}$ is an $\hat{m} \times \hat{m}$ matrix of the form

$$\tilde{F} = Q\overline{F}Q^{-1},$$

where $\overline{F} = diag(Q^T F)$.

Remark 3.3.2.5. Let A represent $\hat{m} \times \hat{m}$ matrix. Then for the HLBPFs vector $\Psi(t)$, it can be written as

$$\Psi^T(t)A\Psi(t) = \hat{A}^T\Psi(t),$$

where $\hat{A}^T = UQ^{-1}$ and $U = diag(Q^TAQ)$ is an $\hat{m}$-vector.

3.3.2.4 *Stochastic operational matrix of HLBPFs*

Theorem 3.3.2.6. *Let* $\Psi(t)$ *be the* $\hat{m}$-*dimensional HLBPFs vector. The integral of this vector is*

$$\int_0^t \Psi(s)ds = QPQ^{-1}\Psi(t) = \Lambda\Psi(t), \qquad (3.12)$$

where Q is introduced in Eq. (3.11) and P is the OM of integration for BPFs derived in Eq. (3.9).

Proof. Let $\Psi(t)$ be the HLBPFs vector.

$$\int_0^t \Psi(s)ds = Q\int_0^t \Phi(s)ds = QP\Phi(t),$$

where $\Psi(t) = Q\Phi(t)$.

Now, Theorem 3.3.2.3 gives

$$\int_0^t \Psi(s)ds = QPQ^{-1}\Psi(t) = \Lambda\Psi(t),$$

where $\Lambda = QPQ^{-1}$.

This completes the proof. $\qquad\qquad\qquad\qquad\square$

Theorem 3.3.2.7. *Let* $\Psi(t)$ *be the* $\hat{m}$-*dimensional HLBPFs vector. The Itô integral of this vector is*

$$\int_0^t \Psi(s)dB(s) = QP_sQ^{-1}\Psi(t) = \Lambda_s\Psi(t), \qquad (3.13)$$

where Λ_s is called the stochastic OM for HLBPFs, Q is introduced in Eq. (3.11) and P_s is the stochastic OM of integration for BPFs in Eq. (3.10).

Proof. Let $\Psi(t)$ be the HLBPFs vector. Using Theorem 3.3.2.3 and Lemma 3.3.2.2, it yields

$$\int_0^t \Psi(s)dB(s) = Q \int_0^t \Phi(s)dB(s) = QP_s\Phi(t).$$

Now, Theorem 3.3.2.3 yields

$$\int_0^t \Psi(s)dB(s) = QP_sQ^{-1}\Psi(t) = \Lambda_s\Psi(t),$$

where $\Lambda_s = QP_sQ^{-1}$. $\qquad\square$

3.3.3 *Approximation of stochastic Volterra–Fredholm integral equation*

In the present analysis, the following SVFIE has been considered:

$$X(t) = f(t) + \int_\alpha^\beta k_1(s,t)X(s)ds + \int_0^t k_2(s,t)X(s)ds$$

$$+ \int_0^t k_3(s,t)X(s)dB(s), t \in [0,T), \tag{3.14}$$

where $X(t)$, $f(t)$ and $k_i(s,t)$, $i = 1,2,3$, are the stochastic processes on $(\Omega, \mathcal{F}, \mathbb{P})$ and $X(t)$ is unknown. Also, $B(t)$ is a Brownian motion process and $\int_0^t k_3(s,x)X(s)dB(s)$ is the Itô integral [105]. Without any loss of generality set $(\alpha, \beta) = (0,1)$. Now, approximating the functions in Eq. (3.14) yields

$$f(t) = F^T\Psi(t) = \Psi^T(t)F,$$

$$X(t) = X^T\Psi(t) = \Psi^T(t)X, \tag{3.15}$$

$$k_i(s,t) = \Psi^T(s)K_i\Psi(t) = \Psi^T(t)K_i\Psi(s), \qquad i = 1,2,3,$$

where X and F are the HLBPFs coefficients vectors and K_i, $i = 1, 2, 3$, are the HLBPFs coefficients matrices defined in Eqs. (3.5) and (3.8). Substituting Eq. (3.15) in Eq. (3.14) yields

$$X^T \Psi(t) = F^T \Psi(t) + X^T \left(\int_0^1 \Psi(s)\Psi^T(t)ds \right) K_1 \Psi(t)$$

$$+ \Psi^T(t)K_2^T \left(\int_0^1 \Psi(s)\Psi^T(t)Xds \right)$$

$$+ \Psi^T(t)K_3^T \left(\int_0^1 \Psi(s)\Psi^T(t)XdB(s) \right).$$

Using relation $\int_0^1 \Psi(s)\Psi^T(t)ds = I_{\hat{m}\times\hat{m}}$ and Remark 3.3.2.4,

$$X^T \Psi(t) = F^T \Psi(t) + X^T K_1 \Psi(t) + \Psi^T(t)K_2^T \left(\int_0^1 \tilde{X}\Psi(s)ds \right)$$

$$+ \Psi^T(t)K_3^T \left(\int_0^1 \tilde{X}\Psi(s)dB(s) \right),$$

where $\tilde{X}$ is an $\hat{m} \times \hat{m}$ matrix. Now, applying OM Λ and Λ_s for HLBPFs derived in Eqs. (3.12) and (3.13) yields

$$X^T \Psi(t) = F^T \Psi(t) + X^T K_1 \Psi(t) + \Psi^T(t)K_2^T \tilde{X}\Lambda\Psi(t)$$

$$+ \Psi^T(t)K_3^T \tilde{X}\Lambda_s \Psi(t).$$

By setting $Y_2 = K_2^T \tilde{X}\Lambda, Y_3 = K_3^T \tilde{X}\Lambda_s$ and using Remark 3.3.2.5, it yields

$$X^T \Psi(t) - X^T K_1 \Psi(t) - \hat{Y}_2^T \Psi(t) - \hat{Y}_3^T \Psi(t) = F^T \Psi(t),$$

in which $\hat{Y}_2$ and $\hat{Y}_3$ are $\hat{m} \times \hat{m}$ matrices and these are linear functions of the vector X for all $t \in [0, 1)$; thus,

$$X^T - X^T K_1 - \hat{Y}_2^T - \hat{Y}_3^T = F^T. \tag{3.16}$$

Since $\hat{Y}_2$ and $\hat{Y}_3$ are linear functions of X, Eq. (3.16) is a linear system of equations. By determining X, an approximate solution of the stochastic Volterra–Fredholm integral equation (3.14) can be obtained by substituting the obtained vector X in Eq. (3.15).

3.3.4 *Convergence and error analysis*

Theorem 3.3.4.1 (convergence analysis [103]). *If a continuous function $f(t) \in L^2(\mathbb{R})$ defined on $[0,1)$ has bounded second derivative $|\frac{\partial^2 f}{\partial t^2}| \leq \mathscr{M}$, the HLBPFs expansion $f(t) = \sum_{n=1}^{N} \sum_{m=0}^{M-1} X(n,m)\psi(n,m,t)$ of $f(t)$ converges uniformly.*

Proof. Let $f(t) \in L^2[0,1]$ and $|\frac{\partial^2 f}{\partial t^2}| \leq \mathscr{M}, \mathscr{M}$ is a positive constant.

The coefficients of HLBP expansion of continuous function $f(t)$ are defined as

$$X(n,m) = \int_0^1 f(t)\psi(n,m,t)dt$$

$$= \int_I f(t)\psi(n,m,t)dt,$$

where $I = \left[\frac{n-1}{N}, \frac{n}{N}\right)$.

Now, substituting $2Nt - 2n + 1 = y$ and $\hat{n} = 2n - 1$

$$X(n,m) = \frac{\sqrt{N(2m+1)}}{2N} \int_{-1}^{1} f\left(\frac{\hat{n}+y}{2N}\right) L_m(y)dy$$

$$= \frac{\sqrt{N(2m+1)}}{2N} \frac{1}{2m+1} \int_{-1}^{1} f\left(\frac{\hat{n}+y}{2N}\right)$$

$$(L'_{m+1}(y) - L'_{m-1}(y))dy,$$

since $(2m+1)L_m(y) = L'_m(y) - L'_{m-1}(y)$.

Now, integrating by parts, it can be obtained as

$$X(n, m) = \frac{\sqrt{N(2m+1)}}{2N(2m+1)} \left(f\left(\frac{\hat{n}+y}{2N}\right) [L_{m+1}(y) - L_{m-1}(y)]_{-1}^{1} \right.$$

$$\left. - \int_{-1}^{1} \frac{\partial f}{\partial y}(L_{m+1}(y) - L_{m-1}(y))dy \right)$$

$$= \frac{-1}{2\sqrt{N(2m+1)}} \int_{-1}^{1} \frac{\partial f}{\partial y}(L_{m+1}(y) - L_{m-1}(y))dy$$

$$= \frac{1}{2\sqrt{N(2m+1)}} \left(\int_{-1}^{1} \frac{\partial^2 f}{\partial y^2} \left(\frac{L_{m+2}(y) - L_m(y)}{2m+3} \right. \right.$$

$$\left. \left. - \frac{L_m(y) - L_m(y) - L_{m-2}(y)}{2m-1} \right) dy \right). \tag{3.17}$$

Now, suppose

$$R_1(y) = (2m-1)L_{m+2}(y) - (2m-1)L_m(y) - (2m+3)L_m(y)$$

$$+(2m+3)L_{m-2}(y) = (2m-1)L_{m+2}(y) - 2(2m+1)L_m(y)$$

$$+(2m+3)L_{m-2}(y). \tag{3.18}$$

Therefore,

$$|X(n, m)| \leq \lambda \int_{-1}^{1} \left| \frac{\partial^2 f}{\partial y^2} \right| |R_1(y)|dy$$

$$\leq \lambda \mathcal{M} \int_{-1}^{1} |R_1(y)|dy, \tag{3.19}$$

where $\lambda = \dfrac{1}{2\sqrt{N(2m+1)(2m+3)(2m-1)}}$ and $\left| \frac{\partial^2 f}{\partial y^2} \right| \leq \mathcal{M}$, by the hypothesis of the theorem.

Now, from Cauchy–Schwarz inequality, it can be obtained as

$$\left(\int_{-1}^{1} |R_1(y)| dy \right)^2 \leq \int_{-1}^{1} 1^2 dy \int_{-1}^{1} \left((2m-1)^2 + (4m+2)^2 \right.$$
$$\left. + (2m+3)^2 \right) \left(L_{m+2}^2(y) + L_m^2(y) + L_{m-2}^2(y) \right) dy.$$

This implies

$$\left(\int_{-1}^{1} |R_1(y)| dy \right)^2 \leq 2 \left((2m-1)^2 + (4m+2)^2 + (2m+3)^2 \right)$$
$$\left(\frac{2}{2m+3} + \frac{2}{2m+1} + \frac{2}{2m-3} \right).$$

Thus,

$$\int_{-1}^{1} |R_1(y)| dy \leq 2 \sqrt{ \frac{ \left((2m-1)^2 + (4m+2)^2 + (2m+3)^2 \right) }{ \left(\frac{1}{2m+3} + \frac{1}{2m+1} + \frac{1}{2m-3} \right) } }. \tag{3.20}$$

Putting Eq. (3.20) in Eq. (3.19),

$$|X(n,m)| \leq 2\mathcal{M} \lambda \eta,$$

where $\eta = \sqrt{ \left((2m-1)^2 + (4m+2)^2 + (2m+3)^2 \right) \left(\frac{1}{2m+3} + \frac{1}{2m+1} + \frac{1}{2m-3} \right) }.$

Therefore, $\sum_{n=1}^{N} \sum_{m=0}^{M-1} X(n,m)$ is absolutely convergent.
Hence, the HLBPFs expansion of $f(t)$ converges uniformly. $\qquad \square$

For error analysis, define $\|X\| = E \left\{ \left(|X(t)|^2 \right)^{1/2} \right\}.$

Theorem 3.3.4.2. *Suppose $X(t)$ is the exact solution of Eq. (3.1) and $X_m(t)$ is its HLBPF approximate solution. Also, it can be assumed that*

$$\|X(t)\| \leq \mu, \ \ t \in [0,1],$$
$$\text{and } \|k_i(s,t)\| \leq M_i, \quad s,t \in [0,1] \times [0,1], \quad i = 1,2,3.$$

Then, for any $\varepsilon > 0$,

$$\|X(t) - X_m(t)\| \leq \frac{\varepsilon[1 + \mu((\beta - \alpha) + \|B(t)\| + 1)]}{1 - [(\beta - \alpha)(M_1 + \varepsilon) + (M_2 + \varepsilon) + \|B(t)\|(M_3 + \varepsilon)]}.$$

Proof. From Eq. (3.1), it yields

$$X(t) - X_m(t) = f(t) - f_m(t) + \int_\alpha^\beta (k_1(s,t)X(s)$$
$$- k_{1m}(s,t)X_m(s))ds$$
$$+ \int_0^t (k_2(s,t)X(s) - k_{2m}(s,t)X_m(s))ds$$
$$+ \int_0^t (k_3(s,t)X(s) - k_{3m}(s,t)X_m(s))dB(s).$$

So, by integral mean value theorem,

$$\|X(t) - X_m(t)\| \le \|f(t) - f_m(t)\| + (\beta - \alpha)\|k_1(\xi_1,t)X(\xi_1)$$
$$- k_{1m}(\xi_1,t)X_m(\xi_1)\|$$
$$+ t\|k_2(\xi_2,t)X(\xi_2) - k_{2m}(\xi_2,t)X_m(\xi_2)\|$$
$$+ \|B(t)\|\|k_3(\xi_3,t)X(\xi_3) - k_{3m}(\xi_3,t)X_m(\xi_3)\|,$$

$$(3.21)$$

where $\xi_1 \in [\alpha,\beta]$ and $\xi_2, \xi_3 \in [0,1]$.

Now, Theorem 3.3.4.1 shows that the HLBPFs expansion for any continuous function $f(t)$ converges uniformly to f. So, for any ε, there exists m such that

$$\|k_i(s,t) - k_{im}(s,t)\| \le \varepsilon, i - 1,2,3,$$
$$\|f(t) - f_m(t)\| \le \varepsilon.$$

Thus, for $i = 1,2,3$, it yields

$$\|k_i(\xi_i,t)X(\xi_i) - k_{im}(\xi_i,t)X_m(\xi_i)\| \le \|k_i(\xi_i,t)\|\|X(\xi_i) - X_m(\xi_i)\|$$
$$+ \|k_i(\xi_i,t) - k_{im}(\xi_i,t)\|\|X(\xi_i)\|$$
$$+ \|k_i(\xi_i,t) - k_{im}(\xi_i,t)\|$$
$$\|X(\xi_i) - X_m(\xi_i)\|$$
$$\le (M_i + \varepsilon)\|X(\xi_i) - X_m(\xi_i)\|$$
$$+ \mu\varepsilon.$$

$$(3.22)$$

Now, substituting Eq. (3.22) in Eq. (3.21) yields

$$
\begin{aligned}
\|X(t) - X_m(t)\| \leq{} & \varepsilon + (\beta - \alpha)\left[(M_1 + \varepsilon)\|X(t) - X_m(t)\| + \mu\varepsilon\right] \\
& + (M_2 + \varepsilon)\|X(t) - X_m(t)\| + \mu\varepsilon \\
& + \|B(t)\|\left[(M_3 + \varepsilon)\|X(t) - X_m(t)\| + \mu\varepsilon\right],
\end{aligned}
$$

replacing ξ_1, ξ_2, ξ_3 by an arbitrary $t \in [0, 1]$.

Hence, the following has been obtained:

$$
\|X(t) - X_m(t)\| \leq \frac{\varepsilon[1 + \mu((\beta - \alpha) + \|B(t)\| + 1)]}{1 - [(\beta - \alpha)(M_1 + \varepsilon) + (M_2 + \varepsilon) + \|B(t)\|(M_3 + \varepsilon)]}.
$$

$\square$

3.3.5 *Numerical examples*

In this section, the efficiency of the HLBPF method has been examined through suitable nontrivial examples.

Example 3.1. Consider the following SVFIE [104]:

$$
X(t) = f(t) + \int_0^1 \cos(s + t)X(s)ds + \int_0^t (s + t)X(s)ds
$$

$$
+ \int_0^t e^{-3(s+t)}X(s)dB(s), t \in [0, 1),
$$

in which

$$
f(t) = t^2 + \sin(1 + t) - 2\cos(1 - t) - 2\sin(t) - \frac{7t^4}{12} + \frac{1}{40}B(t),
$$

where $X(t)$ is an unknown on $(\Omega, \mathcal{F}, \mathbb{P})$ and $B(t)$ is a Brownian motion process. The stochastic OM of HLBPFs and the proposed scheme in Section 3.3.3 are used for solving Example 3.1. A comparison between the numerical results of HLBPFs and second-kind Chebyshev wavelets (CWs) is shown in Table 3.1 and the sample path for $\hat{m} = 64$ has been plotted in Fig. 3.1.

Table 3.1. A comparison between numerical solutions of HLBPFs and CWs for $\hat{m} = 16$, $\hat{m} = 64$ and $\hat{m} = 128$.

	$\hat{m} = 16$		$\hat{m} = 64$		$\hat{m} = 128$	
t	Hybrid Legendre BPF	Second-kind Chebyshev wavelets	Hybrid Legendre BPF	Second-kind Chebyshev wavelets	Hybrid Legendre BPF	Second-kind Chebyshev wavelets
0.1	0.006913	0.006058	0.005348	0.004357	0.011971	0.008284
0.3	0.082015	0.082758	0.073878	0.049951	0.103723	0.089965
0.5	0.228978	0.244836	0.211842	0.155212	0.208927	0.074237
0.7	0.486592	0.461998	0.461905	0.43497	0.503052	0.503779
0.9	0.828383	0.809047	0.806919	0.785316	0.781354	0.822646

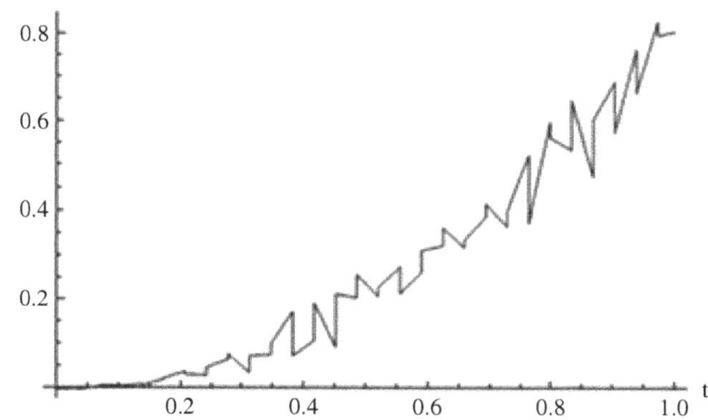

Fig. 3.1. The sample path of Example 3.1 for $\hat{m} = 64$.

Example 3.2. Consider the following SVFIE [104]:

$$X(t) = f(t) + \int_0^1 (s+t)X(s)ds + \int_0^t (s-t)X(s)ds$$

$$+ \frac{1}{125} \int_0^t \sin(s+t)X(s)dB(s), t \in [0,1),$$

in which

$$f(t) = 2 - \cos(1) - (1+t)\sin(1) + \frac{1}{250}\sin(B(t)),$$

Table 3.2. A comparison between numerical solutions of HLBPFs and CWs for $\hat{m} = 16$, $\hat{m} = 32$, $\hat{m} = 64$ and $\hat{m} = 128$.

t	$\hat{m} = 16$		$\hat{m} = 32$		$\hat{m} = 64$		$\hat{m} = 128$	
	Hybrid Legendre BPF	Second-kind Chebyshev wavelets	Hybrid Legendre BPF	Second-kind Chebyshev wavelets	Hybrid Legendre BPF	Second-kind Chebyshev wavelets	Hybrid Legendre BPF	Second-kind Chebyshev wavelets
0.1	0.992653	0.984086	0.963417	0.995998	0.972043	0.963967	0.987766	0.99065
0.3	0.953967	0.942391	0.894796	0.961358	0.927789	0.925242	0.944809	0.942445
0.5	0.872557	0.85293	0.84054	0.868952	0.843569	0.881	0.787147	0.879926
0.7	0.767245	0.734418	0.735131	0.737895	0.745125	0.726265	0.660328	0.564951
0.9	0.632835	0.596789	0.572819	0.621766	0.464226	0.578294	0.514462	0.594356

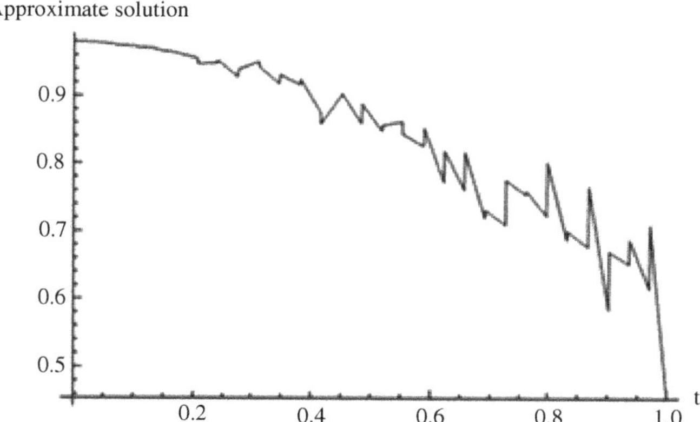

Fig. 3.2. The sample path of Example 3.2 for $\hat{m} = 64$.

where $X(t)$ is an unknown on $(\Omega, \mathcal{F}, \mathbb{P})$ and $B(t)$ is a BM process. The stochastic OM of HLBPFs and the proposed scheme in Section 3.3.3 are used for solving Example 3.2. A comparison between the numerical results of HLBPFs and second-kind CWs is shown in Table 3.2 and the sample path for $\hat{m} = 64$ has been plotted in Fig. 3.2.

3.4 Two-dimensional CWs for Mixed Stochastic Integral Equation

The two-dimensional BPFs, the two-dimensional CWs and their corresponding OM have been discussed in this subsection.

3.4.1 *Two-dimensional BPFs*

An (m_1, m_2) set of two-dimensional-BPFs defined in $t \in [0, T_1)$ and $x \in [0, T_2)$

$$\varphi_{i_1, i_2}(t, x) = \begin{cases} 1, & (i_1 - 1)h_1 \leq t < i_1 h_1 \text{ and } (i_2 - 1)h_2 \leq x < i_2 h_2, \\ 0, & \text{otherwise}, \end{cases}$$

$$(3.23)$$

where $i_1 = 1, 2, \ldots, m_1$ and $i_2 = 1, 2, \ldots, m_2$ with positive integer values for m_1, m_2 and $h_1 = \frac{T_1}{m_1}$, $h_2 = \frac{T_2}{m_2}$. Some properties of two-dimensional-BPFs like orthogonality, disjointness and completeness are similar to one-dimensional-BPFs [117].

3.4.2   Two-dimensional second-kind Chebyshev wavelets

For $m_1 = 0, 1, \ldots, M_1 - 1$, $m_2 = 0, 1, \ldots, M_2 - 1$ and $n_1 = 1, 2, \ldots, 2^{k_1-1}$, $n_2 = 1, 2, \ldots, 2^{k_2-1}$, the two-dimensional CWs on the interval $[0, 1) \times [0, 1)$ are

$$\psi_{n_1,m_1,n_2,m_2}(t, x)$$

$$= \begin{cases} \frac{2}{\pi} 2^{\frac{k_1}{2}} U_{m_1}(2^{k_1}t - 2n_1 + 1) 2^{\frac{k_2}{2}} U_{m_2}(2^{k_2}x - 2n_2 + 1), & \frac{n_1-1}{2^{k_1-1}} \le t < \frac{n_1}{2^{k_1-1}}, \\ 0, & \text{otherwise,} \end{cases}$$

$$\text{and } \frac{n_2 - 1}{2^{k_2-1}} \le x < \frac{n_2}{2^{k_2-1}}. \tag{3.24}$$

Here, two-dimensional CWs [105,122,123] form an orthonormal basis for $L^2_{w_{nk}}[0,1]$ with respect to weight functions $w_{n_1k_1}(t) = w(2^{k_1}t - 2n_1 + 1)$ and $w_{n_2k_2}(x) = w(2^{k_2}x - 2n_2 + 1)$ with $w(t) = \sqrt{1 - t^2}$ and $w(x) = \sqrt{1 - x^{2'}}$.

3.4.2.1   Function approximation

Let $f(t, x)$ over $[0, 1) \times [0, 1)$ satisfy, $f(t, x) \in L^2([0, 1) \times [0, 1))$ can be expanded in terms of two-dimensional-CWs as

$$f(t, x) = \sum_{n_1=1}^{\infty} \sum_{m_1=0}^{\infty} \sum_{n_2=1}^{\infty} \sum_{m_2=0}^{\infty} C(n_1, m_1, n_2, m_2)\psi_{n_1,m_1,n_2,m_2}(t, x).$$

$$\tag{3.25}$$

If Eq. (3.25) is truncated at some values of $2^{k_1-1}, 2^{k_2-1}$ and M_1, M_2, then it can be written as

$$f(t, x) \approx \sum_{n_1=1}^{2^{k_1-1}} \sum_{m_1=0}^{M_1-1} \sum_{n_2=1}^{2^{k_2-1}} \sum_{m_2=0}^{M_2-1} C(n_1, m_1, n_2, m_2)\psi_{n_1,m_1,n_2,m_2}(t, x),$$

$$= X^T \Psi(t \cdot x), \tag{3.26}$$

where C and $\Psi(t,x)$ are $2^{k_1-1}2^{k_2-1}M_1M_2$ column vectors

$$
\begin{aligned}
C = [&C(1,0,1,0),\ldots,C(1,0,1,M_2-1),C(1,0,2,0),\ldots,\\
&C(1,0,2,M_2-1),\ldots,C(1,0,2^{k_2-1},M_2-1),\ldots,\\
&C(1,M_1-1,2^{k_2-1},M_2-1),\ldots,\\
&C(2^{k_1-1},M_1-1,2^{k_2-1},M_2-1)\Big]^T,
\end{aligned}
\tag{3.27}
$$

and

$$
\begin{aligned}
\Psi(t,x) = [&\psi_{1,0,1,0}(t,x),\psi_{1,0,1,M_2-1}(t,x),\psi_{1,0,2,0}(t,x),\ldots,\\
&\psi_{1,0,2,M_2-1}(t,x),\ldots,\psi_{2^{k_1-1},M_1-1,2^{k_2-1},M_2-1}(t,x)\Big]^T.
\end{aligned}
\tag{3.28}
$$

3.4.2.2 *Integration operational matrices*

In this subsection, a stochastic operational matrix (SOM) for two-dimensional-BPFs has been proposed.

Lemma 3.4.2.1. *Let $\Phi(t,x)$ be the two-dimensional-BPFs. The integral can be derived as*

$$
\int_0^t \int_0^1 \Phi(s,y)\,dy\,ds \cong P\Phi(t,x) = [E_{m_1\times m_1}\otimes E_{m_2\times m_2}]\Phi(t,x), \tag{3.29}
$$

where P is the $2^{k_1-1}2^{k_2-1}M_1M_2 \times 2^{k_1-1}2^{k_2-1}M_1M_2$-dimensional integrational OM for two-dimensional-BPFs and

$$
E_{m_1,m_2} = \frac{h}{2}\begin{bmatrix} 1 & 2 & 2 & \cdots & 2 \\ 0 & 1 & 2 & \cdots & 2 \\ 0 & 0 & 1 & \cdots & 2 \\ \vdots & \vdots & \vdots & \ddots & \vdots \\ 0 & 0 & 0 & \cdots & 1 \end{bmatrix}.
$$

Proof. It may be referred to Ref. [117]. $\qquad\qquad\square$

Lemma 3.4.2.2. *Let $\Phi(t,x)$ be the two-dimensional-BPFs. The Itô integral is*

$$\int_0^t \int_0^1 \Phi(s,y)dB(s,y) \cong P_s\Phi(t,x), \qquad (3.30)$$

where P_s is the $2^{k_1-1}2^{k_2-1}M_1M_2 \times 2^{k_1-1}2^{k_2-1}M_1M_2$-dimensional integrational SOM for two-dimensional-BPFs whose rows consist of elements such as

$$(0,\ldots,0,\alpha(i_1,i_2),\beta(i_1,i_2),\ldots,\beta(i_1,i_2),0,\ldots,0,\gamma(i_1,i_2),$$

$$\lambda(i_1,i_2),\ldots,\lambda(i_1,i_2),0,\ldots,0,\gamma(i_1,i_2),\lambda(i_1,i_2),\ldots,\lambda(i_1,i_2)),$$

where $i_1 = 1,2,\ldots,2^{k_1-1}2^{k_2-1}M_1M_2$ and $i_2 = 1,2,\ldots,2^{k_1-1}2^{k_2-1}M_1M_2$

$$\alpha(i_1,i_2) = B((i_1-0.5)h,(i_2-0.5)k) - B((i_1-0.5)h,(i_2-1)k)$$
$$-B((i_1-1)h,(i_2-0.5)k) + B((i_1-1)h,(i_2-1)k),$$

$$\beta(i_1,i_2) = B((i_1-0.5)h,i_2k) - B((i_1-0.5)h,(i_2-1)k)$$
$$-B((i_1-1)h,i_2k) + B((i_1-1)h,(i_2-1)k),$$

$$\gamma(i_1,i_2) = B(i_1h,(i_2-0.5)k) - B(i_1h,(i_2-1)k)$$
$$-B((i_1-1)h,(i_2-0.5)k) + B((i_1-1)h,(i_2-1)k),$$

$$\lambda(i_1,i_2) = B(i_1h,i_2k) - B((i_1-1)h,i_2k) - B(i_1h,(i_2-1)k)$$
$$+B((i_1-1)h,(i_2-1)k).$$

Proof.   It may be further referred to Ref. [117]. $\qquad\qquad\square$

3.4.2.3   *Relation of two-dimensional CWs and two-dimensional BPFs*

Theorem 3.4.2.3. *Let $\Psi(t,x)$ and $\Phi(t,x)$ be $2^{k_1-1}2^{k_2-1}M_1M_2$-dimensional two-dimensional CWs and two-dimensional BPFs vectors. The vector $\Psi(t,x)$ can be expanded as*

$$\Psi(t,x) = Q\Phi(t,x), \qquad (3.31)$$

where Q is a $2^{k_1-1}2^{k_2-1}M_1M_2 \times 2^{k_1-1}2^{k_2-1}M_1M_2$ block matrix.

Proof.   It can be referred to Refs. [117,119]. $\qquad\qquad\square$

Remark 3.4.2.4. For a $2^{k_1-1}2^{k_2-1}M_1M_2$-vector F, then

$$\Psi(t,x)\Psi^T(t.,x)F = \tilde{F}\Psi(t,x),$$

in which $\tilde{F}$ is a $2^{k_1-1}2^{k_2-1}M_1M_2 \times 2^{k_1-1}2^{k_2-1}M_1M_2$ matrix of the form

$$\tilde{F} = Q\bar{F}Q^{-1},$$

where $\bar{F} = diag\left(Q^T F\right)$.

Remark 3.4.2.5. Let A represent a $2^{k_1-1}2^{k_2-1}M_1M_2 \times 2^{k_1-1}2^{k_2-1}M_1M_2$ matrix. Then, for the two-dimensional CWs vector $\Psi(t,x)$,

$$\Psi^T(t,x)A\Psi(t,x) = \hat{A}^T\Psi(t,x),$$

where $\hat{A}^T = UQ^{-1}$ and $U = diag(Q^T AQ)$ is a $2^{k_1-1}2^{k_2-1}M_1M_2$ vector.

3.4.2.4 *Stochastic operational matrix of two-dimensional CWs*

Theorem 3.4.2.6 ([116,117]). *If $\Psi(t,x)$ be the $2^{k_1-1}2^{k_2-1}M_1M_2$-dimensional two-dimensional CWs vector in Eq. (3.28). The integral of this vector is*

$$\int_0^t \int_0^1 \Psi(s,y)dyds = QPQ^{-1}\Psi(t,x) = \Lambda\Psi(t,x), \qquad (3.32)$$

where Q is introduced in Eq. (3.31) and P is the OM of integration for BPFs specified in Eq. (3.29).

Proof. Let $\Psi(t,x)$ imply two-dimensional-second-kind Chebyshev wavelets vector.

$$\int_0^t \int_0^1 \Psi(s,y)dyds = Q\int_0^t \int_0^1 \Phi(s,y)dyds = QP\Phi(t,x),$$

where $\Psi(t,x) = Q\Phi(t,x)$.

Now, from Theorem 3.4.2.3,

$$\int_0^t \int_0^1 \Psi(s.y)dyds = QPQ^{-1}\Psi(t,x) = \Lambda\Psi(t,x),$$

where $\Lambda = QPQ^{-1}$.

This completes the proof. $\qquad\qquad\square$

Theorem 3.4.2.7 ([117,119]). *Let $\Psi(t, x)$ be a $2^{k_1-1}2^{k_2-1}M_1M_2$-dimensional two-dimensional CWs vector in Eq. (3.28). The Itô integral of this vector is*

$$\int_0^t \int_0^1 \Psi(s, y)dB(y, s) = QP_sQ^{-1}\Psi(t, x) = \Lambda_s\Psi(t, x), \qquad (3.33)$$

where Λ_s represents the stochastic operational matrix of two-dimensional CWs, P_s represents the stochastic operational matrix for two-dimensional-BPFs specified in Eq. (3.30) and Q is introduced in Eq. (3.31).

Proof. Let $\Psi(t, x)$ be a two-dimensional CWs vector. Applying Theorem 3.4.2.3 and Lemma 3.4.2.2, it yields

$$\int_0^t \int_0^1 \Psi(s, y)dB(y, s) = Q\int_0^t \int_0^1 \Phi(s, y)dB(y, s) = QP_s\Phi(t, x).$$

Now, from Theorem 3.4.1.1,

$$\int_0^t \int_0^1 \Psi(s.y)dB(y, s) = QP_sQ^{-1}\Psi(t, x) = \Lambda_s\Psi(t, x),$$

where $\Lambda_s = QP_sQ^{-1}$.

This completes the proof. $\qquad\qquad\qquad\qquad\qquad\qquad\qquad$ $\square$

3.4.3 *Approximation of SMVFIE*

Consider the following SMVFIE:

$$u(t, x) = u_0(x, t) + \int_0^t \int_0^1 k_1(t, s, x, y)u(s, y)dyds$$

$$+ \int_0^t \int_0^1 k_2(t, s, x, y)dB(s, y), \quad t, x \in [0, 1], \quad (3.34)$$

where the unknown function $u(t, x)$ is to be obtained for $t, x \in [0, 1]$. Let

$$u(t, x) = U^T\Psi(t) = \Psi^T(t)U,$$

$$u(t, x) = U_0^T\Psi(t) = \Psi^T(t)U_0, \qquad\qquad (3.35)$$

$$k_i(t, x, s, y) = \Psi^T(t, x)K_i\Psi(s, y) = \Psi^T(s, y)K_i^T\Psi(t, x), \quad i = 1, 2,$$

where U and U_0 are the two-dimensional CWs coefficients vectors and $K_i, i = 1, 2$, are the two-dimensional CWs coefficients matrices. The above approximations are substituted in Eq. (3.34), then

$$U^T \Psi(t, x) = U_0^T \Psi(t, x) + \Psi^T(t, x) K_1 \left(\int_0^t \int_0^1 \Psi(s, y) \Psi^T(s, y) U \, dy \, ds \right)$$

$$+ \Psi^T(t, x) K_2 \left(\int_0^t \int_0^1 \Psi(s, y) \Psi^T(s, y) U \, dB(s, y) \right).$$

Using Remark 3.4.2.4,

$$U^T \Psi(t, x) = U_0^T \Psi(t, x) + \Psi^T(t, x) K_1 \left(\int_0^t \int_0^1 \tilde{U} \Psi(s, y) \, dy \, ds \right)$$

$$+ \Psi^T(t, x) K_2 \left(\int_0^t \int_0^1 \tilde{U} \Psi(s, y) \, dB(s, y) \right),$$

where $\tilde{U}$ is a $2^{k_1-1}2^{k_2-1}M_1 M_2 \times 2^{k_1-1}2^{k_2-1}M_1 M_2$ matrix. Now, applying operational matrices Λ, Λ_s for two-dimensional CWs derived in Eqs. (3.32) and (3.33),

$$U^T \Psi(t, x) = U_0^T \Psi(t, x) + \Psi^T(t, x) K_1 \tilde{U} \Lambda \Psi(t, x)$$

$$+ \Psi^T(t, x) K_2 \tilde{U} \Lambda_s \Psi(t, x).$$

By setting $Y_1 = K_1 \tilde{U} \Lambda, Y_2 = K_2 \tilde{U} \Lambda_s$ and using Remark 3.4.2.5, it yields

$$U^T \Psi(t, x) = U_0^T \Psi(t, x) + \hat{Y}_1^T \Psi(t, x) + \hat{Y}_2^T \Psi(t, x),$$

in which $\hat{Y}_0$ and $\hat{Y}_i (i = 1, 2)$ are $2^{k_1-1}2^{k_2-1}M_1 M_2$-dimensional vectors. Thus,

$$U^T = U_0^T + \hat{Y}_1^T + \hat{Y}_2^T. \tag{3.36}$$

Since $\hat{Y}_1^I, \hat{Y}_2^T$ are linear functions of U, Eq. (3.36) is a linear system of equations for U. By determining U, the approximate solution of the mixed stochastic integral equation (3.34) has been determined by substituting the obtained vector U in Eq. (3.35).

3.4.4 *Numerical examples*

The accuracy and efficiency of two-dimensional CWs have been examined through suitable examples of nonlinear mixed stochastic integral equations [117].

Example 3.3.

$$u(t,x) = x^2 e^t - \frac{2}{3}x^3 t^2 + \int_0^t \int_0^1 t^2 e^{-y} u(s,y)\,dy\,ds$$

$$+ \int_0^x \int_0^1 t e^{-y} u(s,y)\,dB(y,s),\, t,x \in [0,1).$$

A comparison between two-dimensional CWs method solution and HLBPF method solution has been shown in Table 3.3 and displayed in Figs. 3.3(a)–(c).

Example 3.4.

$$u(t,x) = \frac{1}{12} + \int_0^t \int_0^1 u(s,y)\,dy\,ds + \int_0^t \int_0^1 u(s,y)\,dB(y,s),\, t,x \in [0,1),$$

with the exact solution $u(t,x) = \frac{1}{12}e^{\frac{1}{2}t + B(t,1)}$. A comparison between two-dimensional CWs method solution, HLBPF method solution and exact solutions has been shown in Table 3.4 and displayed in Figs. 3.4(a)–(c), respectively.

Table 3.3. A comparison between two-dimensional CWs method and HLBPF method solutions for Example 3.3.

x	$k_1=1,k_2=1,$ $M_1=2$ and $M_2=2$ Proposed method solution	HLBPF solution	$k_1=2,k_2=2,$ $M_1=2$ and $M_2=2$ Proposed method solution	HLBPF solution	$k_1=2,k_2=2,$ $M_1=4$ and $M_2=4$ Proposed method solution	HLBPF solution
0.0	0.057112	0.048551	0.040122	0.0394357	0.051663	0.0496333
0.1	0.618895	0.526774	0.548527	0.406777	0.54622	0.5032
0.2	0.943371	1.09931	1.14339	1.04987	1.38895	1.16663
0.3	0.847736	0.75229	0.873394	0.903498	0.108834	0.0838802
0.4	0.661337	0.657734	0.593363	0.771464	0.677127	0.586364
0.5	0.982263	1.67228	1.64119	1.81379	1.48668	1.45337
0.6	0.082275	0.091886	0.090442	0.0807267	0.11374	0.0986566
0.7	1.369935	0.908114	0.698878	0.824468	0.87228	0.796632
0.8	1.81223	1.679122	1.82269	1.65796	2.17553	1.86332

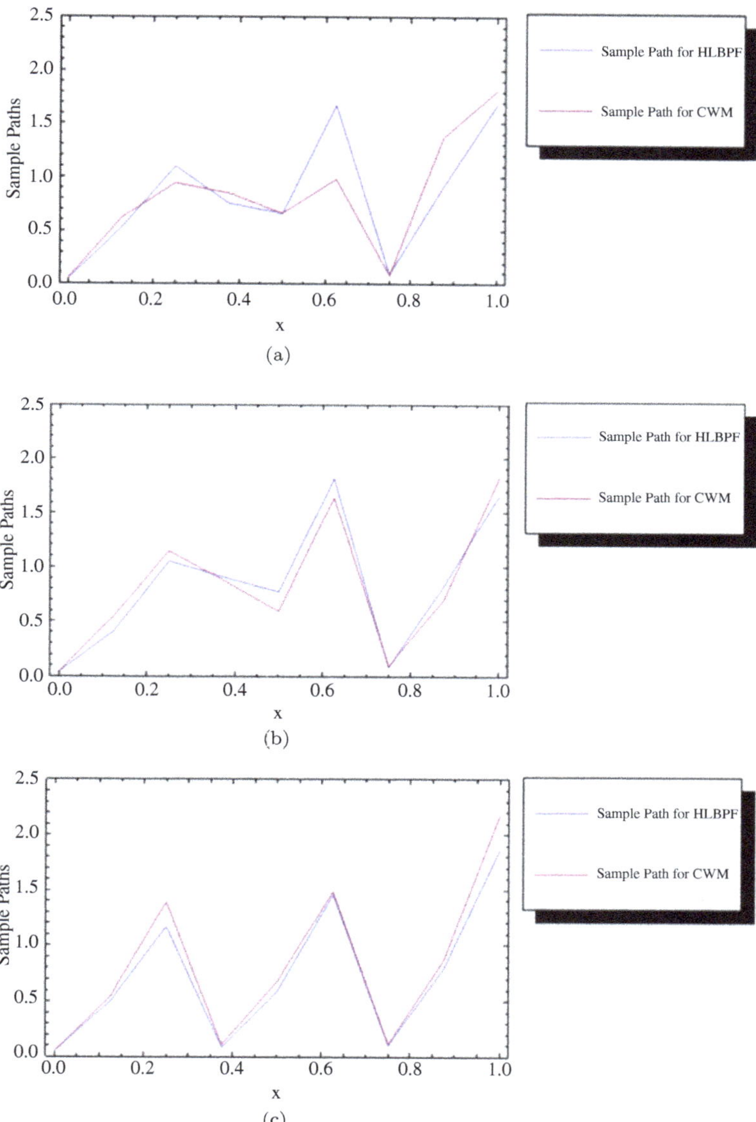

Fig. 3.3. (a) The sample paths of Example 3.3 for $k_1 = 1, k_2 = 1, M_1 = 2$ and $M_2 = 2$. (b) The sample paths of Example 3.3 for $k_1 = 2, k_2 = 2, M_1 = 2$ and $M_2 = 2$. (c) The sample paths of Example 3.3 for $k_1 = 2, k_2 = 2, M_1 = 4$ and $M_2 = 4$.

Table 3.4. A comparison between two-dimensional CWs method, exact and HLBPF method solutions for Example 3.4.

x	$k_1 = 1, k_2 = 1,$ $M_1 = 2$ and $M_2 = 2$			$k_1 = 2, k_2 = 2,$ $M_1 = 2$ and $M_2 = 2$			$k_1 = 2, k_2 = 2,$ $M_1 = 4$ and $M_2 = 4$		
	Proposed method solution	Exact solution	HLBPF solution	Proposed method solution	Exact solution	HLBPF solution	Proposed method solution	Exact solution	HLBPF solution
0.0	0.159113	0.171134	0.14689	0.214051	0.219949	0.248334	0.117291	0.12066	0.141224
0.1	0.166131	0.161503	0.20489	0.157864	0.143199	0.16234	0.17874	0.187729	0.190633
0.2	0.218493	0.231788	0.19963	0.139446	0.143188	0.186638	0.133745	0.139863	0.163021
0.3	0.168326	0.168341	0.17364	0.318598	0.292112	0.248857	0.437386	0.45436	0.604723
0.4	0.194758	0.195762	0.18722	0.045290	0.0488749	0.031599	0.517495	0.492506	0.579328
0.5	0.0798811	0.083615	0.10134	0.22059	0.226348	0.265515	0.105544	0.096084	0.114865
0.6	0.272227	0.254725	0.30478	0.081084	0.0780385	0.071553	0.133635	0.132296	0.171236
0.7	0.406694	0.35271	0.39561	0.30455	0.316297	0.385128	0.040455	0.049236	0.068317
0.8	0.156024	0.102363	0.15669	0.045682	0.0428931	0.050327	0.0452905	0.046253	0.0540967

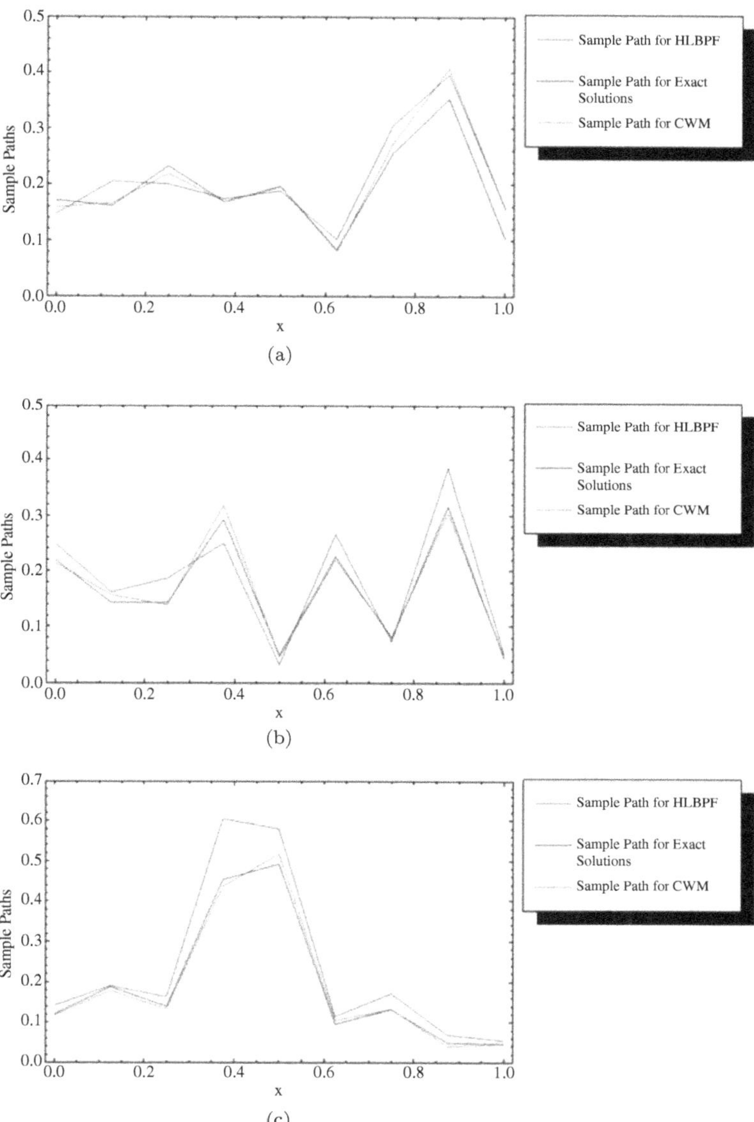

Fig. 3.4. (a) The sample paths of Example 3.4 for $k_1 = 1$, $k_2 = 1$, $M_1 = 2$ and $M_2 = 2$. (b) The sample paths of Example 3.4 for $k_1 = 2$, $k_2 = 2$, $M_1 = 2$ and $M_2 = 2$. (c) The sample paths of Example 3.4 for $k_1 = 2$, $k_2 = 2$, $M_1 = 4$ and $M_2 = 4$.

3.5 Summary

In this chapter, HLBPFs have been applied on the stochastic Volterra–Fredholm integral equation to obtain the numerical results. The stochastic operational matrix of HLBPFs has been used to discretise the stochastic Volterra–Fredholm integral equation. The obtained numerical results have been compared with those obtained from the second-kind CWs. Moreover, the efficiency of the proposed method has been well established by suitable numerical examples.

Also, two-dimensional CWs have been applied to obtain solutions for the nonlinear mixed stochastic integral equation. Operational matrix method based on two-dimensional BPFs and two-dimensional CWs has been used to discretise the mixed stochastic integral equation. Two illustrated examples have been presented to compare the results of two-dimensional CWs method solutions with that obtained by two-dimensional HLBPF method solutions in order to justify the accuracy and applicability of the proposed method.

Chapter 4

Numerical Solutions of Multidimensional Stochastic Integral Equation

4.1 Introduction

In recent years, wavelet theory has been widely applied in different fields of science and engineering [124]. It is legitimate to consider J. Fourier (1768–1830) as the initiator of the theory of integral equations, owing to the fact that he obtained the inversion formula for what is now called the "Fourier transform". One can interpret the inversion formula as providing the inverse operator (an integral operator) of the Fourier integral operator. This interpretation was adopted towards the end of the last century by V. Volterra, who identified the problem of solving integral equations with the problem of finding inverses of certain integral operators [125].

The year 1895 marked a new beginning in the theory of integral equations, due mainly to Volterra [126]. Unlike most of his predecessors, who were aiming at finding the solution of the equation by means of formulas or who dealt with special cases of what is now called "Volterra equations" (the term was introduced by Lalesco who wrote his thesis on this topic with Picard). In the last 15–20 years, one observes a sharp increase in the interest in the theory of Volterra integral equations. The interest in this theory has been stimulated by the steady extension of the volume of applications as well as by the realisation of the fact that the Volterra equations are not only simple special cases of the Fredholm equations but represent a class of equations with their own specific problems.

In the early 1900s, Volterra developed new types of equations termed as integro-differential equations while studying the population growth [16]. In these types of equations, the unknown function occurs on one side as an ordinary derivative and appears on the other side under the integral sign. Integro-differential equations can be modelled from several phenomena occurring in physics and biology. As integral equations arise from distinct origins, such as boundary value problems as in Fredholm equations, and from initial value problems as in Volterra equations, different schemes were implemented to obtain solutions for different integral equations [97].

Random integral equations have been modelled by various researchers to study the nondeterministic behaviour occurring in the general areas of applied science. Wave propagation in random media can be described by stochastic integral equations. According to statistical turbulence theory, stochastic integral equations arise in describing the motion of a point in a continuous fluid in turbulent motion and also arise in the drawbacks of chemical kinetics and metabolising systems [20].

The main characteristic of wavelet methodology is that it reduces these problems to those of solving a system of algebraic equations. CAS wavelet [111], Legendre wavelet [112,113,127], Legendre multiwavelet [114], Haar wavelet [115], Chebyshev wavelet [116] and so on have been widely used in numerical solutions of all kinds of linear or nonlinear integral equations, differential equations and integro-differential equations.

Maleknejad, Khodabin and Rostami obtained numerical solutions of the multidimensional Itô–Volterra integral equations by using BPFs and their stochastic operational matrix (SOM) of integration [125]. Mohammadi proposed Haar wavelets for solving multidimensional stochastic Itô–Volterra integral equations (MDSIVIEs) [127]. Sahu and Saha Ray have applied hybrid Legendre Block-Pulse functions (HLBPFs) for obtaining numerical solutions of a system of nonlinear Fredholm–Hammerstein integral equations [128].

4.2 Outline of Present Study

In Section 4.3, HLBPFs are implemented in determining the approximate solutions for MDSIVIEs. The BPF and the proposed scheme are used for deriving a methodology to obtain stochastic operational matrix. Error and convergence analysis of the scheme is discussed. A brief discussion including numerical examples has been provided to establish the efficiency of the mentioned method.

In Section 4.4, the numerical solutions of MDSIVIEs have been obtained by second-kind Chebyshev wavelets (CWs). The second-kind CWs are orthonormal and have compact support on [0, 1]. The BPFs and their relations to second-kind CWs are employed to derive a general procedure for forming SOM of SKCWs. Convergence and error analysis of the proposed method are also discussed. Furthermore, some examples have been discussed to establish the accuracy of the proposed scheme.

4.2.1 *Multidimensional stochastic Itô–Volterra integral equations*

Consider the MDSIVIE [125,127]

$$X(t) = f(t) + \int_0^t k_0(s,t)X(s)ds + \sum_{i=1}^n \int_0^t k_i(s,t)X(s)dB_i(s),$$

$$t \in [0,T), \tag{4.1}$$

where $X(t), f(t), k_0(s,t)$ and $k_i(s,t), i = 1, 2, \ldots, n$ for $s,t \in [0,T)$, are the stochastic processes defined on $(\Omega, \mathcal{F}, \mathbb{P})$ and $X(t)$ is unknown. Also, $B(t)$ is a BM process and $\int_0^t k_i(s,t)X(s)dB(s)$ is the Itô integral.

4.3   Hybrid Legendre Block-Pulse Functions for MDSIVIEs

4.3.1 *Block-Pulse functions*

A set of Block-Pulse functions [99] $\phi_n(x), n = 1, 2, \ldots, N$ on [0,1), are defined as follows:

$$\phi_n(t) = \begin{cases} 1, & \frac{n-1}{N} \le t < \frac{n}{N}, \\ 0, & \text{otherwise}, \end{cases}$$

with $t \in [0,1), n = 1, 2, \ldots, N$ and $h = \frac{1}{N}$.

The properties of Block-Pulse functions are as follows:

(i) The Block-Pulse functions on the interval $[0,1)$ are disjoint

$$\phi_n(t)\phi_m(t) = \delta_{nm}\phi_n(t),$$

where $n, m = 1, 2, \ldots, N$ and δ_{nm} is Kronecker delta.

(ii) The Block-Pulse functions are orthogonal on the interval $[0,1)$.

$$\int_0^1 \phi_n(t)\phi_m(t)dt = h\delta_{nm}, \quad n, m = 1, 2, \ldots, N.$$

(iii) If $N \to \infty$, then the BPFs set is complete; for every $f \in L^2([0,1))$, Parseval's identity holds,

$$\int_0^1 f^2(t)dt = \sum_{i=1}^{\infty} f_n^2\|\phi_n(t)\|^2,$$

where

$$f_n = \frac{1}{h}\int_0^1 f(t)\phi_n(t)dt.$$

Consider the first N terms of BPFs and write them concisely as N-vector

$$\Phi(t) = (\phi_1(t), \phi_2(t), \ldots, \phi_N(t))^T, \quad t \in [0,1). \tag{4.2}$$

The above representation and disjointness property follows

$$\Phi(t)\Phi^T(t) = \begin{bmatrix} \phi_1(t) & 0 & \cdots & 0 \\ 0 & \phi_2(t) & \cdots & 0 \\ \vdots & \vdots & \ddots & \vdots \\ 0 & 0 & \cdots & \phi_N(t) \end{bmatrix}_{N \times N}.$$

Furthermore, it can be written as $\Phi^T(t)\Phi(t) = 1$ and $\Phi(t)\Phi^T(t)F^T = D_F\Phi(t)$, where D_F usually denotes a diagonal matrix whose diagonal entries are related to a constant vector $F = (f_1, f_2, \ldots, f_N)^T$.

4.3.2 *Hybrid Legendre Block-Pulse functions*

Consider the Legendre polynomials $L_m(x)$ on the interval $[-1, 1]$

$$L_0(x) = 1,$$

$$L_1(x) = x,$$

$$L_m(x) = \tfrac{2m-1}{m} x L_{m-1}(x) - \tfrac{m-1}{m} L_{m-2}(x), m = 2, 3, \ldots$$

The set $\{L_m(x) : m = 0, 1, \ldots\}$ in Hilbert space $L^2[-1, 1]$ is a complete orthogonal set.

For $m = 0, 1, \ldots, M - 1$ and $n = 1, 2, \ldots, N$, the HLBPFs on $[0, 1)$ are defined in $[100, 101]$ as follows:

$$\psi(n, m, t) = \begin{cases} \sqrt{N\,(2m + 1)}\, L_m(2Nt - 2n + 1), & \frac{n-1}{N} \leq t < \frac{n}{N}, \\ 0, & \text{otherwise,} \end{cases}$$

where m and n are the order of the Legendre polynomials and Block-Pulse functions, respectively. The coefficient $\sqrt{N(2m + 1)}$ is for the orthonormality.

4.3.2.1 *Function approximation*

Any function $f(t)$ over the interval $[0,1)$ which is square integrable, i.e., $f(t) \in L^2[0, 1)$, can be expanded in terms of HLBPFs as

$$f(t) = \sum_{n=1}^{\infty} \sum_{m=0}^{\infty} X(n \cdot m) \psi(n, m, t), \tag{4.3}$$

where $X(n, m)\langle f(t), \psi(n, m, t)\rangle$. If the infinite series in Eq. (4.3) is truncated at some values of N and M, then

$$f(t) \approx \sum_{n=1}^{N} \sum_{m=0}^{M-1} X(n, m) \psi(n, m, t) = X^T \Psi(t), \tag{4.4}$$

where X and $\Psi(t)$ are $\hat{m} = NM$ column vectors,

$$X = [X(1, 0), X(1, 1), \ldots, X(1, M - 1), \ldots, X(N, 0),$$

$$X(N, 1), \ldots, X(N, M - 1)]^T, \tag{4.5}$$

and

$$\Psi(t) = [\psi(1,0,t), \psi(1,1,t), \ldots, \psi(1,M-1,t), \ldots$$
$$\psi(N,0,t), \ldots, \psi(N,M-1,t)]^T. \tag{4.6}$$

Since $\int_0^1 \Psi(t)\Psi^T(t)dt = 1$, where the I is an identity matrix of dimension NM, then calculate $X(n,m)$ as

$$X(n,m) = \int_0^1 f(t)\psi(n,m,t)dt, \quad n = 1, 2, \ldots, N$$

$$\text{and } m = 0, 1, \ldots, M-1.$$

Similarly, let $K(s,t) \in L^2([0,1) \times [0,1))$ where

$$K(s,t) \approx \Psi^T(s)K\Psi(t) = \Psi^T(t)K^T\Psi(s), \tag{4.7}$$

where $\Psi(s)$ and $\Psi(t)$ are $\hat{m}$-dimensional HLBP vectors, respectively, and $K = [k_{ij}]_{\hat{m}\times\hat{m}}$ is the $\hat{m} \times \hat{m}$ HLBP coefficient matrix with

$$k_{ij} = \int_0^1 \int_0^1 K(s,t)\Psi_i(s)\Psi_j(t)dtds = \langle \Psi_i(s), \langle K(s,t), \Psi_j(t)\rangle \rangle.$$

4.3.2.2 *Integration operational matrices*

In this section, a SOM for the HLBPFs has been derived.

Lemma 4.3.2.1. *Let $\Phi(t)$ be the $\hat{m}$-dimensional BPFs vector. The integral of this vector is*

$$\int_0^t \Phi(s)ds \cong P\Phi(t), \tag{4.8}$$

where P is called the OM of integration for BPFs and

$$P = \frac{h}{2}\begin{bmatrix} 1 & 2 & 2 & \cdots & 2 \\ 0 & 1 & 2 & \cdots & 2 \\ 0 & 0 & 1 & \vdots & \vdots \\ \vdots & \vdots & \vdots & \ddots & \vdots \\ 0 & 0 & 0 & \cdots & 1 \end{bmatrix}_{\hat{m}\times\hat{m}}.$$

Proof.   It may be referred to Ref. [99]. $\square$

Lemma 4.3.2.2. *Let $\Phi(t)$ be the $\hat{m}$-dimensional BPFs vector. The Itô integral of this vector is*

$$\int_0^t \Phi(s)dB(s) \cong P_s\Phi(t), \tag{4.9}$$

where P_s is called the SOM of integration for BPFs and

$$P_s = \begin{bmatrix} B\left(\frac{h}{2}\right) & B(h) & B(h) & \cdots & B(h) \\ 0 & B\left(\frac{3h}{2}\right) - B(h) & B(2h) - B(h) & \cdots & B(2h) - B(h) \\ 0 & 0 & B\left(\frac{5h}{2}\right) - B(2h) & \cdots & B(3h) - B(2h) \\ \vdots & \vdots & \vdots & \ddots & \vdots \\ 0 & 0 & 0 & \cdots & B\left(\frac{(2\hat{m}-1)h}{2}\right) \\ & & & & -B((\hat{m}-1)h) \end{bmatrix}_{\hat{m}\times\hat{m}}.$$

Proof.   It may be further referred to Ref. [99]. $\square$

4.3.2.3   *Hybrid Legendre Block-Pulse functions and BPFs*

In the following section, the relation between the HLBPFs and BPFs has been reviewed.

Theorem 4.3.2.3. *Let $\Psi(t)$ and $\Phi(t)$ be the $\hat{m}$-dimensional HLBPFs and BPFs vector. The vector $\Psi(t)$ can be expanded by BPFs vector $\Phi(t)$ as*

$$\Psi(t) = Q\Phi(t), \tag{4.10}$$

where Q is an $\hat{m} \times \hat{m}$ block matrix and

$$Q_{ij} = \psi_i\left(\frac{2j-1}{2\hat{m}}\right), \quad i,j = 1, 2, \ldots, \hat{m}.$$

Proof.   Let $\psi_i(t), i = 1, 2, \ldots, \hat{m}$, be the ith element of HLBPFs vector. Expanding $\psi_i(t)$ into an $\hat{m}$-term vector of BPFs, it yields

$$\psi_i(t) = \sum_{j=1}^{\hat{m}} Q_{ij}\phi_j(t), \quad i = 1, 2, \ldots, \hat{m},$$

where Q_{ij} is the $(i,j)^{th}$ element of matrix Q and

$$Q_{ij} = \frac{1}{h} \int_0^1 \psi_i(t)\phi_j(t)dt = \frac{1}{h} \int_{\frac{j-1}{\hat{m}}}^{\frac{j}{\hat{m}}} \psi_i(t)dt$$

$$= \hat{m} \int_{\frac{j-1}{\hat{m}}}^{\frac{j}{\hat{m}}} \psi_i(t)dt, \text{ where } \hat{m} = \frac{1}{h}.$$

Now, using the mean value theorem for integrals in the last equation yields

$$Q_{ij} = \hat{m}\left(\frac{j}{\hat{m}} - \frac{j-1}{\hat{m}}\right)\psi_i(\eta_j) = \psi_i(\eta_j), \quad \eta_j \in \left(\frac{j-1}{\hat{m}}, \frac{j}{\hat{m}}\right).$$

By choosing $\eta_j = \frac{2j-1}{2\hat{m}}$, the above equation becomes

$$Q_{ij} = \psi_i\left(\frac{2j-1}{2\hat{m}}\right), \quad i,j = 1,2,\ldots,\hat{m}.$$

$\square$

Remark 4.3.2.4. For an $\hat{m}$-vector F

$$\Psi(t)\Psi^T(t)F = \tilde{F}\Psi(t),$$

in which $\tilde{F}$ is an $\hat{m} \times \hat{m}$ matrix of the form

$$\tilde{F} = Q\bar{F}Q^{-1},$$

where $\bar{F} = diag(Q^T F)$.

Remark 4.3.2.5. Let A be an arbitrary $\hat{m} \times \hat{m}$ matrix. Then, for the HLBPFs vector $\Psi(t)$,

$$\Psi^T(t)A\Psi(t) = \hat{A}^T\Psi(t),$$

where $\hat{A}^T = UQ^{-1}$ and $U = diag(Q^T AQ)$ is an $\hat{m}$-vector.

4.3.2.4 *Stochastic OM of HLBPFs*

Theorem 4.3.2.6. *Let $\Psi(t)$ be the $\hat{m}$-dimensional HLBPFs vector. The integral of this vector is*

$$\int_0^t \Psi(s)ds = QPQ^{-1}\Psi(t) = \Lambda\Psi(t), \qquad (4.11)$$

where Q is introduced in Eq. (4.10) and is the OM of integration for BPFs in Eq. (4.8).

Proof. Let $\Psi(t)$ be the HLBPFs vector.

$$\int_0^t \Psi(s)ds = Q \int_0^t \Phi(s)ds = QP\Phi(t),$$

where $\Psi(t) = Q\Phi(t)$. $\qquad\qquad\square$

Now, Theorem 4.3.2.3 gives

$$\int_0^t \Psi(s)ds = QPQ^{-1}\Psi(t) = \Lambda\Psi(t),$$

where $\Lambda = QPQ^{-1}$.

Theorem 4.3.2.7. *Let $\Psi(t)$ be the $\hat{m}$-dimensional HLBPFs vector. The Itô integral of this vector is*

$$\int_0^t \Psi(s)dB(s) = QP_sQ^{-1}\Psi(t) = \Lambda_s\Psi(t), \qquad (4.12)$$

where Λ_s is called stochastic OM for HLBPFs, Q is introduced in Eq. (4.10) and P_s is the SOM of integration for BPFs in Eq. (4.9).

Proof. Let $\Psi(t)$ be the HLBPFs vector. Using Theorem 4.3.2.3 and Lemma 4.3.2.2 yields

$$\int_0^t \Psi(s)dB(s) = Q \int_0^t \Phi(s)dB(s) = QP_s\Phi(t).$$

Now, Theorem 4.3.2.3 yields

$$\int_0^t \Psi(s)dB(s) = QP_sQ^{-1}\Psi(t) = \Lambda_s\Psi(t),$$

where $\Lambda_s = QP_sQ^{-1}$. $\qquad\qquad\square$

4.3.3   *Approximation of MDSIVIE*

Consider the following MDSIVIE:

$$X(t) = f(t) + \int_0^t k_0(s,t)X(s)ds + \sum_{i=1}^{n} \int_0^t k_i(s,t)X(s)dB_i(s),$$

$$t \in [0, T), \tag{4.13}$$

where $X(t)$, $f(t)$ and $k_i(s,t)$, $i = 0, 1, 2, \ldots, n$, are the stochastic processes defined on $(\Omega, \mathcal{F}, \mathbb{P})$ and $X(t)$ is unknown. Also, $B(t) = (B_1(t), B_2(t), \ldots, B_n(t))$ is a multidimensional BM process and $\int_0^t k_i(s,x)X(s)dB_i(s)$, $i = 0, 1, 2, \ldots, n$, is the Itô integral [105]. Now, the functions in Eq. (4.13) have been approximated as

$$f(t) = F^T \Psi(t) = \Psi^T(t)F,$$

$$X(t) = X^T \Psi(t) = \Psi^T(t)X, \tag{4.14}$$

$$k_i(s,t) = \Psi^T(s)k_i\Psi(t) = \Psi^T(t)K_i\Psi(s), \quad i = 0, 1, 2, \ldots, n,$$

where X and F are the HLBPFs coefficients vector and $K_i, i = 0, 1, 2, \ldots, n$, are the HLBPFs coefficients matrices defined in Eqs. (4.5) and (4.7). Substituting Eq. (4.14) in Eq. (4.13) yields

$$X^T \Psi(t) = F^T \Psi(t) + \Psi^T(t)K_0^T \left(\int_0^t \Psi(s)\Psi^T(s)X ds \right)$$

$$+ \sum_{i=1}^{n} \Psi^T(t)K_i^T \left(\int_0^t \Psi(s)\Psi^T(s)X dB_i(s) \right).$$

Using Remark 4.3.2.1,

$$X^T \Psi(t) = F^T \Psi(t) + \Psi^T(t)K_0^T \left(\int_0^t \tilde{X}\Psi(s)ds \right)$$

$$+ \sum_{i=1}^{n} \Psi^T(t)K_i^T \left(\int_0^t \tilde{X}\Psi(s)dB_i(s) \right),$$

where $\tilde{X}$ is an $\hat{m} \times \hat{m}$ matrix. Implementing Λ and Λ_{s_i} for HLBPFs derived in Eqs. (4.11) and (4.12) yields

$$X^T \Psi(t) = F^T \Psi(t) + \Psi^T(t) K_0^T \tilde{X} \Lambda \Psi(t) + \sum_{i=1}^{n} \Psi^T(t) K_i^T \tilde{X} \Lambda_{s_i} \Psi(t).$$

By setting $Y_0 = K_0^T \tilde{X} \Lambda$, $Y_i = K_i^T \tilde{X} \Lambda_{s_i}$ and using Remark 4.3.2.2,

$$X^T \Psi(t) - \hat{Y}_0^T \Psi(t) - \sum_{i=1}^{n} \hat{Y}_i^T \Psi(t) = F^T \Psi(t),$$

in which $\hat{Y}_0$ and $\hat{Y}_i$ are $\hat{m}$-dimensional vectors. Thus,

$$X^T - \hat{Y}_0^T - \sum_{i=1}^{n} \hat{Y}_i^T = F^T. \tag{4.15}$$

Since $\hat{y}_0$ and $\hat{Y}_i$ are linear functions of X, Eq. (4.15) is a linear system of equations for X. By determining X, the solution of the multidimensional stochastic Itô–Volterra integral equation (4.13) can be determined by substituting the obtained vector X in Eq. (4.14).

4.3.4 *Convergence and error analysis*

Theorem 4.3.4.1 (convergence analysis [103]). If a continuous function $f(t) \in L^2(\mathbb{R})$ on $[0,1]$ has bounded second derivative $|\frac{\partial^2 f}{\partial t^2}| \leq \mathcal{M}$, the hybrid Legendre Block-Pulse functions expansion $f(t) = \sum_{n=1}^{N} \sum_{m=0}^{M-1} X(n,m)\psi(n,m,t)$ of $f(t)$ converges uniformly.

Proof. It has been proved in Theorem 3.3.4.1 of Chapter 3.

For error analysis, define $\|X\| = E\left\{ \left(|X(t)|^2\right)^{\frac{1}{2}} \right\}$. $\qquad \square$

Theorem 4.3.4.2 (error analysis). Suppose $X(t)$ is the exact solution of Eq. (4.1) and $X_m(t)$ is its HLBPF approximate solution. Also, assume that

$$\|X(t)\| \leq \mu, t \in [0,1],$$
$$\text{and} \quad \|k_i(s,t)\| \leq M_i, (s,t) \in [0,1] \times [0,1], i = 1,2,3.$$

For any $\varepsilon > 0$,

$$\|X(t) - X_m(t)\| \leq \frac{\left[\delta + \mu\varepsilon + \mu\sum_{i=1}^{n}\sup_{t\in[0,1)}|B_i(t)|\varepsilon_i\right]}{1 - \left[(M_0 + \varepsilon_0) + \sum_{i=1}^{n}\sup_{t\in[0,1)}|B_i(t)|(M_i + \varepsilon_i)\right]}.$$

Proof.   From Eq. (4.1), it yields

$$X(t) - X_m(t) = f(t) - f_m(t) + \int_0^t (k_0(s,t)X(s) - k_{0m}(s,t)X_m(s))ds$$

$$+ \sum_{i=1}^{n}\int_0^t (k_i(s,t)X(s) - k_{im}(s,t)X_m(s))dB(s).$$

So, by integral mean value theorem,

$$\|X(t) - X_m(t)\| \leq \|f(t) - f_m(t)\| + t\|k_0(\xi_0,t)X(\xi_0)$$

$$- k_{0m}(\xi_0,t)X_m(\xi_0)\|$$

$$+ \sum_{i=1}^{n}|B_i(t)|\,\|k_i(\xi_i,t)X(\xi_i)$$

$$- k_{im}(\xi_i,t)X_m(\xi_i)\|, \tag{4.16}$$

where $\xi_i \in [0,1], i = 0,1,2,\ldots,n$.

Now, Theorem 4.3.4.1 shows that the HLBPFs expansion for any continuous function $f(t)$ converges uniformly to f. So, for any ε, there exists m such that

$$\|k_i(s,t) - k_{im}(s,t)\| \leq \varepsilon_i, i = 0,1,2,3,\ldots,n,$$

$$\|f(t) - f_m(t)\| \leq \delta.$$

Thus, for $i = 0,1,2,3\ldots,n$, it can be written as

$$\|k_i(\xi_i,t)X(\xi_i) - k_{im}(\xi_i,t)X_m(\xi_i)\| \leq \|k_i(\xi_i,t)\|\,\|X(\xi_i) - X_m(\xi_i)\|$$

$$+ \|k_i(\xi_i,t) - k_{im}(\xi_i,t)\|$$

$$\|X(\xi_i)\| + \|k_i(\xi_i,t) - k_{im}(\xi_i,t)\|$$

$$\|X(\xi_i) - X_m(\xi_i)\|$$

$$\leq (M_i + \varepsilon_i)\|X(\xi_i) - X_m(\xi_i)\| + \mu\varepsilon_i. \tag{4.17}$$

Now, substituting Eq. (4.17) in Eq. (4.16) yields

$$\|X(t) - X_m(t)\| \leq \delta + t\left[(M_0 + \varepsilon_0)\|X(t) - X_m(t)\| + \mu\varepsilon_0\right]$$

$$+ \sum_{i=1}^{n} |B_i(t)|\left[(M_i + \varepsilon_i)\|X(t) - X_m(t)\| + \mu\varepsilon_i\right]$$

$$\leq \delta + \left[(M_0 + \varepsilon_0)\|X(t) - X_m(t)\| + \mu\varepsilon_0\right]$$

$$+ \sum_{i=1}^{n} \sup_{t\in[0,1)} |B_i(t)|[(M_i + \varepsilon_i)$$

$$\times \|X(t) - X_m(t)\| + \mu\varepsilon_i].$$

Replacing $\xi_i, i = 0, 1, 2, \ldots, n$, by an arbitrary $t \in [0, 1)$.
Hence,

$$\|X(t) - X_m(t)\| \leq \frac{\left[\delta + \mu\varepsilon + \mu\sum_{i=1}^{n} \sup_{t\in[0,1)} |B_i(t)|\varepsilon_i\right]}{1 - \left[(M_0 + \varepsilon_0) + \sum_{i=1}^{n} \sup_{t\in[0,1)} |B_i(t)|(M_i + \varepsilon_i)\right]}.$$

$$\square$$

4.3.5   *Numerical examples*

Some suitable nontrivial examples have been solved using HLBPFs in this section.

Example 4.1 Consider the following MDSIVIE [127]:

$$X(t) = f(t) + \int_0^t r(s)X(s)ds + \sum_{i=1}^{3} \int_0^t \alpha_i(s)X(s)dB(s), t \in [0, 1),$$

in which $f(t) = \frac{1}{12}$ with $r(s) = s^2, \alpha_1(s) = \sin(s), \alpha_2(s) = \cos(s)$ and $\alpha_3(s) = s$. The exact solution of this MDSIVIE is

$$X(t) = \frac{1}{12} \exp\left(\int_0^t \left(r(s) - \frac{1}{2}\sum_{i=1}^{3} \alpha_i^2(s)\right)ds + \sum_{i=1}^{3} \int_0^t \alpha_i(s)dB_i(s)\right),$$

where $X(t)$ is unknown on $(\Omega, \mathcal{F}, \mathbb{P})$ and $B(t) = (B_1(t), B_2(t), B_3(t))$ is a multidimensional BM process. The stochastic OM for HLBPFs

Table 4.1. A comparison between the approximate solutions based on HLBPFs, exact solutions and approximate solutions based on second-kind CWs for Example 4.1.

| | $\hat{m} = 5$ | | | $\hat{m} = 6$ | | |
| | Approximate solutions (HLBPF) | Exact solutions for Example 4.1 | Approximate solutions for second-kind Chebyshev wavelets [102] | Approximate solutions (HLBPF) | Exact solutions for Example 4.1 | Approximate solutions for second-kind Chebyshev wavelets [102] |
t						
0.1	0.061066	0.054984	0.063669	0.050083	0.040785	0.082868
0.3	0.093887	0.081219	0.08859	0.083105	0.091071	0.079014
0.5	0.110224	0.095005	0.138552	0.074281	0.034843	0.051892
0.7	0.028255	0.190394	0.142213	0.006996	0.000957	0.073193
0.9	0.028792	0.027482	0.218212	0.503987	0.398823	0.010641

presented in Section 4.3.2 and the proposed method described in Section 4.3.3 are used for solving Example 4.1. A comparison between the approximate and exact solutions of the above problem is shown in Table 4.1.

Example 4.2 Consider the following MDSIVIE [127]:

$$X(t) = f(t) + \frac{1}{20} \int_0^t X(s)ds + \sum_{i=1}^4 \int_0^t \alpha_i X(s)dB(s), t \in [0,1),$$

in which $f(t) = \frac{1}{200}$ with $\alpha_1 = \frac{1}{50}, \alpha_2 = \frac{2}{50}, \alpha_3 = \frac{4}{50}$ and $\alpha_4 = \frac{9}{50}$. The exact solution of this MDSIVIE is

$$X(t) = \frac{1}{200} \exp\left(\left(\frac{1}{20} - \frac{1}{2}\sum_{i=1}^4 \alpha_i^2 \right) t + \sum_{i=1}^4 \alpha_i B_i(t) \right),$$

where $X(t)$ is an unknown stochastic process on the probability space $(\Omega, \mathcal{F}, \mathbb{P})$ and $B(t) = (B_1(t), B_2(t), B_3(t), B_4(t))$ is a multidimensional Brownian motion process. The stochastic operational matrix for HLBPFs presented in Section 4.3.2 and the proposed method described in Section 4.3.3 are used for solving Example 4.2. A comparison between the approximate and exact solutions of the above problem is shown in Table 4.2.

Table 4.2. A comparison between the approximate solutions based on HLBPFs, exact solutions and approximate solutions based on second-kind CWs for Example 4.2.

	$\hat{m} = 5$			$\hat{m} = 6$		
t	Approximate solutions (HLBPF)	Exact solutions for Example 4.1	Approximate solutions for second-kind Chebyshev wavelets [102]	Approximate solutions (HLBPF)	Exact solutions for Example 4.1	Approximate solutions for second-kind Chebyshev wavelets [102]
0.1	0.004697	0.005381	0.004883	0.005456	0.004750	0.005346
0.3	0.004677	0.005211	0.005938	0.005389	0.005445	0.004834
0.5	0.004523	0.005171	0.004398	0.002185	0.005254	0.004443
0.7	0.004680	0.004882	0.005897	0.004811	0.004199	0.004036
0.9	0.003628	0.004822	0.003765	0.004760	0.004145	0.006616

4.4   Second-Kind CWs for MDSIVIEs

4.4.1   *Second-kind CWs*

Consider the second-kind Chebyshev polynomials $U_m(t)$ of degree m on the interval $[0, 1]$

$$U_m(t) = \frac{\sin((m+1)\theta)}{\sin\theta}, t = \cos\theta.$$

For $m = 0, 1, \ldots, M - 1$ and $n = 1, 2, \ldots, 2^{k-1}$, the second-kind CWs on the interval $[0,1)$ are defined in Refs. [102] as follows:

$$\psi(n, m, t) = \begin{cases} \sqrt{\dfrac{2}{\pi}} 2^{\frac{k}{2}} U_m(2^k t - 2n + 1), & \frac{n-1}{2^{k-1}} \le t < \frac{n}{2^{k-1}}, \\ 0, & \text{otherwise.} \end{cases}$$

Here, second-kind CWs form an orthonormal basis for $L^2_{w_{nk}}[0,1]$ with respect to weight functions

$$w_{nk}(t) = w(2^k t - 2n + 1),$$

in which $w(t) = \sqrt{1 - t^2}$.

4.4.1.1 *Function approximation*

Any function $f(t)$ over $[0,1]$ which is square integrable, i.e., $f(t) \in L^2_{w_{nk}}[0,1)$, can be expanded in terms of second-kind CWs as

$$f(t) = \sum_{n=1}^{\infty} \sum_{m=0}^{\infty} X(n \cdot m)\psi(n,m,t), \tag{4.18}$$

where $X(n,m) = \langle f(t), \psi(n,m,t)\rangle_{w_{nk}}$. If Eq. (4.18) is truncated at some values of 2^{k-1} and M, then

$$f(t) \approx \sum_{n=1}^{2^{k-1}} \sum_{m=0}^{M-1} X(n,m)\psi(n,m,t) = X^T\Psi(t), \tag{4.19}$$

where X and $\Psi(t)$ are $\hat{m} = 2^{k-1}M$ column vectors,

$$X = \Big[X(1,0), X(1,1), \dots, X(1,M-1), \dots, X(2^{k-1},0),$$

$$X(2^{k-1},1), \dots, X(2^{k-1},M-1)\Big]^T, \tag{4.20}$$

and

$$\Psi(t) = \Big[\psi(1,0,t), \psi(1,1,t), \dots, \psi(1,M-1,t), \dots,$$

$$\psi(2^{k-1},0,t), \dots, \psi(2^{k-1},M-1,t)\Big]^T. \tag{4.21}$$

Since $\langle \Psi(t), \Psi^T(t)\rangle_{w_{nk}} = I$, where the I is an identity matrix of dimension $2^{k-1}M$, then calculate $X(n,m)$ as

$$X(n,m) = \langle f(t), \psi(n,m,t)\rangle_{w_{nk}}, \quad n = 1,2,\dots,2^{k-1}$$

$$\text{and} \quad m = 0,1,\dots,M-1.$$

Similarly, let $K(s,t) \in L^2_{\mathbf{w}\otimes\mathbf{w}}([0,1) \times [0,1))$, where $\mathbf{w}(t) = [w_{nk}(t)]$ for $\frac{n-1}{2^{k-1}} \leq t < \frac{n}{2^{k-1}}$. Then,

$$K(s,t) \approx \Psi^T(s)K\Psi(t) = \Psi^T(t)K^T\Psi(s), \tag{4.22}$$

where $\Psi(s)$ and $\Psi(t)$ are $\hat{m}$-dimensional second-kind CWs vectors and $K = [k_{ij}]_{\hat{m}\times\hat{m}}$ is the $\hat{m} \times \hat{m}$ second-kind CWs coefficient matrix with

$$k_{ij} = \Big\langle \Psi_i(s), \langle K(s,t), \Psi_j(t)\rangle_{w_{nk}}\Big\rangle_{w_{nk}}.$$

4.4.1.2   *Integration operational matrices*

Here, a stochastic OM for the second-kind CWs has been derived.

Lemma 4.4.1.1. *Let $\Phi(t)$ be the $\hat{m}$-dimensional BPFs vector. The integral of this vector is*

$$\int_0^t \Phi(s)ds \cong P\Phi(t), \tag{4.23}$$

where P is called the OM of integration for BPFs and

$$P = \frac{h}{2}\begin{bmatrix} 1 & 2 & 2 & \cdots & 2 \\ 0 & 1 & 2 & \cdots & 2 \\ 0 & 0 & 1 & \vdots & \vdots \\ \vdots & \vdots & \vdots & \ddots & 2 \\ 0 & 0 & 0 & \cdots & 1 \end{bmatrix}_{\hat{m}\times\hat{m}}.$$

Proof. It may be referred to Ref. [99]. $\square$

Lemma 4.4.1.2. *Let $\Phi(t)$ be the $\hat{m}$-dimensional BPFs vector. The Itô integral of this vector is*

$$\int_0^t \Phi(s)dB(s) \cong P_s\Phi(t), \tag{4.24}$$

where P_s is called the SOM of integration for BPFs and

$$P_s = \begin{bmatrix} B\left(\frac{h}{2}\right) & B(h) & B(h) & \cdots & B(h) \\ 0 & B\left(\frac{3h}{2}\right)-B(h) & B(2h)-B(h) & \cdots & B(2h)-B(h) \\ 0 & 0 & B\left(\frac{5h}{2}\right)-B(2h) & \cdots & B(3h)-B(2h) \\ \vdots & \vdots & \vdots & \ddots & \vdots \\ 0 & 0 & 0 & \cdots & \begin{matrix}B\left(\frac{(2\hat{m}-1)h}{2}\right)\\ -B((\hat{m}-1)h)\end{matrix} \end{bmatrix}_{\hat{m}\times\hat{m}}.$$

Proof. It may be further referred to Ref. [99]. $\square$

4.4.1.3   *Second-kind CWs and BPFs*

In the following section, the relation between the second-kind CWs and BPFs has been reviewed.

Theorem 4.4.1.3. *Let $\Psi(t)$ and $\Phi(t)$ be the $\hat{m}$-dimensional second-kind CWs and BPFs vector. The vector $\Psi(t)$ can be expanded by BPFs vector $\Phi(t)$ as*

$$\Psi(t) = Q\Phi(t), \tag{4.25}$$

where Q is an $\hat{m} \times \hat{m}$ block matrix and

$$Q_{ij} = \psi_i\left(\frac{2j-1}{2\hat{m}}\right), \quad i, j = 1, 2, \ldots, \hat{m}.$$

Proof.   Let $\psi_i(t), i = 1, 2, \ldots, \hat{m}$, be the ith element of second-kind CWs vector. Expanding $\psi_i(t)$ into an $\hat{m}$-term vector of BPFs, it can be written as

$$\psi_i(t) = \sum_{j=1}^{\hat{m}} Q_{ij}\phi_j(t), \quad i = 1, 2, \ldots, \hat{m},$$

where Q_{ij} is the $(i, j)^{th}$ element of matrix Q and

$$Q_{ij} = \frac{1}{h}\int_0^1 \psi_i(t)\phi_j(t)dt = \frac{1}{h}\int_{\frac{j-1}{\hat{m}}}^{\frac{j}{\hat{m}}} \psi_i(t)dt$$

$$= \hat{m}\int_{\frac{j-1}{\hat{m}}}^{\frac{j}{\hat{m}}} \psi_i(t)dt, \quad \text{where } \hat{m} = \frac{1}{h}.$$

Now, using the mean value theorem for integrals in the last equation yields

$$Q_{ij} = \hat{m}\left(\frac{j}{\hat{m}} - \frac{j-1}{\hat{m}}\right)\psi_i(\eta_j) = \psi_i(\eta_j), \quad \eta_j \in \left(\frac{j-1}{\hat{m}}, \frac{j}{\hat{m}}\right).$$

By choosing $\eta_j = \frac{2j-1}{2\hat{m}}$, it yields

$$Q_{ij} = \psi_i\left(\frac{2j-1}{2\hat{m}}\right), \quad i, j = 1, 2, \ldots, \hat{m}.$$

This completes the proof of the theorem. $\square$

Remark 4.4.1.4. For an $\hat{m}$-vector F

$$\Psi(t)\Psi^T(t)F = \tilde{F}\Psi(t),$$

in which $\tilde{F}$ is an $\hat{m} \times \hat{m}$ matrix of the form

$$\tilde{F} = Q\bar{F}Q^{-1},$$

where $\bar{F} = diag(Q^T F)$.

Remark 4.4.1.5. Let A represents an $\hat{m} \times \hat{m}$ matrix. Then, for the second-kind CWs vector $\Psi(t)$

$$\Psi^T(t)A\Psi(t) = \hat{A}^T\Psi(t),$$

where $\hat{A}^T = UQ^{-1}$ and $U = diag(Q^T AQ)$ is an $\hat{m}$-vector.

4.4.1.4   *Stochastic OM of second-kind CWs*

Theorem 4.4.1.6. *Let $\Psi(t)$ be the $\hat{m}$-dimensional second-kind CWs vector. The integral of this vector is*

$$\int_0^t \Psi(s)ds = QPQ^{-1}\Psi(t) = \Lambda\Psi(t), \qquad (4.26)$$

where Q is introduced in Eq. (4.25) and P is the OM of integration for BPFs specified in Eq. (4.23).

Proof.   Let $\Psi(t)$ be the second-kind CWs vector.

$$\int_0^t \Psi(s)ds = Q\int_0^t \Phi(s)ds = QP\Phi(t),$$

where $\Psi(t) = Q\Phi(t)$.
 Now, Theorem 4.4.1.3 gives

$$\int_0^t \Psi(s)ds = QPQ^{-1}\Psi(t) = \Lambda\Psi(t),$$

where $\Lambda = QPQ^{-1}$.
 This completes the proof. $\square$

Theorem 4.4.1.7. *Let $\Psi(t)$ be the $\hat{m}$-dimensional second-kind CWs vector. The Itô integral of this vector is*

$$\int_0^t \Psi(s)dB(s) = QP_sQ^{-1}\Psi(t) = \Lambda_s\Psi(t), \qquad (4.27)$$

where Λ_s is called SOM for second-kind CWs, Q is introduced in Eq. (4.25) and P_s is the stochastic OM of integration for BPFs specified in Eq. (4.24).

Proof. Let $\Psi(t)$ be the second-kind CWs vector. Using Theorem 4.4.1.3 and Lemma 4.4.1.2, it yields

$$\int_0^t \Psi(s)dB(s) = Q\int_0^t \Phi(s)dB(s) = QP_s\Phi(t).$$

Now, Theorem 4.4.1.3 yields

$$\int_0^t \Psi(s)dB(s) = QP_sQ^{-1}\Psi(t) = \Lambda_s\Psi(t),$$

where $\Lambda_s = QP_sQ^{-1}$.

This completes the proof of the theorem. $\qquad\square$

4.4.2 *Approximation of multidimensional stochastic Itô–Volterra integral equation*

In the present analysis, MDIVIEs have been considered as

$$X(t) = f(t) + \int_0^t k_0(s,t)X(s)ds$$

$$+ \sum_{i=1}^n \int_0^t k_i(s,t)X(s)dB_i(s)s, \quad t \in [0,T), \qquad (4.28)$$

where $X(t)$, $f(t)$ and $k_i(s,t)$, $i = 0,1,2,\ldots,n$, are the stochastic processes defined on $(\Omega, \mathcal{F}, \mathbb{P})$ and $X(t)$ is unknown. Also, $B(t) = (B_1(t), B_2(t), \ldots, B_n(t))$ is a multidimensional BM process and $\int_0^t k_i(s,x)X(s)dB_i(s)$, $i = 0,1,2,\ldots,n$, is the Itô integral [105].

Now, the functions in Eq. (4.28) are approximated as

$$f(t) = F^T \Psi(t) = \Psi^T(t)F,$$
$$X(t) = X^T \Psi(t) = \Psi^T(t)X, \tag{4.29}$$
$$k_i(s,t) = \Psi^T(s)K_i\Psi(t) = \Psi^T(t)K_i\Psi(s), \quad i = 0, 1, 2, \ldots, n,$$

where X and F are the second-kind CWs vector and K_i, $i = 0, 1, 2, \ldots, n$, are the second-kind CWs coefficients matrices defined in Eqs. (4.20) and (4.22). Substituting Eq. (4.29) in Eq. (4.28),

$$X^T \Psi(t) = F^T \Psi(t) + \Psi^T(t)K_0^T \left(\int_0^t \Psi(s)\Psi^T(s)X ds \right)$$
$$+ \sum_{i=1}^n \Psi^T(t)K_i^T \left(\int_0^t \Psi(s)\Psi^T(s)X dB_i(s) \right).$$

Using relation $\langle \Psi(s), \Psi^T(s) \rangle_{W_{nk}} = I_{\hat{m} \times \hat{m}}$ and Remark 4.4.1.4, it can be written as

$$X^T \Psi(t) = F^T \Psi(t) + \Psi^T(t)K_0^T \left(\int_0^t \tilde{X}\Psi(s) ds \right)$$
$$+ \sum_{i=1}^n \Psi^T(t)K_i^T \left(\int_0^t \tilde{X}\Psi(s) dB_i(s) \right),$$

where $\tilde{X}$ is an $\hat{m} \times \hat{m}$ matrix. Exerting Λ and Λ_{s_i} for second-kind CWs derived in Eqs. (4.26) and (4.27), it can be written as

$$X^T \Psi(t) = F^T \Psi(t) + \Psi^T(t)K_0^T \tilde{X}\Lambda\Psi(t) + \sum_{i=1}^n \Psi^T(t)K_i^T \tilde{X}\Lambda_{s_i}\Psi(t).$$

By setting $Y_0 = K_0^T \tilde{X}\Lambda, Y_i = K_i^T \tilde{X}\Lambda_{s_i}$ and using Remark 4.4.1.5 yield

$$X^T \Psi(t) - \hat{Y}_0^T \Psi(t) - \sum_{i=1}^n \hat{Y}_i^T \Psi(t) = F^T \Psi(t),$$

in which $\hat{Y}_0$ and $\hat{y}_i$ are $\hat{m}$-dimensional vectors. This equation holds for all $t \in [0, 1)$, thus

$$X^T - \hat{Y}_0^T - \sum_{i=1}^{n} \hat{Y}_i^T = F^T. \tag{4.30}$$

Since $\hat{Y}_0$ and $\hat{Y}_i$ are linear functions of X, Eq. (4.30) is a linear system of equations for X. By determining X, the approximate solution of the multidimensional stochastic Itô–Volterra integral equation (4.28) can be determined by substituting the obtained vector X in Eq. (4.29).

4.4.3   *Convergence and error analysis*

Theorem 4.4.3.1 (convergence analysis [103]). *If a continuous function $f(t) \in L^2_{w_{nk}}[0, 1]$ has bounded second derivative $|\frac{\partial^2 f}{\partial t^2}| \leq \mathcal{M}$, the second-kind CWs expansion $f(t) = \sum_{n=1}^{2^{k-1}} \sum_{m=0}^{M-1} X(n, m)\psi(n, m, t)$ of $f(t)$ converges uniformly.*

Proof.   Let $f(t) \in L^2_{\mathbf{w}_{nk}}[0, 1]$ and $|\frac{\partial^2 f}{\partial t^2}| \leq \mathcal{M}$, where $\mathcal{M}$ is a positive constant.

The coefficients of second-kind CWs of continuous function $f(t)$ are defined as

$$X(n, m) = \int_0^1 f(t)\psi(n, m, t)dt$$

$$= \int_I f(t)\psi(n, m, t)dt,$$

where $I = \left[\frac{n-1}{2^{k-1}}, \frac{n}{2^{k-1}}\right)$.

Now, substituting $2^k t - 2n + 1 = y$ and $\hat{n} = 2n - 1$ yields

$$X(n, m) = \frac{\sqrt{\frac{2}{\pi}}2^k}{2^k} \int_{-1}^1 f\left(\frac{\hat{n} + y}{2^k}\right) U_m(y)dy,$$

$$= \frac{1}{\sqrt{2^{k-1}\pi}} \frac{1}{2m + 1} \int_{-1}^1 f\left(\frac{\hat{n} + y}{2^k}\right) (U'_{m+1}(y) - U'_{m-1}(y))dy,$$

since $(2m + 1)U_m(y) = U'_m(y) - U'_{m-1}(y)$.

Now, integrating by parts, it yields

$$
X(n,m) = \frac{1}{\sqrt{2^{k-1}}\pi(2m+1)} \left(f\left(\frac{\hat{n}+y}{2^k}\right) [U_{m+1}(y) - U_{m-1}(y)]_{-1}^{1} \right.
$$

$$
\left. - \int_{-1}^{1} \frac{\partial f}{\partial y}(U_{m+1}(y) - U_{m-1}(y))dy \right)
$$

$$
= \frac{1}{\sqrt{2^{k-1}}\pi(2m+1)} \left(\int_{-1}^{1} \frac{\partial f}{\partial y}(U_{m+1}(y) - U_{m-1}(y))dy \right)
$$

$$
= \frac{1}{\sqrt{2^{k-1}}\pi(2m+1)} \left(\int_{-1}^{1} \frac{\partial^2 f}{\partial y^2}\left(\frac{U_{m+2}(y) - U_m(y)}{2m+3} \right. \right.
$$

$$
\left. \left. - \frac{U_m(y) - U_{m-2}(y)}{2m-1} \right)dy \right).
$$

$$(4.31)$$

Now, suppose

$$
R_1(y) = (2m-1)U_{m+2}(y) - (2m-1)U_m(y)
$$

$$
-(2m+3)U_m(y) + (2m+3)U_{m-2}(y)
$$

$$
= (2m-1)U_{m+2}(y) - 2(2m+1)U_m(y) + (2m+3)U_{m-2}(y).
$$

$$(4.32)$$

Therefore,

$$
|X(n,m)| \le \lambda \int_{-1}^{1} \left| \frac{\partial^2 f}{\partial y^2} \right| |R_1(y)| dy
$$

$$
\le \lambda \mathcal{M} \int_{-1}^{1} |R_1(y)| dy,
$$

$$(4.33)$$

where $\lambda = \dfrac{1}{2\sqrt{2^{k-1}}\pi(2m+1)(2m+3)(2m-1)}$ and $|\frac{\partial^2 f}{\partial y^2}| \le \mathcal{M}$, by the hypothesis of the theorem.

Now, from Cauchy–Schwarz inequality, it becomes

$$
\left(\int_{-1}^{1} |R_1(y)| dy \right)^2 \le \int_{-1}^{1} 1^2 dy \int_{-1}^{1} ((2m-1)^2 + (4m+2)^2
$$

$$
+ (2m+3)^2) \left(U_{m+2}^2(y) + U_m^2(y) + U_{m-2}^2(y) \right) dy.
$$

This implies

$$\left(\int_{-1}^{1} |R_1(y)|dy\right)^2 \leq 2\left((2m-1)^2 + (4m+2)^2 + (2m+3)^2\right)$$

$$\times \left(\frac{2}{2m+3} + \frac{2}{2m+1} + \frac{2}{2m-3}\right).$$

Thus,

$$\int_{-1}^{1} |R_1(y)|dy$$

$$\leq 2\sqrt{\begin{array}{c}((2m-1)^2 + (4m+2)^2 + (2m+3)^2) \\ \times \left(\frac{1}{2m+3} + \frac{1}{2m+1} + \frac{1}{2m-3}\right)\end{array}}. \tag{4.34}$$

Plugging Eq. (4.34) in Eq. (4.33), it can be obtained as

$$|X(n,m)| \leq 2\mathcal{M}\lambda\eta,$$

where $\eta = \sqrt{((2m-1)^2 + (4m+2)^2 + (2m+3)^2)\left(\frac{1}{2m+3} + \frac{1}{2m+1} + \frac{1}{2m-3}\right)}.$

Therefore, $\sum_{n=1}^{2^{k-1}} \sum_{m=0}^{M-1} X(n,m)$ is absolutely convergent.

Hence, the second-kind CWs expansion of $f(t)$ converges uniformly. $\qquad\square$

For error analysis, define $\|X\| = E\{(|X(t)|^2)^{\frac{1}{2}}\}$.

Theorem 4.4.3.2. (error analysis). *Suppose $X(t)$ is the exact solution of Eq. (4.1) and $X_m(t)$ is its second-kind CWs solution. Also, assume that*

$$\|X(t)\| \leq \mu, t \in [0,1],$$
$$and \quad \|k_i(s,t)\| \leq M_i, \quad (s,t) \in [0,1] \times [0,1], \quad i = 1,2,3.$$

Then, for any $\varepsilon > 0$,

$$\|X(t) - X_m(t)\| \leq \frac{\left[\delta + \mu\varepsilon + \mu\sum_{i=1}^{n} \sup_{t \in [0,1)} |B_i(t)|\varepsilon_i\right]}{1 - \left[(M_0 + \varepsilon_0) + \sum_{i=1}^{n} \sup_{t \in [0,1)} |B_i(t)|(M_i + \varepsilon_i)\right]}.$$

Proof. From Eq. (4.1), it yields

$$X(t) - X_m(t) = f(t) - f_m(t) + \int_0^t \left(k_0(s,t)X(s) - k_{0m}(s,t)X_m(s) \right) ds$$

$$+ \sum_{i=1}^n \int_0^t \left(k_i(s,t)X(s) - k_{im}(s,t)X_m(s) \right) dB(s).$$

So, by integral mean value theorem,

$$\|X(t) - X_m(t)\| \leq \|f(t) - f_m(t)\|$$
$$+ t\|k_0(\xi_0,t)X(\xi_0) - k_{0m}(\xi_0,t)X_m(\xi_0)\|$$
$$+ \sum_{i=1}^n |B_i(t)|$$
$$\times \|k_i(\xi_i,t)X(\xi_i) - k_{im}(\xi_i,t)X_m(\xi_i)\|, \quad (4.35)$$

where $\xi_i \in [0,1], i = 0,1,2,\dots,n$.

Now, Theorem 4.4.3.2 shows that the second-kind CWs expansion for any continuous function $f(t)$ converges uniformly to f. So, for any ε, there exists m such that

$$\|k_i(s,t) - k_{im}(s,t)\| \leq \varepsilon_i, i = 0,1,2,3,\dots,n,$$
$$\|f(t) - f_m(t)\| \leq \delta.$$

Thus, for $i = 0,1,2,3,\dots,n$,

$$\|k_i(\xi_i,t)X(\xi_i) - k_{im}(\xi_i,t)X_m(\xi_i)\| \leq \|k_i(\xi_i,t)\|\|X(\xi_i) - X_m(\xi_i)\|$$
$$+ \|k_i(\xi_i,t) - k_{im}(\xi_i,t)\|\|X(\xi_i)\|$$
$$+ \|k_i(\xi_i,t) - k_{im}(\xi_i,t)\|$$
$$\|X(\xi_i) - X_m(\xi_i)\|$$
$$\leq (M_i + \varepsilon_i)\|X(\xi_i) - X_m(\xi_i)\|$$
$$+ \mu\varepsilon_i. \quad (4.36)$$

Now, substituting Eq. (4.36) in Eq. (4.35),

$$\|X(t) - X_m(t)\| \leq \delta + t\left[(M_0 + \varepsilon_0)\|X(t) - X_m(t)\| + \mu\varepsilon_0\right]$$

$$+ \sum_{i=1}^{n} |B_i(t)|\left[(M_i + \varepsilon_i)\|X(t) - X_m(t)\| + \mu\varepsilon_i\right]$$

$$\leq \delta + \left[(M_0 + \varepsilon_0)\|X(t) - X_m(t)\| + \mu\varepsilon_0\right]$$

$$+ \sum_{i=1}^{n} \sup_{t\in[0,1)} |B_i(t)|\left[(M_i + \varepsilon_i)\|X(t) - X_m(t)\| + \mu\varepsilon_i\right].$$

Replacing $\xi_i, i = 0, 1, 2, \ldots, n$, by an arbitrary $t \in [0, 1)$.
  Hence,

$$\|X(t) - X_m(t)\| \leq \frac{\left[\delta + \mu\varepsilon + \mu\sum_{i=1}^{n}\sup_{t\in[0,1)}|B_i(t)|\varepsilon_i\right]}{1 - \left[(M_0 + \varepsilon_0) + \sum_{i=1}^{n}\sup_{t\in[0,1)}|B_i(t)|(M_i + \varepsilon_i)\right]}.$$

$$\square$$

4.4.4 *Numerical examples*

In this section, the accuracy of the proposed second-kind CWs method has been examined through suitable nontrivial examples.

Example 4.3. Consider the following MDSIVIE [127]:

$$X(t) = f(t) + \int_0^t r(s)X(s)ds + \sum_{i=1}^{3} \int_0^t \alpha_i(s)X(s)dB(s), t \in [0, 1),$$

in which $f(t) = \frac{1}{12}$ with $r(s) = s^2, \alpha_1(s) = \sin(s), \alpha_2(s) = \cos(s)$ and $\alpha_3(s) = s$. The exact solution of this MDSIVIE is

$$X(t) = \frac{1}{12}\exp\left(\int_0^t \left(r(s) - \frac{1}{2}\sum_{i=1}^{3}\alpha_i^2(s)\right)ds + \sum_{i=1}^{3}\int_0^t \alpha_i(s)dB_i(s)\right),$$

where $X(t)$ is an unknown on $(\Omega, \mathcal{F}, \mathbb{P})$ and $B(t) = (B_1(t), B_2(t),$

Table 4.3. A comparison between the approximate solutions based on second-kind CWs and exact solutions for $\hat{m} = 5$ and $\hat{m} = 6$ for Example 4.3.

	$\hat{m} = 5$		$\hat{m} = 6$	
t	Second-kind CWs solutions for Example 4.3	Exact solutions of Example 4.3	Second-kind CWs solutions for Example 4.3	Exact solutions of Example 4.3
0.1	0.063669	0.058134	0.082868	0.102891
0.3	0.08859	0.080269	0.079014	0.056993
0.5	0.138552	0.150826	0.051892	0.044815
0.7	0.142213	0.101087	0.073193	0.023806
0.9	0.218212	0.074673	0.010641	0.003749

$B_3(t))$ is a multidimensional BM process. The SOM for second-kind CWs presented in Section 4.4.1 and the proposed scheme described in Section 4.4.2 are used for solving Example 4.3. A comparison between the approximate and exact solutions of the above problem is shown in Table 4.3.

Example 4.4 Consider the following MDSIVIE [127]:

$$X(t) = f(t) + \frac{1}{20} \int_0^t X(s)ds + \sum_{i=1}^4 \int_0^t \alpha_i X(s)dB(s), t \in [0, 1),$$

in which $f(t) = \frac{1}{200}$ with $\alpha_1 = \frac{1}{50}, \alpha_2 = \frac{2}{50}, \alpha_3 = \frac{4}{50}$ and $\alpha_1 = \frac{9}{50}$. The exact solution of this MDSIVIE

$$X(t) = \frac{1}{200} \exp\left(\left(\frac{1}{20} - \frac{1}{2}\sum_{i=1}^4 \alpha_i^2\right) t + \sum_{i=1}^4 \alpha_i B_i(t)\right),$$

where $X(t)$ is an unknown on $(\Omega, \mathcal{F}, \mathbb{P})$ and $B(t) = (B_1(t), B_2(t), B_3(t), B_4(t))$ is a multidimensional Brownian motion process. The stochastic operational matrix for second-kind CWs presented in Section 4.4.1 and the proposed method described in Section 4.4.2 are used for solving Example 4.4. A comparison between the approximate and exact solutions of the above problem is shown in Table 4.4.

Table 4.4. A comparison between the approximate solutions based on second-kind CWs and exact solutions for $\hat{m} = 5$ and $\hat{m} = 6$ for Example 4.4.

t	$\hat{m} = 5$		$\hat{m} = 6$	
	Second-kind CWs solutions for Example 4.4	Exact solutions of Example 4.4	Second-kind CWs solutions for Example 4.4	Exact solutions of Example 4.4
0.1	0.004883	0.004862	0.005346	0.005683
0.3	0.005938	0.003732	0.004834	0.004819
0.5	0.004398	0.005245	0.004443	0.005526
0.7	0.005897	0.005115	0.004036	0.004764
0.9	0.003765	0.006589	0.006616	0.004333

4.5   Summary

In this chapter, HLBPFs and second-kind CWs have been successfully applied on the of MDSIVIEs to obtain the numerical results. The BPFs and their relations to HLBPFs and second-kind CWs are employed to derive a general procedure for forming SOM for both the schemes. These stochastic operational matrices have been used to discretise the multidimensional stochastic Volterra–Fredholm integral equation and then the system of integral equations has been reduced to a system of nonlinear algebraic equations using the proposed approximating methods. The obtained numerical solutions have been compared with the exact solutions to establish the accuracy of the proposed methods.

Chapter 5

Numerical Solutions of Stochastic Integral Equations with Fractional Brownian Motion

5.1 Introduction

Fractional Brownian motion (FBM) was first introduced in 1940 by Andrei Nikolaevich Kolmogorov [129], who was studying spiral curves in Hilbert space. It was considered by Richard Allen Hunt [130] in the context of random Fourier transforms and by Akiva Moiseevich Yaglom [131], who studied the correlation structure of processes that have stationary nth order increments. However, it is undoubtedly the seminal paper of Mandelbrot and Van Ness which put the focus on FBM and gave it its name. The term FBM was coined by Mandelbrot and Van Ness in the now classical paper [18].

In recent years, many researchers such as CAS wavelet [111], Legendre wavelet [112,113,124], Legendre multiwavelet [114], Haar wavelet [115], Chebyshev wavelet [116] and so on have been widely used in the numerical solution of all kinds of linear or nonlinear integral equations, differential equations and integro-differential equations. Mirzaee and Hamzeh obtained solutions of stochastic differential equations (SDEs) based on FBM [132]. Maleknejad, Khodabin and Rostami obtained numerical solutions of the MDSIVIEs by using BPFs and their stochastic OM of integration [125]. Mohammadi proposed Haar wavelets for solving MDSIVIEs [127]. Mohammadi proposed second-kind CW for solving stochastic Itô–Volterra integral equations [57]. Gupta and Saha

143

Ray obtained numerical solutions of fractional fifth-order Sawada–Kotera equation using second-kind CW method [133]. Sahu and Saha Ray have applied CW method for numerical solutions of integro-differential form of Lane–Emden type differential equations [134]. Asgari *et al.* [135] and Maleknejad *et al.* [136] obtained approximate solutions of nonlinear Volterra–Fredholm–Hammerstein integral equations by applying Bernstein OM of integration. Mirzaee and Samadyar obtained approximate results of nonlinear stochastic Itô–Volterra integral equations having FBM [119]. Sahu and Saha Ray have applied Bernstein collocation for solving Fredholm integral equations [120].

5.2   Outline of Present Study

In Section 5.3, fractional Brownian motion and its applications to stochastic integral equations have been studied. Second-kind CWs have been used to obtain the numerical solutions of fractional stochastic Itô–Volterra integral equation (FSIVIE). These functions are orthonormal and have compact support on [0, 1]. The proposed method reduces the system of integral equations to a system of linear algebraic equations. Convergence and error analysis of the proposed method have been discussed. Furthermore, some examples have been discussed to establish the accuracy of the proposed scheme.

In Section 5.4, fractional Brownian motion and its applications to nonlinear stochastic integral equations have been discussed. Approximate solutions of nonlinear fractional stochastic Itô–Volterra integral equation (NLFSIVIE) have been obtained by using Bernstein polynomials. Error and convergence analysis of the Bernstein polynomials method have been mentioned. Furthermore, few examples have been discussed to establish the accuracy of Bernstein polynomials.

5.2.1   *Fractional stochastic integral equations*

Consider FSIVIE [119] and fix $\frac{1}{2} < H < 1$.

$$X(t) = f(t) + \int_0^t k_0(s,t)X(s)ds + \sum_{i=1}^{n} \int_0^t k_i(s,t)X(s)dB_i(s)$$

$$+ \int_0^t g(s,t)X(s)dB^H(s), \ t \in [0,T], \tag{5.1}$$

where $X(t)$, $f(t)$, $k_0(s,t)$, $k_i(s,t)$, $i = 1, 2, \ldots, n$ and $g(s,t)$ for $s, t \in [0, T)$ are the stochastic processes defined on the same probability space $\Omega, \mathcal{F}, \mathbb{P}$ and $X(t)$ is unknown. Also, $B(t)$ is a BM process and $B^H(t)$ is fractional Brownian motion with $\int_0^t k_i(s,t)X(s)dB(s)$ being the Itô integral.

5.2.2 *Fractional stochastic nonlinear integral equations*

5.2.2.1 *Derivation of a stochastic integral equation driven by fractional Brownian motion*

Let $(B^H(\tau))_{\tau \geq 0}$ be a fractional Brownian motion with Hurst parameter H such that $H > \frac{1}{2}$. Stochastic differential equation of the form has been investigated

$$dY(\tau) = b(\tau, Y(\tau))d\tau + \sigma(\tau, Y(\tau))dB^H(\tau), \tag{5.2}$$

$$Y(\tau_0) = Y_0,$$

where $\tau_0 \in (0, T]$, Y_0 is a random vector in $\mathbb{R}^n$ and the random functions b and σ satisfy with probability 1, i.e., a.s. the following conditions:

1. $b \in C([0, T] \times \mathbb{R}^n, \mathbb{R}^n)$, $\quad \sigma \in C^1([0, T] \times \mathbb{R}^n, \mathbb{R}^n)$,
2. for each $\tau \in [0, T]$, the functions $b(\tau, .)$, $\frac{\partial \sigma(\tau, \cdot)}{\partial y^i}$, $\frac{\partial \sigma(\tau, \cdot)}{\partial \tau}$ are locally Lipschitz for each $i \in \{1, \ldots, n\}$.

Consider the pathwise auxiliary partial differential equation on $[0, T] \times \mathbb{R}^n \times \mathbb{R}$,

$$\frac{\partial \Psi}{\partial z}(\tau, w, z) = \sigma(\tau, \Psi(\tau, w, z)), \tag{5.3}$$

$$\Psi(\tau_0, W_0, Z_0) = Y_0,$$

where W_0 is an arbitrary random vector in $\mathbb{R}^n$ and Z_0 an arbitrary random variable in $\mathbb{R}$. From the theory of differential equations, it follows that a.s. there exists a local solution $\Psi \in C^1([0, T] \times \mathbb{R}^n \times \mathbb{R}, \mathbb{R}^n)$ in a neighbourhood N of (τ_0, W_0, Z_0) with partial derivatives being Lipschitz in the variable w and

$$\det \left(\frac{\partial \Psi^i}{\partial w^j}(\tau, w, z) \right)_{1 \leq i,j \leq n} \neq 0.$$

Consider the pathwise differential equation (in matrix representation) on $[0, T]$,

$$dU(\tau) = \left(\frac{\partial \Psi}{\partial u}(\tau, U(\tau), B^H(\tau)) \right)^{-1}$$

$$\times \left[b(\tau, \Psi(\tau, U(\tau), B^H(\tau))) - \frac{\partial \Psi}{\partial \tau}(\tau, U(\tau), B^H(\tau)) \right] d\tau,$$

$$U(\tau_0) = U_0.$$

which has a unique local solution on a maximal interval $(\tau_0^1, \tau_0^2) \subseteq [0, T]$ with $\tau_0 \in (\tau_0^1, \tau_0^2)$ (Theorem 6.1 of Ref. [137]).

Now, stochastic Itô formula (see Ref. [138], Theorem 5.10) has been applied to the random function $X(\tau, z) = \Psi(\tau, U(\tau), z)$ and the fractional Brownian motion $B^H(\tau)$ to obtain

$$\Psi(\tau, U(\tau), z) - \Psi(\tau_0, U(\tau_0), z)$$

$$= \sum_{j=1}^{n} \int_{\tau_0}^{\tau} \frac{\partial \Psi}{\partial u^j} \left(s, U(s), B^H(s) \right) dU^j(s)$$

$$+ \int_{\tau_0}^{\tau} \frac{\partial \Psi}{\partial z} \left(s, U(s), B^H(s) \right) dB^H(s)$$

$$+ \int_{\tau_0}^{\tau} \frac{\partial \Psi}{\partial \tau} \left(s, U(s), B^H(s) \right) ds$$

$$= \int_{\tau_0}^{\tau} b \left(s, \Psi \left(s, U(s), B^H(s) \right) \right) ds$$

$$+ \int_{\tau_0}^{\tau} \sigma \left(s, \Psi \left(s, U(s), B^H(s) \right) \right) dB^H(s).$$

Therefore, $Y(\tau) := \Psi\left(\tau, U(\tau), B^H(\tau)\right)$ satisfies

$$Y(\tau) = Y_0 + \int_{\tau_0}^{\tau} b(s, Y(s))ds + \int_{\tau_0}^{\tau} \sigma(s, Y(s))dB^H(s).$$

Consider the NLFSIVIE equation [135] and fix $\frac{1}{2} < H < 1$

$$f(x) = g(x) + \lambda_1 \int_0^x b(y, f(y))dy + \lambda_2 \int_0^x \sigma(y, f(y))dB^H(y),$$
$$x \in [0, T), \tag{5.4}$$

where λ_1, λ_2 are parameters and $f(x)$, $g(x)$, $b(x, f(x))$ and $\sigma(x, f(x))$, where $x \in [0, T)$, are the stochastic processes defined on $\Omega, \mathcal{F}, \mathbb{P}$ and $f(x)$ is to be obtained. $B(x)$ is the BM process and $B^H(x)$ is FBM with $\int_0^x \sigma(y, f(y))dB^H(y)$ being the fractional Itô integral.

5.3 Second-Kind CWs for Fractional Stochastic Integral Equations

In this section, the Block-Pulse functions (BPFs), second-kind CW function and the corresponding operational matrices have been briefly introduced.

5.3.1 *Block-Pulse functions*

A set of BPFs [99] $\phi_n(x)$, $n = 1, 2, \ldots, N$ on the interval $[0, 1)$ are defined as follows:

$$\phi_n(t) = \begin{cases} 1, & \dfrac{n-1}{N} \leq t < \dfrac{n}{N}, \\ 0, & \text{otherwise,} \end{cases}$$

with $t \in [0, 1)$, $n = 1, 2, \ldots, N$ and $h = \frac{1}{N}$.

The properties of BPFs are as follows:

(i) The BPFs on the interval $[0, 1)$ are disjoint.

$$\phi_n(t)\phi_m(t) = \delta_{nm}\phi_n(t),$$

where $n, m = 1, 2, \ldots, N$ and δ_{nm} is Kronecker delta.

(ii) The BPFs are orthogonal on the interval $[0, 1)$.

$$\int_0^1 \phi_n(t)\phi_m(t)dt = h\delta_{nm}, \quad n, m = 1, 2, \ldots, N.$$

(iii) If $N \to \infty$, then the BPFs' set is complete; for every $f \in L^2([0,1))$, Parseval's identity holds,

$$\int_0^1 f^2(t)dt = \sum_{i=1}^{\infty} f_n^2 \|\phi_n(t)\|^2,$$

where

$$f_n = \frac{1}{h}\int_0^1 f(t)\phi_n(t)dt.$$

Consider the first N terms of BPFs and write them concisely as N-vector

$$\Phi(t) = (\phi_1(t), \phi_2(t), \ldots, \phi_N(t))^T, \quad t \in [0, 1). \tag{5.5}$$

The above representation and disjointness property follows

$$\Phi(t)\Phi^T(t) = \begin{pmatrix} \phi_1(t) & 0 & \cdots & 0 \\ 0 & \phi_2(t) & \cdots & 0 \\ \vdots & \vdots & \ddots & \vdots \\ 0 & 0 & \cdots & \phi_N(t) \end{pmatrix}_{N \times N}.$$

Furthermore, $\Phi^T(t)\Phi(t) = 1$ and $\Phi(t)\Phi^T(t)F^T = D_F\Phi(t)$, where D_F usually denotes a diagonal matrix whose diagonal entries are related to a constant vector $F = (f_1, f_2, \ldots, f_N)^T$.

5.3.2   *Second-kind Chebyshev wavelets*

Consider the second-kind Chebyshev polynomials $U_m(t)$ of degree m on the interval $[0, 1]$.

$$U_m(t) = \frac{\sin((m+1)\theta)}{\sin\theta}, \quad t = \cos\theta.$$

For $m = 0, 1, \ldots, M-1$ and $n = 1, 2, \ldots, 2^{k-1}$, the second-kind CWs on the interval $[0,1)$ are defined in Refs. [102,133] as follows:

$$\psi(n, m, t) = \begin{cases} \sqrt{\dfrac{2}{\pi}} 2^{\frac{k}{2}} U_m(2^k t - 2n + 1), & \dfrac{n-1}{2^{k-1}} \leq t < \dfrac{n}{2^{k-1}}, \\ 0, & \text{otherwise.} \end{cases}$$

Here, second-kind Chebyshev wavelets form an orthonormal basis for $L^2_{w_{nk}}$ with respect to weight functions

$$w_{nk}(t) = w(2^k t - 2n + 1),$$

in which $w(t) = \sqrt{1 - t^2}$.

5.3.2.1 *Function approximation*

Any function $f(t)$ over the interval $[0, 1)$ which is square integrable, i.e., $f(t) \in L^2_{w_{nk}}[0, 1)$, can be expanded in terms of second-kind CWs as

$$f(t) = \sum_{n=1}^{\infty} \sum_{m=0}^{\infty} X(n.m)\psi(n, m, t), \tag{5.6}$$

where $X(n, m) = \langle f(t), \psi(n, m, t) \rangle_{w_{nk}}$. If Eq. (5.4) is truncated at some values of 2^{k-1} and M, then it can be written as

$$f(t) \approx \sum_{n=1}^{2^{k-1}} \sum_{m=0}^{M-1} X(n, m)\psi(n, m, t) = X^T \Psi(t), \tag{5.7}$$

where X and $\Psi(t)$ are $\hat{m} = 2^{k-1}M$ column vectors given by

$$X = [X(1,0), X(1,1), \ldots, X(1, M-1), \ldots, X(2^{k-1}, 0),$$
$$X(2^{k-1}, l), \ldots, X(2^{k-1}, M-1)]^T, \tag{5.8}$$

and

$$\Psi(t) = [\psi(1, 0, t), \psi(1, 1, t), \ldots, \psi(1, M-1, t), \ldots,$$
$$\psi(2^{k-1}, 0, t), \ldots, \psi(2^{k-1}, M-1, t)]^T. \tag{5.9}$$

Since $\langle \Psi(t), \Psi^T(t) \rangle_{w_{nk}} = I$, where the I is an identity matrix of dimension $2^{k-1}M$, then calculate $X(n, m)$ as

$$X(n, m) = \langle f(t), \psi(n, m, t) \rangle_{w_{nk}},$$

$$n = 1, 2, \ldots, 2^{k-1} \quad \text{and} \quad m = 0, 1, \ldots, M - 1.$$

Similarly, let $K(s, t) \in L^2_{\mathbf{w} \otimes \mathbf{w}}([0, 1) \times [0, 1))$, where

$$K(s, t) \approx \Psi^T(s) K \Psi(t) = \Psi^T(t) K^T \Psi(s), \tag{5.10}$$

where $\Psi(s)$ and $\Psi(t)$ are $\hat{m}$-dimensional second-kind Chebyshev wavelets vectors, respectively, and $K = [k_{ij}]_{\hat{m} \times \hat{m}}$ is the $\hat{m} \times \hat{m}$ second-kind Chebyshev wavelets coefficient matrix with

$$k_{ij} = \langle \Psi_i(s), \langle K(s, t), \Psi_j(t) \rangle_{w_{nk}} \rangle_{w_{nk}}.$$

5.3.2.2 *Integration operational matrices*

In this section, a stochastic OM for the second-kind CWs has been derived.

Lemma 5.3.2.1. *Let $\Phi(t)$ be the $\hat{m}$-dimensional BPFs vector. The integral of this vector is*

$$\int_0^t \Phi(s)ds \cong P\Phi(t), \tag{5.11}$$

where P is called the OM of integration for BPFs and

$$P = \frac{h}{2} \begin{bmatrix} 1 & 2 & 2 & \cdots & 2 \\ 0 & 1 & 2 & \cdots & 2 \\ 0 & 0 & 1 & \vdots & \vdots \\ \vdots & \vdots & \vdots & \ddots & 2 \\ 0 & 0 & 0 & \cdots & 1 \end{bmatrix}_{\hat{m} \times \hat{m}}.$$

Proof. It may be referred to Ref. [99]. $\qquad\qquad\square$

Lemma 5.3.2.2. *Let $\Phi(t)$ be the $\hat{m}$-dimensional BPFs. The Itô integral of this vector is*

$$\int_0^t \Phi(s)dB(s) \cong P_S\Phi(t), \tag{5.12}$$

where P_S is called the stochastic OM of integration for BPFs and

$$\begin{bmatrix} B\left(\frac{h}{2}\right) & B(h) & B(h) & \dots & B(h) \\ 0 & B\left(\frac{3h}{2}\right) - B(h) & B(2h) - B(h) & \dots & B(2h) - B(h) \\ 0 & 0 & B\left(\frac{5h}{2}\right) - B(2h) & \dots & B(3h) - B(2h) \\ \vdots & \vdots & \vdots & \ddots & \vdots \\ 0 & 0 & 0 & \dots & B\left(\frac{(2\hat{m}-1)h}{2}\right) \\ & & & & -B((\hat{m}-1)h) \end{bmatrix}_{\hat{m}\times\hat{m}}.$$

Proof. It may be further referred to Ref. [99]. $\qquad\square$

Lemma 5.3.2.3. *Let $\Phi(t)$ be the $\hat{m}$-dimensional BPFs vector. The Itô integral of this vector is*

$$\int_0^t \Phi(s)dB^H(s) \cong P_S^H\Phi(t), \tag{5.13}$$

where P_S^H is called the fractional stochastic OM of integration for BPFs and

$$P_s = \begin{bmatrix} B^H\left(\frac{h}{2}\right) & B^H(h) & B^H(h) & \dots & B^H(h) \\ 0 & B^H\left(\frac{3h}{2}\right) - B^H(h) & B^H(2h) - B^H(h) & \dots & B^H(2h) - B^H(h) \\ 0 & 0 & B^H\left(\frac{5h}{2}\right) - B^H(2h) & \dots & B^H(3h) - B^H(2h) \\ \vdots & \vdots & \vdots & \ddots & \vdots \\ 0 & 0 & 0 & \dots & B^H\left(\frac{(2\hat{m}-1)h}{2}\right) \\ & & & & -B^H((\hat{m}-1)h) \end{bmatrix}_{\hat{m}\times\hat{m}}.$$

Proof. It may be further referred to Ref. [132]. $\qquad\square$

5.3.2.3 *Second-kind CWs and BPFs*

In the following section, the relation between the second-kind CWs and BPFs has been reviewed.

Theorem 5.3.2.4. *Let $\Psi(t)$ and $\Phi(t)$ be the $\hat{m}$-dimensional second-kind CWs and BPFs vector. The vector $\Psi(t)$ can be expanded by*

BPFs vector $\Phi(t)$ as

$$\Psi(t) = Q\Phi(t), \qquad\qquad (5.14)$$

where Q is an $\hat{m} \times \hat{m}$ block matrix and

$$Q_{ij} = \psi_i\left(\frac{2j-1}{2\hat{m}}\right), \quad i,j = 1, 2, \ldots, \hat{m}.$$

Proof.   Let $\psi_i(t)$, $i = 1, 2, \ldots, \hat{m}$, be the ith element of second-kind CWs vector. Expanding $\psi_i(t)$ into an $\hat{m}$-term vector of BPFs yields

$$\psi_i(t) = \sum_{j=1}^{\hat{m}} Q_{ij}\phi_j(t), \quad i = 1, 2, \ldots, \hat{m},$$

where Q_{ij} is the $(i,j)^{th}$ element of matrix Q and

$$Q_{ij} = \frac{1}{h}\int_0^1 \psi_i(t)\phi_j(t)dt = \frac{1}{h}\int_{\frac{j-1}{\hat{m}}}^{\frac{j}{\hat{m}}} \psi_i(t)dt = \hat{m}\int_{\frac{j-1}{\hat{m}}}^{\frac{j}{\hat{m}}} \psi_i(t)dt,$$

$$\text{where } \hat{m} = \frac{1}{h}.$$

Now, using the mean value theorem for integrals in the last equation yields

$$Q_{ij} = \hat{m}\left(\frac{j}{\hat{m}} - \frac{j-1}{\hat{m}}\right)\psi_i(\eta_j) = \psi_i(\eta_j), \quad \eta_j \in \left(\frac{j-1}{\hat{m}}, \frac{j}{\hat{m}}\right).$$

By choosing $\eta_j = \frac{2j-1}{2\hat{m}}$, it yields

$$Q_{ij} = \psi_i\left(\frac{2j-1}{2\hat{m}}\right), \quad i,j = 1, 2, \ldots, \hat{m}.$$

$\qquad\qquad\qquad\qquad\qquad\qquad\qquad\qquad\qquad\qquad\qquad\qquad\qquad\square$

Remark 5.3.2.5. For an $\hat{m}$-vector F

$$\Psi(t)\Psi^T(t)F = \tilde{F}\Psi(t),$$

in which $\tilde{F}$ is an $\hat{m} \times \hat{m}$ matrix of the form

$$\tilde{F} = Q\bar{F}Q^{-1},$$

where $\bar{F} = diag(Q^T F)$.

Remark 5.3.2.6. Let A represent an $\hat{m} \times \hat{m}$ matrix. Then, for the SKCWs vector $\Psi(t)$,

$$\Psi^T(t)A\Psi(t) = \hat{A}^T\Psi(t),$$

where $\hat{A}^T = UQ^{-1}$ and $U = diag(Q^T AQ)$ is an $\hat{m}$-vector.

5.3.2.4 *Stochastic OM of second-kind CWs*

Theorem 5.3.2.7. *Let $\Psi(t)$ be the $\hat{m}$-dimensional second-kind CWs vector. The integral of this vector is*

$$\int_0^t \Psi(s)ds = QPQ^{-1}\Psi(t) = \Lambda\Psi(t), \tag{5.15}$$

where Q is introduced in Eq. (5.14) and P is the OM of integration for BPFs specified in Eq. (5.11).

Proof.   Let $\Psi(t)$ be the second-kind CWs vector.

$$\int_0^t \Psi(s)ds = Q\int_0^t \Phi(s)ds = QP\Phi(t),$$

where $\Psi(t) = Q\Phi(t)$.

Now, Theorem 5.3.2.4 gives

$$\int_0^t \Psi(s)ds = QPQ^{-1}\Psi(t) = \Lambda\Psi(t),$$

where $\lambda = QPQ^{-1}$. $\qquad\qquad\qquad\qquad\qquad\qquad\qquad\square$

Theorem 5.3.2.8. *Let $\Psi(t)$ be the $\hat{m}$-dimensional SKCWs vector. The Itô integral of this vector is*

$$\int_0^t \Psi(s)dB(s) = QP_sQ^{-1}\Psi(t) = \Lambda_s\Psi(t), \tag{5.16}$$

where Λ_s is called SOM for second-kind CWs, Q is introduced in Eq. (5.12) and P_s is the SOM of integration for BPFs specified in Eq. (5.14).

Proof. Let $\Psi(t)$ be the SKCWs vector. Using Theorem 5.3.2.4 and Lemma 5.3.2.2 yields

$$\int_0^t \Psi(s)dB(s) = Q \int_0^t \Phi(s)dB(s) = QP_s\Phi(t).$$

Now, Theorem 5.3.2.4 yields

$$\int_0^t \Psi(s)dB(s) = QP_sQ^{-1}\Psi(t) = \Lambda_s\Psi(t),$$

where $\Lambda_s = QP_sQ^{-1}$. $\qquad\square$

Theorem 5.3.2.9. *Let $\Psi(t)$ be the $\hat{m}$-dimensional SKCWs vector. The Itô integral of this vector is*

$$\int_0^t \Psi(s)dB^H(s) = QP_S^HQ^{-1}\Psi(t) = \Lambda_S^H\Psi(t), \qquad (5.17)$$

where Λ_S^H is called fractional SOM for SKCWs, Q is introduced in Eq. (5.14) and P_S^H is the fractional SOM of integration for BPFs specified in Eq. (5.13).

Proof. Let $\Psi(t)$ be the SKCWs vector. Using Theorem 5.3.2.4 and Lemma 5.3.2.2 yields

$$\int_0^t \Psi(s)dB^H(s) = Q \int_0^t \Phi(s)dB^H(s) = QP_S^H\Phi(t).$$

Now, Theorem 5.3.2.4 yields

$$\int_0^t \Psi(s)dB^H(s) = QP_S^HQ^{-1}\Psi(t) = \Lambda_S^H\Psi(t),$$

where $\Lambda_S^H = QP_S^HQ^{-1}$. $\qquad\square$

5.3.3 *Approximation of fractional stochastic integral*

Consider the fractional stochastic integral equation for $\frac{1}{2} < H < 1$ [132].

$$X(t) = f(t) + \int_0^t k_0(s,t)X(s)ds + \sum_{i=1}^n \int_0^t k_i(s,t)X(s)dB_i(s)$$

$$+ \int_0^t g(s,t)X(s)dB^H(s), \quad t \in [0,T), \tag{5.18}$$

where $X(t)$, $f(t)$, $k_0(s,t)$, $k_i(s,t)$, $i = 1,2,\ldots,n$ and $g(s,t)$ for $s,t \in [0,T)$ are the stochastic processes on $(\Omega, \mathcal{F}, \mathbb{P})$ and $X(t)$ is unknown. Also, $B(t) = (B_1(t), B_2(t), \ldots, B_n(t))$ is a multidimensional BM process and $B^H(t)$ is the fractional BM [18]. Now, the functions in Eq. (5.16) have been approximated as

$$f(t) = F^T\Psi(t) = \Psi^T(t)F,$$

$$X(t) = X^T\Psi(t) = \Psi^T(t)X, \tag{5.19}$$

$$vk_i(s,t) = \Psi^T(s)K_i\Psi(t) = \Psi^T(t)K_i^T(s), \quad i = 0,1,2,\ldots,n,$$

$$g(s,t) = \Psi^T(s)G\Psi(t) = \Psi^T(t)G^T\Psi(s),$$

where X and F are the second-kind CWs coefficients vector and K_i, $i = 0,1,2,\ldots,n$, and G are the second-kind CWs coefficients matrices defined in Eqs. (5.8) and (5.10). Substituting Eq. (5.19) in Eq. (5.18) yields

$$X^T\Psi(t) = F^T\Psi(t) + \Psi^T(t)K_0^T\left(\int_0^t \Psi(s)\Psi^T(s)Xds\right)$$

$$+ \sum_{i=1}^n \Psi^T(t)K_i^T\left(\int_0^t \Psi(s)\Psi^T(s)XdB_i(s)\right)$$

$$+ \Psi^TG^T\left(\int_0^t \Psi(s)\Psi^T(s)XdB^H(s)\right).$$

Using relation $\langle \Psi(s), \Psi^T(s)\rangle_{w_{nk}} = I_{\hat{m}\times\hat{m}}$ and Remark 5.3.2.5, it yields

$$X^T\Psi(t) = F^T\Psi(t) + \Psi^T(t)K_0^T\left(\int_0^t \tilde{X}\Psi(s)ds\right)$$

$$+ \sum_{i=1}^n \Psi^T(t)K_i^T\left(\int_0^t \tilde{X}\Psi(s)dB_i(s)\right)$$

$$+ \Psi^T(t)G^T\left(\int_0^t \tilde{X}\Psi(s)dB^H(s)\right),$$

where $\tilde{X}$ is an $\hat{m}\times\hat{m}$. matrix. Exerting Λ, Λ_{s_i} and Λ_S^H for second-kind CWs derived in Eqs. (5.15), (5.16) and (5.17) yields

$$X^T\Psi(t) = F^T\Psi(t) + \Psi^T(t)K_0^T\tilde{X}\Lambda\Psi(t)$$

$$+ \sum_{i=1}^n \Psi^T(t)K_i^T\tilde{X}\Lambda_{s_i}\Psi(t)$$

$$+ \Psi^t(t)G^T\tilde{X}\lambda_S^H\Psi(t).$$

By setting $y_0 = K_0^T\tilde{X}\lambda$, $Y_i = K_i^T\tilde{X}\Lambda_{s_i}$, $Y^H = G^T\tilde{X}\Lambda_S^H$ and using Remark 5.3.2.6

$$X^T\Psi(t) - \hat{Y}_0^T\Psi(t) - \sum_{i=1}^n \hat{Y}_i^T\Psi(t) - \hat{Y}^H\Psi(t) = F^T\Psi(t),$$

where $\hat{y}_0$, $\hat{Y}_i$ and $\hat{Y}^H$ are $\hat{m}$-dimensional vectors. This equation holds for all $t \in [0,1)$, thus

$$X^T - \hat{Y}_0^T - \sum_{i=1}^n \hat{Y}_i^T - \hat{Y}^H = F^T. \tag{5.20}$$

Since $\hat{Y}_0$, $\hat{Y}_i$ and $\hat{Y}^H$ are linear functions of X, Eq. (5.20) is a linear system of equations. By determining X, the approximate solution of the fractional stochastic Itô–Volterra integral equation (5.18) can be determined by substituting the obtained vector X in Eq. (5.19).

5.3.4 Convergence and error analysis

Theorem 5.3.4.1 (convergence analysis [103]). *If a continuous function $f(t) \in L^2_{\mathbf{w}_{nk}}[0,1]$ has bounded second derivative $|\frac{\partial^2 f}{\partial t^2}| \leq \mathcal{M}$, the second-kind CWs expansion $f(t) \sum_{n=1}^{2^{k-1}} \sum_{m=0}^{M-1} X(n,m)\psi(n,m,t)$ of $f(t)$ converges uniformly.*

Proof. It has been proved in Theorem 4.4.3.1 of Chapter 4.

For error analysis, define $\|X\| = E\{(|X(t)|^2)^{\frac{1}{2}}\}$.

Theorem 5.3.4.2 (error analysis). *Suppose $X(t)$ is the exact solution of Eq. (5.1) and $X_m(t)$ is its second-kind CWs approximate solution. Also, assume that*

$$\|X(t)\| \leq \mu, t \in [0,1],$$

$$\|k_i(s,t)\| \leq M_i, (s,t) \in [0,1] \times [0,1], \quad i = 1,2,3,$$

$$and \quad \|g(s,t)\| \leq \Omega, (s,t) \in [0,1] \times [0,1].$$

Then, for any $\varepsilon > 0$,

$$\|X(t) - X_m(t)\|$$

$$\leq \frac{\left[\delta + \mu\varepsilon_0 + \mu\sum_{i=1}^{n}\sup_{t\in[0,1)}|B_i(t)|\varepsilon_i + B^H(t)\mu\varsigma\right]}{1 - \left[\begin{array}{l}(M_0 + \varepsilon_0) + \sum_{i=1}^{n}\sup_{t\in[0,1)}|B_i(t)|(M_i + \varepsilon_i) \\ +B^H(t)\mu(\Omega + \varsigma)\end{array}\right]}.$$

$$(5.21)$$

Proof. From Eq. (5.1), it yields

$$X(t) - X_m(t) = f(t) - f_m(t) + \int_0^t (k_0(s,t)X(s) - k_{0m}(s,t)X_m(s))ds$$

$$+ \sum_{i=1}^{n}\int_0^t (k_i(s,t)X(s) - k_{im}(s,t)X_m(s))dB(s)$$

$$+ \int_0^t (g(s,t)X(s) - g_m(s,t)X_m(s))dB^H(s). \quad (5.22)$$

$\square$

Implementing the integral mean value theorem,

$$\|X(t) - X_m(t)\| \leq \|f(t) - f_m(t)\|$$
$$+ t\|k_0(\xi_0, t)X(\xi_0) - k_{0m}(\xi_0, t)X_m(\xi_0)\|$$
$$+ \sum_{i=1}^{n} |B_i(t)| \|k_i(\xi_i, t)X(\xi_i) - k_{im}(\xi_i, t)X_m(\xi_i)\|$$
$$+ |B^H(t)| \|g(\xi, t)X(\xi) - g_m(\xi, t)X_m(\xi)\|, \quad (5.23)$$

where $\xi_i \in [0, 1]$, $i = 0, 1, 2, \ldots, n$.

Now, Theorem 5.3.4.1 shows that the second-kind CWs expansion for any continuous function $f(t)$ converges uniformly to f. So, for any ε, there exists m such that

$$\|k_i(s, t) - k_{im}(s, t)\| \leq \varepsilon_i, \; i = 0, 1, 2, 3, \ldots, n,$$
$$g(\xi, t - g_m)(\xi, t)\| \leq \varsigma,$$
$$\|f(t) - f_m(t)\| \leq \delta.$$

Thus, for $i = 0, 1, 2, 3, \ldots, n$,

$$\|k_i(\xi_i, t)X(\xi_i) - k_{im}(\xi_i, t)X_m(\xi_i)\| \leq \|k_i(\xi_i, t)\|$$
$$\|X(\xi_i) - X_m(\xi_i)\| + \|k_i(\xi t) - k_{im}(\xi_i, t)\| \|X(\xi_i)\|$$
$$+ \|k_i(\xi_i, t) - k_{im}(\xi_i, t)\| \|X(\xi_i) - X_m(\xi_i)\|$$
$$\leq (M_i + \varepsilon_i)\|X(\xi_i) - X_m(\xi_i)\| + \mu\varepsilon_i. \quad (5.24)$$

Similarly,

$$\|g(\xi, t)X(\xi) - g_m(\xi, t)X_m(\xi)\| \leq \|g(\xi, t)\| \|X(\xi) - X_m(\xi)\|$$
$$+ \|g(\xi, t) - g_m(\xi, t)\| \|X(\xi)\| \|$$
$$+ \|g(\xi, t) - g_m(\xi, t)\| \|X(\xi) - X_m(\xi)\|$$
$$\leq (\Omega + \varsigma)\|X(\xi) - X_m(\xi)\| + \mu\varsigma. \quad (5.25)$$

Now, substituting Eqs. (5.24) and (5.25) in Eq. (5.23) yields

$$\|X(t) - X_m(t)\| \le \delta + t[(M_0 + \varepsilon_0)\|X(t) - X_m(t)\| + \mu\varepsilon_0]$$

$$+ \sum_{i=1}^{n} |B_i(t)|[(M_i + \varepsilon_i)\|X(t) - X_m(t)\| + \mu\varepsilon_i]$$

$$+ B^H(t)[(\Omega + \varsigma)\|X(t) - X_m(t)\| + \mu\varsigma]$$

$$\le \delta + [(M_0 + \varepsilon_0)\|X(t) - X_m(t)\| + \mu\varepsilon_0]$$

$$+ \sum_{i=1}^{n} \sup_{t\in[0,1)} |B_i(t)|[(M_i + \varepsilon_i)\|X(t) - X_m(t)\| + \mu\varepsilon_i]$$

$$+ B^H(t)[(\Omega + \varsigma)\|X(t) - X_m(t)\| + \mu\varsigma].$$

Replacing ξ_i, $i = 0, 1, 2, \ldots, n$, by an arbitrary $t \in [0,1)$.
Hence,

$$\|X(t) - X_m(t)\|$$
$$\le \frac{\left[\delta + \mu\varepsilon_0 + \mu\sum_{i=1}^{n}\sup_{t\in[0,1)}|B_i(t)|\varepsilon_i + B^H(t)\mu\varsigma\right]}{1 - \left[(M_0 + \varepsilon_0) + \sum_{i-1}^{n}\sup_{t\in[0,1)}|B_i(t)|(M_i + \varepsilon_i) + B^H(t)(\Omega + \varsigma)\right]}.$$

$\square$

5.3.5 *Numerical examples*

In this section, the accuracy and efficiency of the proposed second-kind Chebyshev wavelets method have been examined through suitable nontrivial examples.

Example 5.1. Consider the following fractional SDE [132]:

$$dx(s) = -\frac{1}{5}s^2 x(s)ds - \frac{1}{10}x(s)dB^H(s)$$

$$-\frac{1}{6}x(s)dB_1(s) - \frac{1}{30}x(ds)dB_2(s), s \in [0,1), T < 1.$$

Table 5.1. A comparison between the approximate solutions based on second-kind CWs and exact solutions for $\hat{m} = 6$ for $H = \frac{2}{3}$ and $\hat{m} = 6$ for $H = \frac{3}{4}$ for Example 5.1.

	$\hat{m} = 6$ for $H = \frac{2}{3}$		$\hat{m} = 6$ for $H = \frac{3}{4}$	
t	Proposed method solution for Example 5.1	Exact solution for Example 5.1	Proposed method solution for Example 5.1	Exact solution for Example 5.1
0.05	0.0356694	0.035661	0.031425	0.0309469
0.1	0.032542	0.033468	0.0319611	0.0320497
0.15	0.0328696	0.0328048	0.0377915	0.0378633
0.2	0.0333409	0.0322541	0.0298987	0.0307432

The exact solution of this FSIVIE is

$$X(t) = \frac{1}{30} \exp\left(-\frac{1}{10} B^H(t) - \frac{1}{15} t^3 - \frac{1}{200} t^{2H} - \frac{1}{6} B_1(t) - \frac{1}{72} t \right.$$

$$\left. - \frac{1}{30} B_2(t) - \frac{1}{1800} t \right),$$

where $X(t)$ is an unknown on $\Omega, \mathcal{F}, \mathbb{P}$. The stochastic operational matrix for second-kind CWs presented in Section 5.3.2.4 and the proposed method described in Section 5.3.3 are used for solving Example 5.1. A comparison between the approximate and exact solutions for $T = 0.2$ of the above problem is shown in Table 5.1 and the sample paths of Example 5.1 for $\hat{m} = 6$ with Hurst index $H = \frac{2}{3}$ and $H = \frac{3}{4}$ have been shown in Figs. 5.1(a) and 5.1(b), respectively.

Example 5.2. Consider the following fractional SDE [132]:

$$dx(s) = -\frac{1}{6} s^2 x(s) ds - \frac{1}{30} x(s) dB^H(s) - \frac{1}{10(1-s)} x(s) dB_1(s)$$

$$- \frac{1}{30} x(s) dB_2(s), \quad s \in [0, 1), \quad T < 1.$$

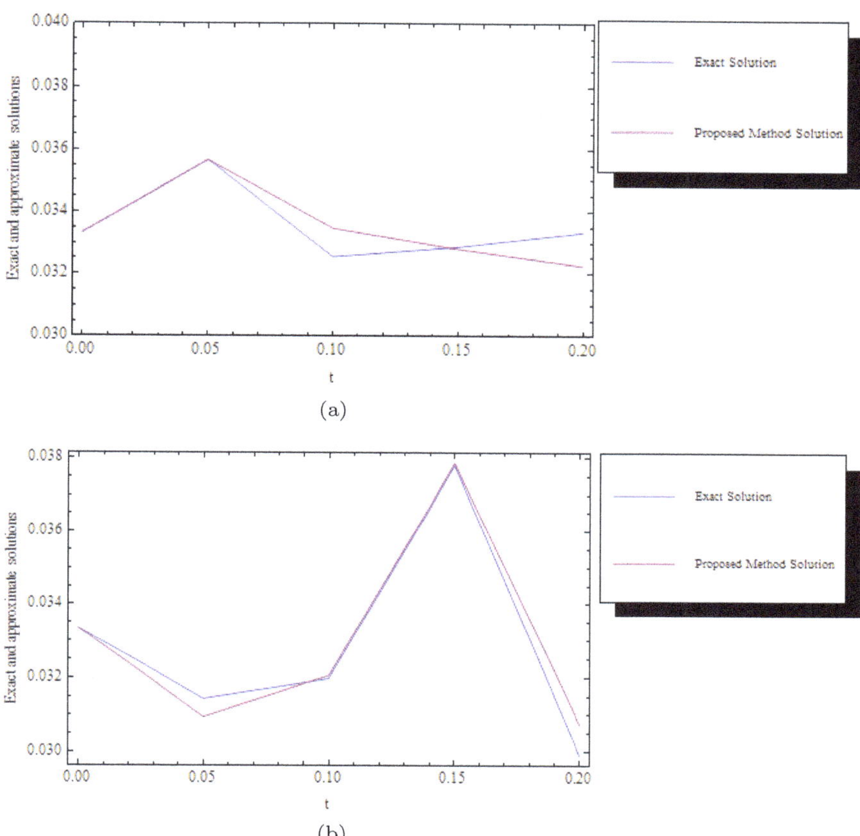

Fig. 5.1. (a) The sample paths of Example 5.1 for $\hat{m} = 6$ for $H = \frac{2}{3}$. (b) The sample paths of Example 5.1 for $\hat{m} = 6$ for $H = \frac{3}{4}$.

The exact solution of this FSIVIE is

$$
X(t) = \frac{1}{12} \exp\left(-\frac{1}{30} B^H(t) - \frac{1}{18} t^3 - \frac{1}{1800} t^{2H} - \int_0^t \frac{(1-s)}{10} dB_1(s) \right.
$$
$$
\left. -\frac{1}{1800} t - \frac{1}{30} B_2(t) + \frac{1}{200(1-t)} \right),
$$

Table 5.2. A comparison between the approximate solutions based on second-kind CWs and exact solutions for $\hat{m} = 6$ for $H = \frac{2}{3}$ and $\hat{m} = 6$ for $H = \frac{3}{4}$ for Example 5.2.

	$\hat{m} = 6$ for $H = \frac{2}{3}$		$\hat{m} = 6$ for $H = \frac{3}{4}$	
t	Proposed method solution for Example 5.2	Exact solution for Example 5.2	Proposed method solution for Example 5.2	Exact solution for Example 5.2
0.05	0.0800834	0.0808709	0.0831644	0.0849975
0.1	0.0860038	0.084305	0.080416	0.077087
0.15	0.0916945	0.0941429	0.0979782	0.0986258
0.2	0.0910696	0.0858352	0.0882836	0.090564

where $X(t)$ is an unknown on $(\Omega, \mathcal{F}, \mathbb{P})$. The stochastic operational matrix for second-kind CWs presented in Section 5.3.2.4 and the proposed method described in Section 5.3.3 are used for solving Example 5.2. A comparison between the approximate and exact solutions for $T = 0.2$ of the above problem is shown in Table 5.2, and the sample paths of Example 5.2 for $\hat{m} = 6$ with Hurst index $H = \frac{2}{3}$ and $H = \frac{3}{4}$ have been shown in Figs. 5.2(a) and 5.2(b), respectively.

5.4 Bernstein Polynomials for Fractional Stochastic Nonlinear Integral Equations

In this section, a brief discussion regarding Bernstein polynomials and their corresponding operational matrices has been provided.

5.4.1 *Bernstein polynomials*

Bernstein polynomials of mth degree over $[0, 1]$ are defined [120] as

$$B_{i,m} = \binom{m}{i} x^i (1 - x)^{m-i}, \quad i = 0, 1, \ldots, m, \qquad (5.26)$$

where $\binom{m}{i} = \frac{m!}{i!(m-1)!}$.

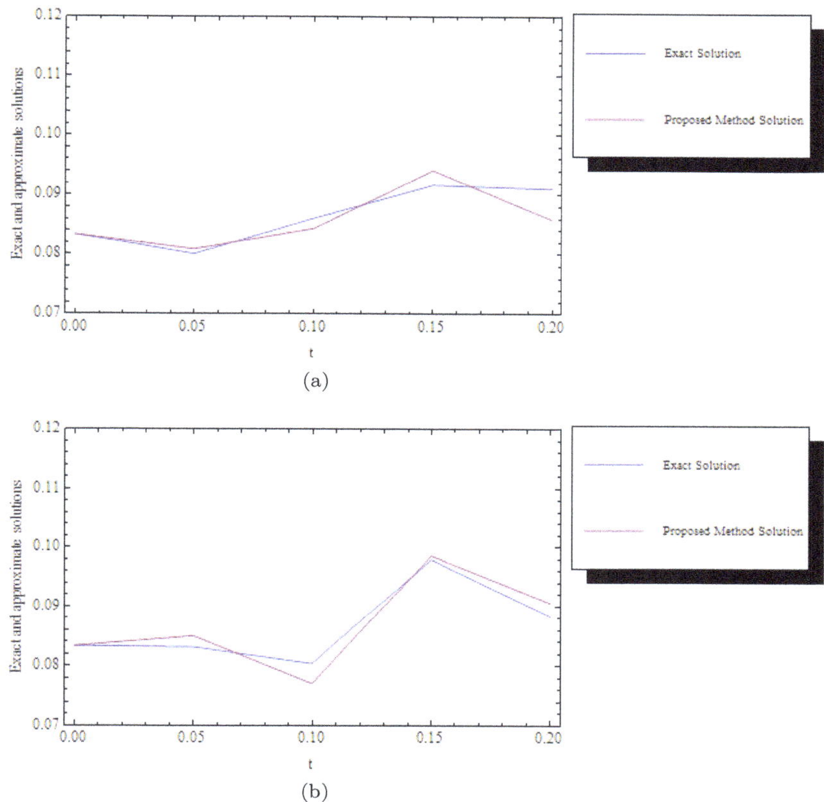

Fig. 5.2. (a) The sample paths of Example 5.2 for $\hat{m} = 6$ for $H = \frac{2}{3}$. (b) The sample paths of Example 5.2 for $\hat{m} = 6$ for $h = \frac{3}{4}$.

These $m + 1$ polynomials having degree m have the following conditions:

(i) $B_{i,m}(x) = 0$, if $i < 0$ or $i > m$,

(ii) $B_{i,m}(0) = B_{i,m}(1) = 0$, for $1 \leq i \leq m - 1$,

(iii) $\sum_{i=0}^{m} B_{i,m}(x) = 1$.

5.4.1.1 *Function approximation*

Here, $\hat{f}(x)$ over $[0, 1)$ which is square integrable, i.e., $\hat{f}(x) \in L^2[0, 1)$, is expanded in accordance to Bernstein polynomials of degree m as

$$\hat{f}(x) = \sum_{i=0}^{m} c_i B_{i,m}(x) = C^T \phi(x), \tag{5.27}$$

where $B(x)$ and C are column vectors given by

$$C = [c_0, c_1, \ldots, c_m]^T, \tag{5.28}$$

and

$$\Phi(x) = [B_{0,m}(x), B_{l,m}(x), \ldots, B_{m,m}(x)]^T, \tag{5.29}$$

where

$$\Phi(x) = A T_m(x), \tag{5.30}$$

and

$$A = \begin{bmatrix} (-1)^0 \binom{m}{0} & (-1)^1 \binom{m}{0}\binom{m-0}{1} & \cdots & (-1)^{m-0}\binom{m}{0}\binom{m-0}{m-0} \\ & \ddots & & \vdots \\ 0 & (-1)^0\binom{m}{i} & \cdots & (-1)^{m-1}\binom{m}{i}\binom{m-i}{m-i} \\ \vdots & \ddots & \ddots & \vdots \\ 0 & \cdots & 0 & (-1)^0\binom{m}{m} \end{bmatrix},$$

$$T_n(x) = \begin{bmatrix} 1 \\ x \\ \vdots \\ x^m \end{bmatrix}. \tag{5.31}$$

5.4.1.2 *Integration operational matrices*

Lemma 5.4.1.1. *Let $B(x)$ be the $m+1$-dimensional Bernstein polynomials. The integral can be derived as*

$$\int_0^x \Phi(y)dy \cong P\Phi(x), \tag{5.32}$$

where P is the $(m+1) \times (m+1)$-dimensional integrational OM for Bernstein polynomials.

Proof. Now from Eq. (5.30), it can be written as

$$\int_0^x \Phi(y)dy = \int_0^x AT_m(y)dy = A\int_0^x T_m(y)dy = A\begin{bmatrix} x \\ \frac{1}{2}x^2 \\ \vdots \\ \frac{1}{m+1}x^{m+1} \end{bmatrix}$$

$$= A_p X_p,$$

where A_p is an $(m+1) \times (m+1)$ matrix.

$$A_p = A\begin{bmatrix} 1 & 0 & \cdots & 0 \\ 0 & \frac{1}{2} & \cdots & 0 \\ \vdots & \ddots & \ddots & \vdots \\ 0 & \cdots & 0 & \frac{1}{m+1} \end{bmatrix} \quad \text{and} \quad X_p = \begin{bmatrix} x \\ x^2 \\ \vdots \\ x^{m+1} \end{bmatrix}.$$

Now, the elements of vector X_p in terms of $\Phi(x)$ have been approximated. From Eq. (5.30), $T_m(x) = A^{-1}\Phi(x)$, then for $k = 0, 1, \ldots, m$,

$$x^k = A_{(k+1)}^{-1}\Phi(x),$$

where $A_{(k+1)}^{-1}$ is $k+1$th row of A^{-1} for $k = 0, 1, \ldots, m$. Its needed to approximate $x^{m+1} \cong c_{m+1}^T \Phi(x)$.

Now,

$$c_{m+1} = \mathbf{D}^{-1}\int_0^1 x^{m+1}\Phi(x)dx = \mathbf{D}^{-1}\begin{bmatrix} \int_0^1 x^{m+1}B_{0,m}(x)dx \\ \int_0^1 x^{m+1}B_{1,m}(x)dx \\ \vdots \\ \int_0^1 x^{m+1}B_{m,m}(x)dx \end{bmatrix}$$

$$= \frac{\mathbf{D}^{-1}}{2m+2} \begin{bmatrix} \dfrac{\dbinom{m}{0}}{\dbinom{2m+1}{m+1}} \\[1.2em] \dfrac{\dbinom{m}{1}}{\dbinom{2m+1}{m+2}} \\[1.2em] \vdots \\[1.2em] \dfrac{\dbinom{m}{m}}{\dbinom{2m+1}{2m+1}} \end{bmatrix},$$

where $\mathbf{D} = \langle \Phi(x), \Phi(x) \rangle$.

Let $B = \begin{bmatrix} A_{[2]}^{-1} \\ A_{[3]}^{-1} \\ \\ A_{[m+1]}^{-1} \\ c_{m+1}^{T} \end{bmatrix}$, then $X_p \cong B\Phi(x)$. Therefore, the operational matrix of integration has been obtained as $P = A_p B$. $\qquad\square$

Lemma 5.4.1.2. *Let $\Phi(t)$ be the $m+1$-dimensional Bernstein polynomials. The Itô integral can be derived as*

$$\int_0^x \Phi(y) dB^H(y) \cong P_s^H \Phi(x), \tag{5.33}$$

where P_S^H is the $(m+1) \times (m+1)$-dimensional fractional integrational stochastic operational matrix for Bernstein polynomials using Eq. (5.31) is given by $P_S^H = A D_S^H A^{-1}$. Also, $B^H(x)$ and $B^H\left(\frac{x}{2}\right)$ for $0 \leq t \leq 1$ have been approximated by $B^H(0.5)$ and $B^H(0.25)$, respectively. So,

$$D_S^H = \begin{bmatrix} \dfrac{(t/2)^{H-\frac{1}{2}}}{\Gamma\left(H+\frac{1}{2}\right)B^H(0.5)} & 0 & 0 \\[2em] 0 & \dfrac{(t/2)^{H-\frac{1}{2}}}{\Gamma\left(H+\frac{1}{2}\right)}\left(\frac{3}{4}B^H(0.5)-\frac{1}{2}B^H(0.25)\right) & \cdots & 0 \\[1em] \vdots & \vdots & \ddots & \vdots \\[1em] 0 & 0 & \cdots & \begin{array}{l}\dfrac{(t/2)^{H-\frac{1}{2}}}{\Gamma\left(H+\frac{1}{2}\right)}\left(\left(1-\frac{m}{4}\right)\right.\\[0.5em] \times B^H(0.5)\\[0.3em] \left.-\frac{m}{2^m}B^H(0.25)\right)\end{array} \end{bmatrix}.$$

Proof.   It may be further referred to Ref. [132]. $\square$

5.4.2   *Approximation of fractional nonlinear stochastic integral*

In the present analysis, the fractional nonlinear stochastic integral equation for Hurst index, $\frac{1}{2} < H < 1$, has been considered.

$$f(x) = g(x) + \lambda_1 \int_0^x b(y, f(y))dy + \lambda_2 \int_0^x \sigma(y, f(y))dB^H(y),$$

$$x \in [0, T), \tag{5.34}$$

where λ_1, λ_2 are parameters and $f(x)$, $g(x)$, $b(x, f(x))$, $\sigma(x, f(x))$ where $x \in [0, T)$, are defined on $(\Omega, \mathcal{F}, \mathbb{P})$, and $f(x)$ is unknown. Also, $B^H(x)$ is the FBM defined in Ref. [18].

$$z_1(x) = b(x, f(x)), z_2(x) = \sigma(x, f(x)). \tag{5.35}$$

Now, Eq. (5.35) is substituted in Eq. (5.34),

$$\begin{cases} z_1(x) = b(x, g(x) + \lambda_1 \int_0^x z_1(y)dy + \lambda_2 \int_0^x z_2(y)dB^H(y)), \\[0.8em] z_2(x) = \sigma(x, g(x) + \lambda_1 \int_0^x z_1(y)dy + \lambda_2 \int_0^x z_2(y)dB^H(y)). \end{cases} \tag{5.36}$$

Now, the Bernstein polynomial approximation $z_1(x)$ and $z_2(x)$ can be written as

$$z_1(x) = Z_1^T \Phi(x), z_2(x) = Z_2^T \Phi(x). \tag{5.37}$$

where Z_1 and Z_2 are defined in Eq. (5.28).

Now, using Eqs. (5.37), (5.32) and (5.33),

$$\int_0^x z_1(y)dy = Z_1^T \int_y^x \Phi(y)dy = Z_1^T P\Phi(x), \tag{5.38}$$

and

$$\int_0^x z_2(y)dB^H(y) = Z_2^T \int_0^x \Phi(y)dB^H(y) = Z_2^T P_S^H \Phi(x). \tag{5.39}$$

Substituting Eqs. (5.37), (5.38) and (5.39) in Eq. (5.36),

$$\begin{cases} Z_1^T \Phi(x) = b(x, g(x)) + \lambda_1 Z_1^T P\Phi(x) + \lambda_2 Z_2^T P_S^H \Phi(x)), \\ Z_2^T \Phi(x) = \sigma(x, g(x) + \lambda_1 Z_1^T P\Phi(x) + \lambda_2 Z_2^T P_S^H \Phi(x)). \end{cases} \tag{5.40}$$

Now, collocating Eq. (5.40) in $m+1$ Newton–Cotes nodes, $x_i = \frac{(2i-1)}{2(m+1)}$, $i = 1, 2, \ldots, m+1$. Then, Eq. (5.40) is of the form

$$\begin{cases} Z_1^T \Phi(x_i) = b(x_i, g(x_i) + \lambda_1 Z_1^T P\Phi(x_i) + \lambda_2 Z_2^T P_S^H \Phi(x_i)), \\ Z_2^T \Phi(x_i) = \sigma(x, g(x_i) + \lambda_1 Z_1^T P\Phi(x_i) + \lambda_2 Z_2^T P_S^H \Phi(x_i)). \end{cases} \tag{5.41}$$

Then, the nonlinear system Eq. (5.41) has been solved for Z_1 and Z_2. Then, Eq. (5.34) can be approximated to be

$$f(x) = g(x) + \lambda_1 Z_1^T P\Phi(x) + \lambda_2 Z_2^T P_S^H \Phi(x).$$

5.4.3 *Convergence and error analysis*

Theorem 5.4.3.1 (convergence analysis). *If $f(x) \in L^2[0,1]$ is a continuous function, the Bernstein polynomial expansion $f(x) = \sum_{m=0}^m c_i B_{i,m}(x)$ of $f(x)$ converges uniformly.*

Proof.   Refer to Refs. [119,139]. $\square$

Theorem 5.4.3.2 (error analysis). *Let $f(x) \in L^2[0,1]$ imply the exact solution of Eq. (5.2) and $f_n(x) \in L^2[0,1]$ be its approximate solution. Also, suppose that $\hat{z}_i(x), i = 1, 2$, are the approximate forms of $z_i(x)$, $i = 1, 2$, by Bernstein polynomial approximations,*

$$\hat{z}_i(x) = B_n(b(x, f(x))), \hat{z}_2(x) = B_n(\sigma(x, f(x))),$$

and

$$z_{1n}(x) = b(x, f_n(x)), z_{2n}(x) = \sigma(x, f_n(x)).$$

Also using Lipschitz conditions assume that [119,139]

$$\|b(x, f(x)) - b_n(x, f(x))\| \le L_1 \|f(x) - f_n(x)\|,$$
$$\|\sigma(x, f(x)) - \sigma_n(x, f(x))\| \le L_2 \|f(x) - f_n(x)\|,$$

and

$$\|g(x) - g_n(x)\| \le \delta, \quad x \in [0, 1).$$

Then,

$$\|f(x) - f_n(x)\|^2 \le [(\lambda_1^2 + \lambda_2^2)\varepsilon + 12\delta^2][1 + 12(\lambda_1^2 L_1^2 + \lambda_2^2 L_2^2)$$
$$\int_0^x e^{12(\lambda_1^2 L_1^2 + \lambda_2^2 L_2^2)(x-y)} dy], \quad x \in [0, 1),$$

where $\|x\| = (E|x|^2)^{\frac{1}{2}}$.

Proof. In Ref. [128], it has been defined that

$$E\left[\left(\int_\alpha^\beta f(x) dB^H(x)\right)^2\right] = \int_\alpha^\beta E[(f(x))^2](dx)^{2H}$$
$$= 2H(2H - 1) \int_\alpha^\beta \int_\alpha^\gamma$$
$$\times (y - x) E(f(x))^2] dx dy.$$

From Eq. (5.35),

$$z_1(x) = b(x, f(x)), z_2(x) = \sigma(x, f(x)).$$

The approximate solution of Eq. (5.2) is

$$f_n(x) = g_n(x) + \lambda_1 \int_0^x b_n(y, f(y))dy + \lambda_2 \int_0^x \sigma_n(y, f(y))dB^H(y).$$

Now,

$$f(x) - f_n(x) = g(x) - g_n(x) + \lambda_1 \int_0^x (b(y, f(y)) - b_n(y, f(y)))dy$$

$$+ \lambda_2 \int_0^x (\sigma(y, f(y)) - \sigma_n(y, f(y)))dB^H(y).$$

So, applying $\left(\sum_{i=1}^n x_i\right)^2 \le n \sum_{i=1}^n x_i^2$ yields

$$\|f(x) - f_n(x)\|^2 \le 3\left(\|g(x) - g_n(x)\|^2 + \lambda_1^2 \left\|\int_0^x (z_1(y) - z_{1n}(y))dy\right\|^2\right.$$

$$\left. + \lambda_2^2 \left\|\int_0^x (z_2(y) - z_{2n}(y))dB^H(y)\right\|^2\right). \qquad (5.42)$$

Again,

$$\left\|\int_0^x (z_2(y) - z_{2n}(y))dB^H(y)\right\|^2$$

$$= E\left[\left(\int_0^x (z_2(y) - z_{2n}(y))dB^H(y)\right)^2\right]$$

$$= \int_0^x E[z_2(y) - z_{2n}(y))]^2 (dy)^{2H}$$

$$= 2H(2H - 1)\int_0^x \int_0^t (t - y)^{2H-2} E[(z_2(y) - z_{2n}(y))^2]dtdy$$

$$= 2H(2H - 1)\int_0^x \int_y^x (t - y)^{2H-2} E[(z_2(y) - z_{2n}(y))^2]dtdy$$

$$= 2H \int_0^x E[(z_2(y) - z_{2n}(y))^2](x - y)^{2H-1}dy.$$

As $0 < y < x < 1$ and $\frac{1}{2} < H < 1$; so, $0 < (x-y)^{2H-1} < 1$,

$$E\left[\left(\int_0^x (z_2(y) - z_{2n}(y))dB^H(y)\right)^2\right] \le 2\int_0^x E[(z_2(y) - z_{2n}(y)^2]dy.$$

$$\|f(x) - f_n(x)\|^2 \le 3(\|g(x) - g_n(x)\|^2$$

$$+ |\lambda_1|^2 \int_0^x \|z_1(y) - z_{1n}(y)\|^2 dy + 2|\lambda_2|^2$$

$$\times \int_0^x \|z_2(y) - \hat{z}_{2n}(y)\|^2 dy)$$

$$\le 6\left(\|g(x) - g_n(x)\|^2 + |\lambda_1|^2 \int_0^x \|(z_1(y) - \hat{z}_1(y) - \hat{z}_{1n}(y))\|^2 dy\right.$$

$$\left. + |\lambda_2|^2 \int_0^x \|(z_2(y) - \hat{z}_2(y) + \hat{z}_2(y) - z_{2n}(y))\|^2 dy\right)$$

$$\le 12\left(\|g(x) - g_n(x)\|^2 + |\lambda_1|^2 \int_0^x \|z_1(y) - \hat{z}_1(y)\|^2 dy + |\lambda_1|^2\right.$$

$$\times \int_0^x \|\hat{z}_1(y) - z_{1n}(y)\|^2 dy + |\lambda_2|^2 \int_0^x \|z_2(y) - \hat{z}_2(y)\|^2 dy + |\lambda_2|^2$$

$$\left. \times \int_0^x \|\hat{z}_2(y) - z_{2n}(y)\|^2 dy\right).$$

From Theorem 5.4.1.1, for any ε, there exists $n > 0$ such that

$$\|\hat{z}_k(y) - z_{kn}(y)\|^2 \le \frac{\varepsilon}{12|\lambda_k|^2}, \quad k = 1, 2.$$

Now, from Ref. [128], it can be seen that the Bernstein expansion for $f(x)$ converges uniformly and using the Lipschitz condition,

$$\|b(x, f(x)) - b_n(x, f(x))\| \le L_1\|f(x) - f_n(x)\|,$$

$$\|\sigma(x, f(x)) - \sigma_n(x, f(x))\| \le L_2\|f(x) - f_n(x)\|,$$

$$\|g(x) - g_n(x)\| \le \delta.$$

Thus, it yields

$$\|f(x) - f_n(x)\|^2 \le (\lambda_1^2 + \lambda_2^2)\varepsilon + 12\delta^2 + 12(\lambda_1^2 L_1^2 + \lambda_2^2 L_2^2)$$
$$\int_0^x \|f(y) - f_n(y)\|^2 dy.$$

Now, from Grönwall inequality,

$$\|f(x) - f_n(x)\|^2 \le [(\lambda_1^2 + \lambda_2^2)\varepsilon + 12\delta^2][1 + 12(\lambda_1^2 L_1^2 + \lambda_2^2 L_2^2)$$
$$\int_0^x e^{12(\lambda_1^2 L_1^2 + \lambda_2^2 L_2^2)(x-y)} dy].$$

Hence, it is proved. $\qquad\square$

5.4.4 Numerical examples

The accuracy and efficiency of Bernstein polynomials have been examined by solving the following nonlinear fractional stochastic integral equations [119].

Example 5.3.

$$f(x) = f_0 - 2Ha^2 \int_0^x y^{2H-1} f(y)(1 - f^2(y)) dy$$
$$+ a \int_0^x (1 - f^2(y)) dB^H(y), \quad x \in [0, 1),$$

where $f(x) = \tanh(ab^H(x) + \operatorname{arctanh}(f_0))$ is the exact solution and $f(x)$ is an unknown stochastic process on $(\Omega, \mathcal{F}, \mathbb{P})$. Stochastic operational matrix for Bernstein polynomials presented in Section 5.4.1.2 implemented in solving the above equation. Approximate, exact solutions and results from modifications of hat functions (MHFs) have been compared for $a = \frac{1}{30}$ and $f_0 = \frac{1}{10}$ of the above problem are shown in Tables 5.3–5.6, and the sample paths of Example 5.3 for $m = 6$, $m = 8$, $m = 10$ and $m = 20$ with Hurst index $H = 0.5$ and $H = 0.8$ have been shown in Figs. 5.3(a)–5.3(h).

Table 5.3. A comparison between the approximate solutions based on Bernstein polynomials, exact solutions and MHFs for $m = 6$ for $H = 0.5$ and $m = 6$ for $H = 0.8$ of Example 5.3.

	$m = 6$ for $H = 0.5$			$m = 6$ for $H = 0.8$		
x	Proposed method solution	Exact solution	MHFs	Proposed method solution	Exact solution	MHFs
0.0	0.100745	0.109317	0.0921149	0.0471007	0.0558465	0.0567325
0.1	0.111883	0.101518	0.129321	0.0841747	0.089354	0.104105
0.2	0.13812	0.14454	0.154111	0.16786	0.167261	0.17111
0.3	0.147757	0.132744	0.127451	0.0214639	0.0192107	0.0251751
0.4	0.168337	0.167697	0.158146	0.134057	0.149098	0.14459
0.5	0.108863	0.0787976	0.0741869	0.108197	0.105359	0.0916713

Table 5.4. A comparison between the approximate solutions based on Bernstein polynomials, exact solutions and MHFs for $m = 8$ for $H = 0.5$ and $m = 8$ for $H = 0.8$ of Example 5.3.

	$m = 8$ for $H = 0.5$			$m = 8$ for $H = 0.8$		
x	Proposed method solution	Exact solution	MHFs	Proposed method solution	Exact solution	MHFs
0.0	0.105274	0.115814	0.115592	0.0784226	0.0716548	0.0741488
0.1	0.0350798	0.0370297	0.0377588	0.164295	0.170581	0.158714
0.2	0.0320345	0.0369338	0.0391641	0.162202	0.161501	0.144197
0.3	0.0938113	0.0924733	0.109382	0.0970816	0.0963307	0.0999462
0.4	0.145425	0.125328	0.159424	0.0300571	0.0331714	0.0281943
0.5	0.118111	0.125814	0.136774	0.0341472	0.0331541	0.0363138
0.6	0.0668849	0.0698239	0.0710771	0.131853	0.135597	0.155293
0.7	0.0191287	0.0186108	0.0204277	0.108687	0.0960146	0.112864

Table 5.5. A comparison between the approximate solutions based on Bernstein polynomials, exact solutions and MHFs for $m = 10$ for $H = 0.5$ and $m = 10$ for $H = 0.8$ of Example 5.3.

	$m = 10$ for $H = 0.5$			$m = 10$ for $H = 0.8$		
x	Proposed method solution	Exact solution	MHF	Proposed method solution	Exact solution	MHF
0.0	0.0994867	0.100878	0.118941	0.106485	0.105583	0.0779137
0.1	0.115453	0.113305	0.11892	0.0857393	0.0895879	0.842485
0.2	0.108356	0.107051	0.118915	0.107079	0.106915	0.100871
0.3	0.109253	0.108101	0.118902	0.109474	0.107777	0.100663
0.4	0.110416	0.113347	0.118889	0.101145	0.102257	0.100039
0.5	0.124175	0.124792	0.118876	0.105427	0.107064	0.116865
0.6	0.11273	0.113268	0.118863	0.114952	0.111771	0.122021
0.7	0.109629	0.110772	0.11885	0.10285	0.101906	0.112468
0.8	0.11038	0.109643	0.118837	0.0949449	0.0945244	0.119099
0.9	0.114076	0.113218	0.118824	0.115552	0.114408	0.116075

Table 5.6. A comparison between the approximate solutions based on Bernstein polynomials, exact solutions and MHFs for $m = 20$ for $H = 0.5$ and $m = 20$ for $H = 0.8$ of Example 5.3.

	$m = 20$ for $H = 0.5$		$m = 20$ for $H = 0.8$	
x	Proposed method solution	Exact solution	Proposed method solution	Exact solution
0.0	0.0990621	0.0999297	0.0936294	0.11227
0.1	0.0209905	0.0252601	0.136276	0.145514
0.2	0.0890391	0.0872236	0.083625	0.08496
0.3	0.0629903	0.0622027	0.0191622	0.018879
0.4	0.129902	0.13452	0.0536205	0.0570127
0.5	0.01699	0.0168559	0.0293628	0.0292542
0.6	0.139899	0.137547	0.0236787	0.0251135
0.7	0.15099	0.142733	0.387841	0.338056
0.8	0.169903	0.167838	0.209454	0.202529
0.9	0.159203	0.162729	0.246707	0.247399

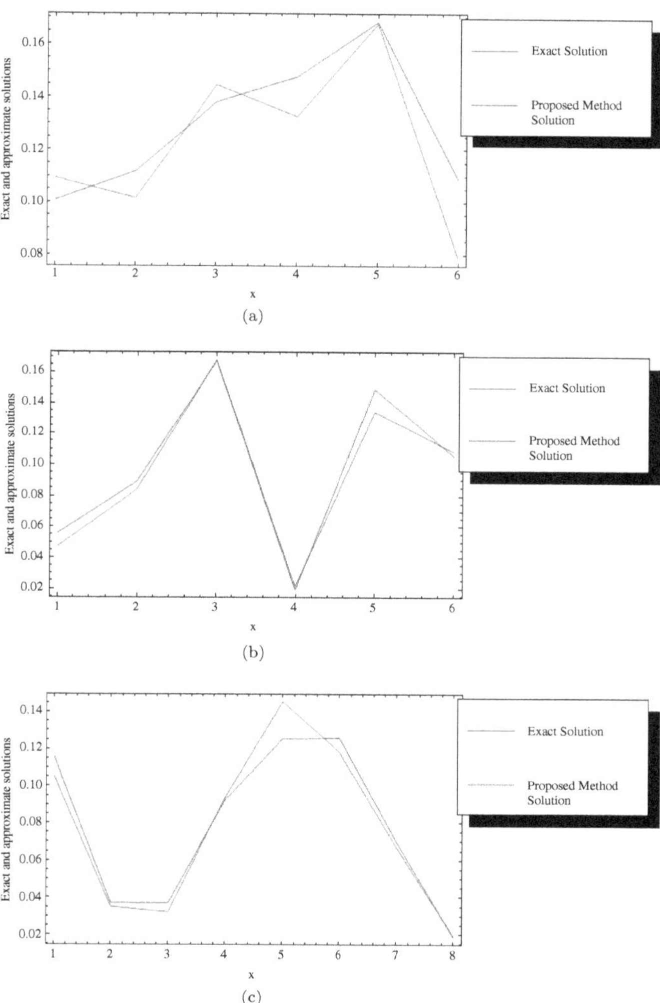

Fig. 5.3. (a) The sample paths of Example 5.3 for $m = 6$ for $H = 0.5$. (b) The sample paths of Example 5.3 for $m = 6$ for $H = 0.8$. (c) The sample paths of Example 5.3 for $m = 8$ for $H = 0.5$. (d) The sample paths of Example 5.3 for $m = 8$ for $H = 0.8$. (e) The sample paths of Example 5.3 for $m = 10$ for $H = 0.5$. (f) The sample paths of Example 5.3 for $m = 10$ for $H = 0.8$. (g) The sample paths of Example 5.3 for $m = 20$ for $H = 0.5$. (h) The sample paths of Example 5.3 for $m = 20$ for $H = 0.8$.

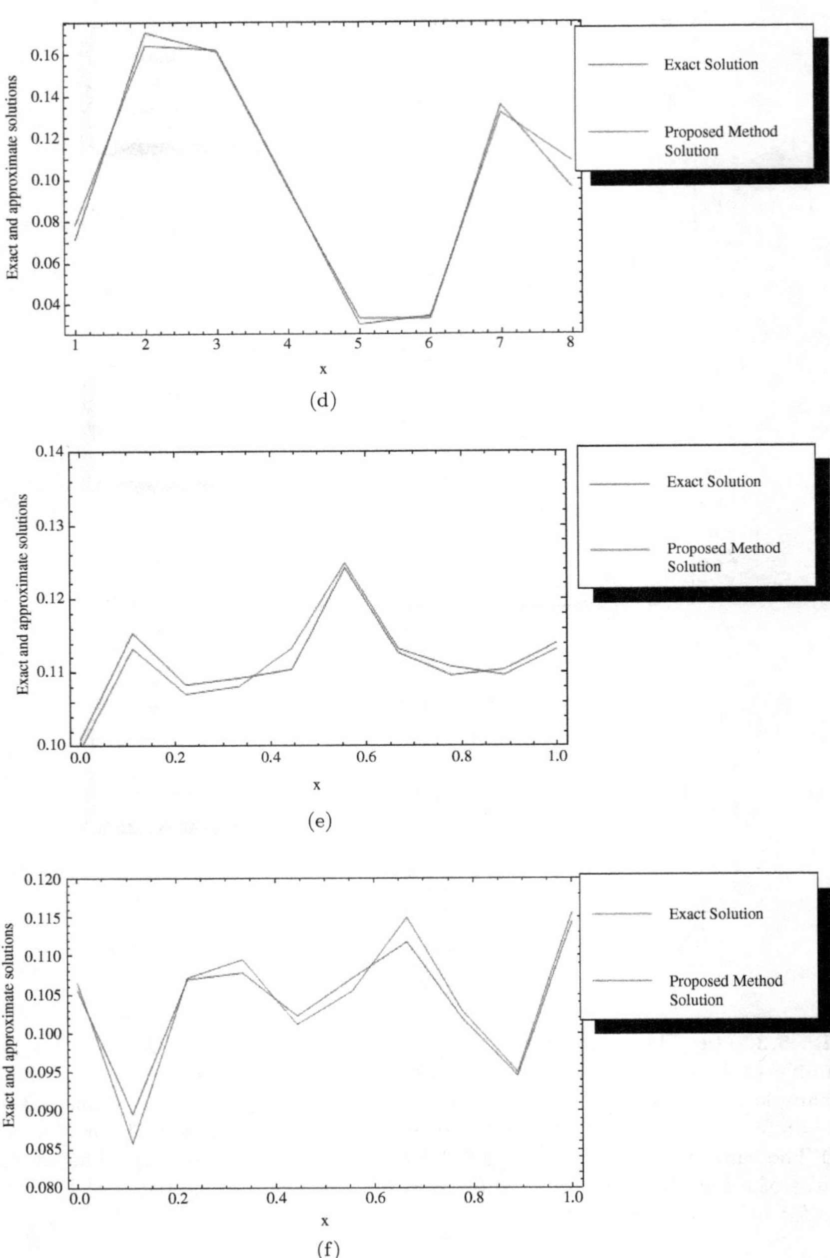

Fig. 5.3. (*Continued*)

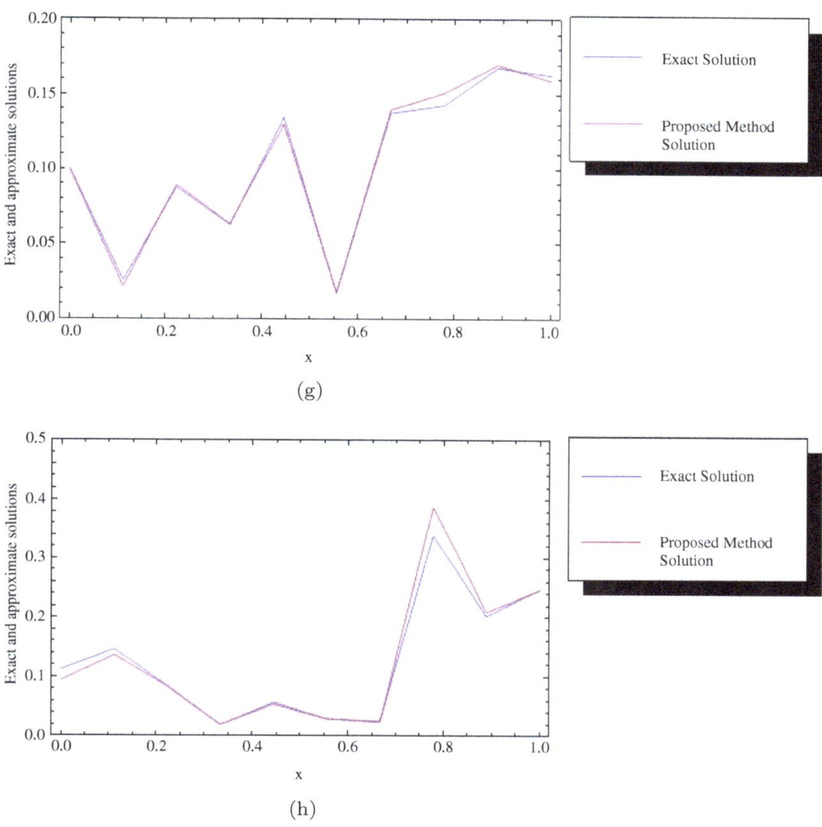

(g)

(h)

Fig. 5.3. *(Continued)*

Example 5.4.

$$f(x) = f_0 - Ha^2 \int_0^x \tanh(f(y))y^{2H-1}\mathrm{sech}^2(f(y))dy$$

$$+a \int_0^x \mathrm{sech}(f(y))dB^H(y), \ x \in [0,1),$$

where $f(x) = \arcsin h(aB^H(x) + \sinh(f_0))$ is the exact solution. Stochastic operational matrix for Bernstein polynomials presented in Section 5.4.1.2 has been implemented in solving the above stochastic integral equation. Approximate, exact solutions and results from

Table 5.7. A comparison between the approximate solutions based on Bernstein polynomials, exact solutions and MHFs for $m = 6$ for $H = 0.5$ and $m = 6$ for $H = 0.8$ of Example 5.4.

| | $m = 6$ for $H = 0.5$ | | | $m = 6$ for $H = 0.8$ | | |
x	Proposed method solution	Exact solution	MHF	Proposed method solution	Exact solution	MHF
0.0	0.136296	0.121052	0.129587	0.0564155	0.059831	0.0564155
0.1	0.199861	0.188489	0.211449	0.184224	0.190082	0.184224
0.2	0.119397	0.119106	0.126462	0.165725	0.172966	0.165725
0.3	0.196856	0.181701	0.191363	0.0619012	0.0568834	0.0619012
0.4	0.0972546	0.089781	0.090527	0.20768	0.210487	0.20768
0.5	0.0677775	0.0734895	0.0638727	0.0896471	0.0920498	0.0896471

Table 5.8. A comparison between the approximate solutions based on Bernstein polynomials, exact solutions and MHFs for $m = 8$ for $H = 0.5$ and $m = 8$ for $H = 0.8$ of Example 5.4.

| | $m = 8$ for $H = 0.5$ | | | $m = 8$ for $H = 0.8$ | | |
x	Proposed method solution	Exact solution	MHF	Proposed method solution	Exact solution	MHF
0.0	0.114698	0.110376	0.0926994	0.143848	0.148215	0.149018
0.1	0.00995975	0.0094432	0.0092349	0.13301	0.154958	0.150194
0.2	0.197377	0.193317	0.225483	0.199056	0.174105	1.90111
0.3	0.0795201	0.0715207	0.0754659	0.0583327	0.0439418	0.049512
0.4	0.123504	0.127348	0.143068	0.230787	0.254958	0.28959
0.5	0.180005	0.188916	0.161357	0.282547	0.214125	0.262094
0.6	0.0836435	0.0800012	0.0836531	0.0151967	0.0128406	0.0110424
0.7	0.326271	0.318033	0.303168	0.129832	0.118988	0.111254

modifications of hat functions (MHF) have been compared for $a = \frac{1}{30}$ and $f_0 = \frac{1}{10}$ of the above problem as shown in Tables 5.7–5.10 with the sample paths of Example 5.4 for $m = 6$, $m = 8$, $m = 10$ and $m = 20$ with Hurst index $H = 0.5$ and $H = 0.8$ have been shown in Figs. 5.4(a)–5.4(h).

Table 5.9. A comparison between the approximate solutions based on Bernstein polynomials, exact solutions and MHFs for $m = 10$ for $H = 0.5$ and $m = 10$ for $H = 0.8$ of Example 5.4.

	$m = 10$ for $H = 0.5$			$m = 10$ for $H = 0.8$		
x	Proposed method solution	Exact solution	MHF	Proposed method solution	Exact solution	MHF
0.0	0.0966096	0.0928758	0.102099	0.114698	0.080478	0.0812128
0.1	0.0958346	0.0928712	0.102087	0.0821148	0.0837401	0.0812082
0.2	0.103978	0.10014	0.102076	0.0872368	0.0884184	0.081202
0.3	0.0967571	0.10036	0.102065	0.0953329	0.0950098	0.0891944
0.4	0.09964	0.101296	0.102054	0.106211	0.104354	0.101185
0.5	0.108739	0.111025	0.102043	0.106931	0.104747	0.111175
0.6	0.109005	0.112032	0.102031	0.120658	0.12674	0.121164
0.7	0.119924	0.119284	0.10202	0.117604	0.114352	0.102115
0.8	0.11985	0.119289	0.102009	0.113085	0.119229	0.118114
0.9	0.140604	0.143203	0.101998	0.137808	0.132231	0.121125

Table 5.10. A comparison between the approximate solutions based on Bernstein polynomials, exact solutions and MHFs for $m = 20$ for $H = 0.5$ and $m = 20$ for $H = 0.8$ of Example 5.4.

	$m = 20$ for $H = 0.5$		$m = 20$ for $H = 0.8$	
x	Proposed method solution	Exact solution	Proposed method solution	Exact solution
0.0	0.104654	0.119795	0.10101	0.0885571
0.1	0.0246481	0.0255236	0.0510078	0.0579506
0.2	0.104642	0.0978306	0.101005	0.0981575
0.3	0.0446365	0.0445441	0.299958	0.297614
0.4	0.124631	0.122905	0.18098	0.189373
0.5	0.204626	0.209107	0.130896	0.127706
0.6	0.146226	0.140939	0.200557	0.201344
0.7	0.0804628	0.0777828	0.0493446	0.0445414
0.8	0.220466	0.225909	0.33954	0.317499
0.9	0.224779	0.22931	0.0934974	0.107631

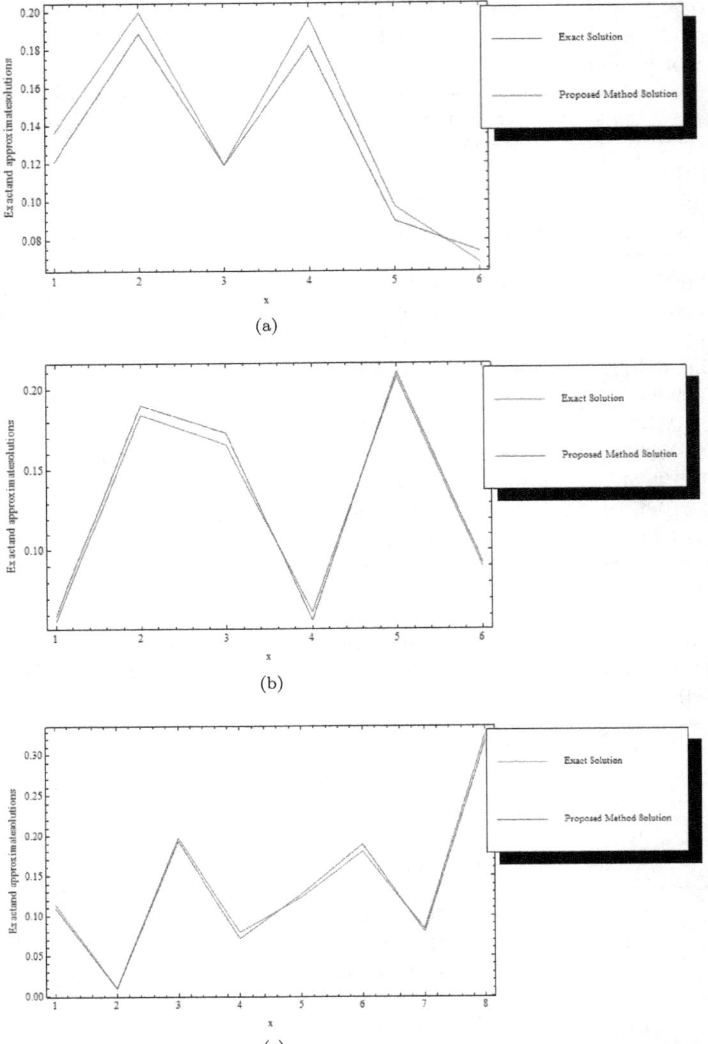

Fig. 5.4. (a) The sample paths of Example 5.4 for $m = 6$ for $H = 0.5$. (b) The sample paths of Example 5.4 for $m = 6$ for $H = 0.8$. (c) The sample paths of Example 5.4 for $m = 0.8$ for $H = 0.5$. (d) The sample paths of Example 5.4 for $m = 0.8$ for $H = 0.8$. (e) The sample paths of Example 5.4 for $m = 10$ for $H = 0.5$. (f) The sample paths of Example 5.4 for $m = 10$ for $H = 0.8$. (g) The sample paths for Example 5.4 for $m = 20$ for $H = 0.5$. (h) The sample paths for Example 5.4 for $m = 20$ for $H = 0.8$.

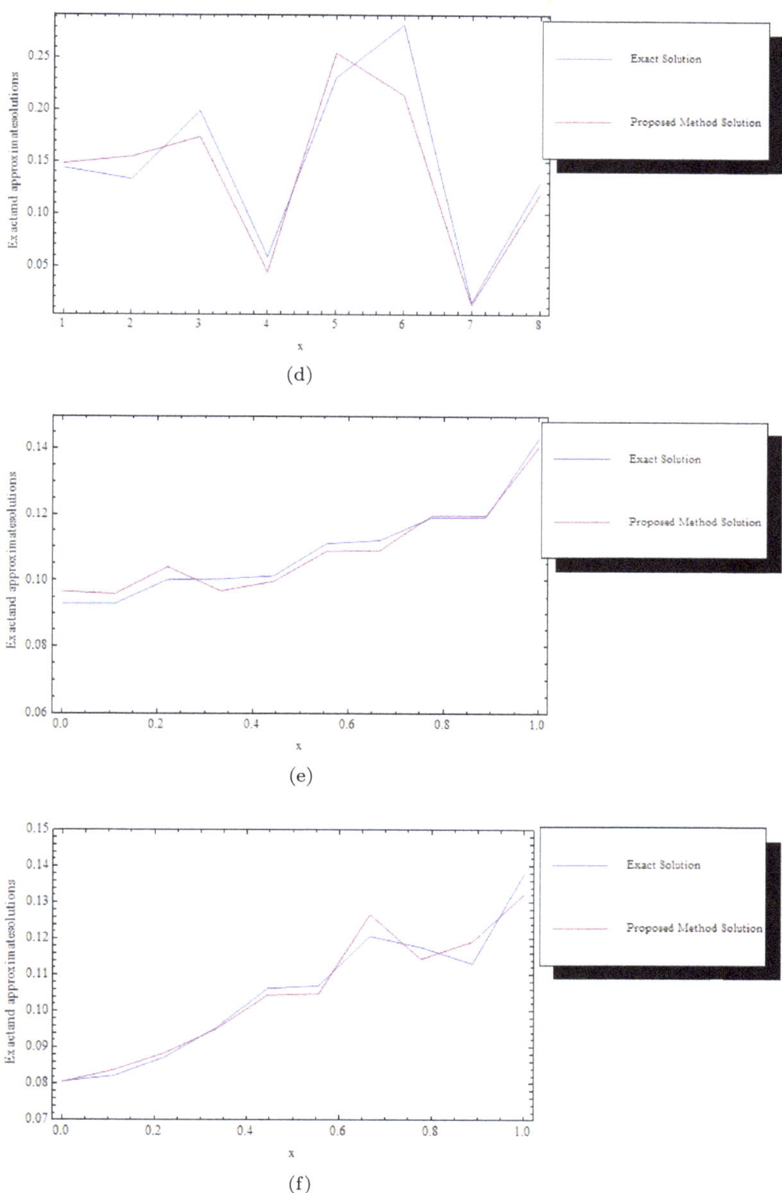

Fig. 5.4. (*Continued*)

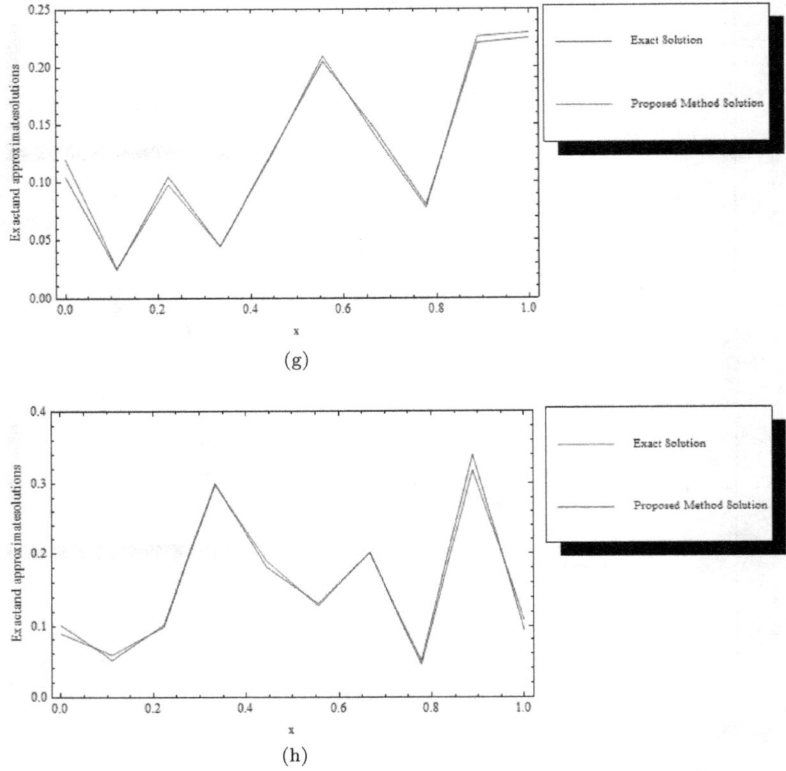

(g)

(h)

Fig. 5.4. (*Continued*)

Example 5.5. The stochastic model driven with fBm for stochastic volatilities, stock prices and electricity prices can be expressed similar to Ref. [140] by

$$f(x) = f_0 + \int_0^x \kappa(\mu - \ln(f(y)))f(y)dy + \int_0^x \sigma f(y)dB^H(y), \quad x \in [0,1),$$

where $f(x) = \exp(\exp(-\kappa x)[\int_0^x (\kappa\mu - \sigma^2 H y^{2H-1})\exp(\kappa y)dy + \sigma \int_0^x \exp(\kappa y)dB^H(y) + \ln(f_0)])$ is the exact solution. Stochastic operational matrix for Bernstein polynomials presented in Section 5.4.1.2 has been implemented in solving the above stochastic integral equation. Approximate and exact solutions have been

Table 5.11. A comparison between the approximate solutions based on Bernstein polynomials and exact solutions for $m = 6$ for $H = 0.5$ and $m = 6$ for $H = 0.8$ of Example 5.5.

| x | $m = 6$ for $H = 0.5$ | | $m = 6$ for $H = 0.8$ | |
	Proposed method solution	Exact solution	Proposed method solution	Exact solution
0.0	0.714629	0.58846	0.830034	0.803626
0.1	1.20327	1.11987	1.05582	0.914427
0.2	1.23895	1.25399	0.884578	0.884183
0.3	0.87849	0.872974	0.691632	0.763762
0.4	1.13219	1.16494	1.25104	1.312
0.5	1.68921	1.77669	1.28869	1.1749

Table 5.12. A comparison between the approximate solutions based on Bernstein polynomials and exact solutions for $m = 8$ for $H = 0.5$ and $m = 8$ for $H = 0.8$ of Example 5.5.

| x | $m = 8$ for $H = 0.5$ | | $m = 8'$ for $H = 0.8$ | |
	Proposed method solution	Exact solution	Proposed method solution	Exact solution
0.0	1.03612	1.0448	1.22856	1.08992
0.1	1.2503	1.45321	1.42048	1.22222
0.2	1.68458	1.78305	0.629685	0.529631
0.3	0.739006	0.758814	1.18564	1.23459
0.4	0.61342	0.631252	1.00633	1.00013
0.5	0.43075	0.495891	0.892542	0.834812
0.6	1.76208	1.74672	1.26414	1.2577
0.7	1.29526	1.40199	0.937474	0.9148

compared for $\kappa = 0.4$, $\mu = 1$, $\sigma = \frac{1}{5}$ and $f_0 = \frac{1}{10}$ of the above problem as shown in Tables 5.11–5.14, also the sample paths of Example 5.5 for $m = 6$, $m = 8$, $m = 10$ and $m = 20$ with Hurst index $H = 0.5$ and $H = 0.8$ have been shown in Figs. 5.5(a)–5.5(h).

Table 5.13. A comparison between the approximate solutions based on Bernstein polynomials and exact solutions for $m = 10$ for $H = 0.5$ and $m = 10$ for $h = 0.8$ of Example 5.5.

| | $m = 10$ for $H = 0.5$ | | $m = 10$ for $H = 0.8$ | |
| | Proposed method solution | Exact solution | Proposed method solution | Exact solution |
x				
0.0	1.29604	1.2745	0.741304	0.75281
0.1	0.907367	0.902923	1.4742	1.33356
0.2	1.59853	1.3396	1.54469	1.52283
0.3	0.76361	0.700684	0.848487	0.965605
0.4	0.608214	0.681133	1.0563	1.01827
0.5	0.912316	0.946004	0.779223	0.828499
0.6	0.671256	0.621984	1.25133	1.20906
0.7	0.982652	0.547559	0.762308	0.739747
0.8	1.31039	1.27161	1.03025	1.03186
0.9	0.739543	0.703352	1.0633	1.08796

Table 5.14. A comparison between the approximate solutions based on Bernstein polynomials and exact solutions for $m = 20$ for $H = 0.5$ and $m = 20$ for $H = 0.8$ of Example 5.5.

| | $m = 20$ for $H = 0.5$ | | $m = 20$ for $H = 0.8$ | |
| | Proposed method solution | Exact solution | Proposed method solution | Exact solution |
x				
0.0	0.836049	0.842673	1.00515	1.06546
0.1	1.00525	1.06126	1.07434	1.32807
0.2	1.07444	0.906346	1.14354	1.27049
0.3	1.14364	1.21712	1.21274	1.29104
0.4	1.21284	1.4185	1.28194	1.11486
0.5	1.28203	1.72973	1.35113	1.39427
0.6	1.35123	1.53003	1.4203	1.16327
0.7	1.82043	2.33322	0.894129	0.833906
0.8	0.478962	0.597707	1.55837	1.44064
0.9	0.879238	0.93496	1.1627	1.14096

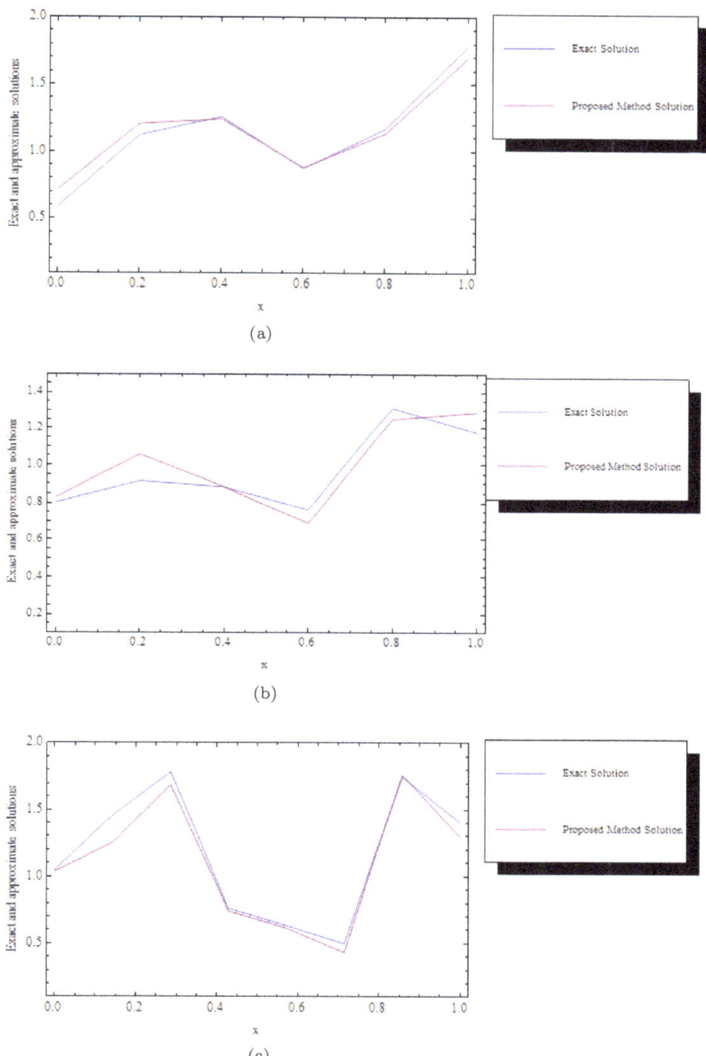

Fig. 5.5. (a) The sample paths for Example 5.5 for $m = 6$ for $H = 0.5$. (b) The sample paths for Example 5.5 for $m = 6$ for $H = 0.8$. (c) The sample paths for Example 5.5 for $m = 8$ for $H = 0.5$. (d) The sample paths for Example 5.5 for $m = 8$ for $H = 0.8$. (e) The sample paths for Example 5.5 for $m = 10$ for $H = 0.5$. (f) The sample paths of Example 5.5 for $m = 10$ for $H = 0.8$. (g) The sample paths of Example 5.5 for $m = 20$ for $H = 0.5$. (h) The sample paths for Example 5.5 for $m = 20$ for $H = 0.8$.

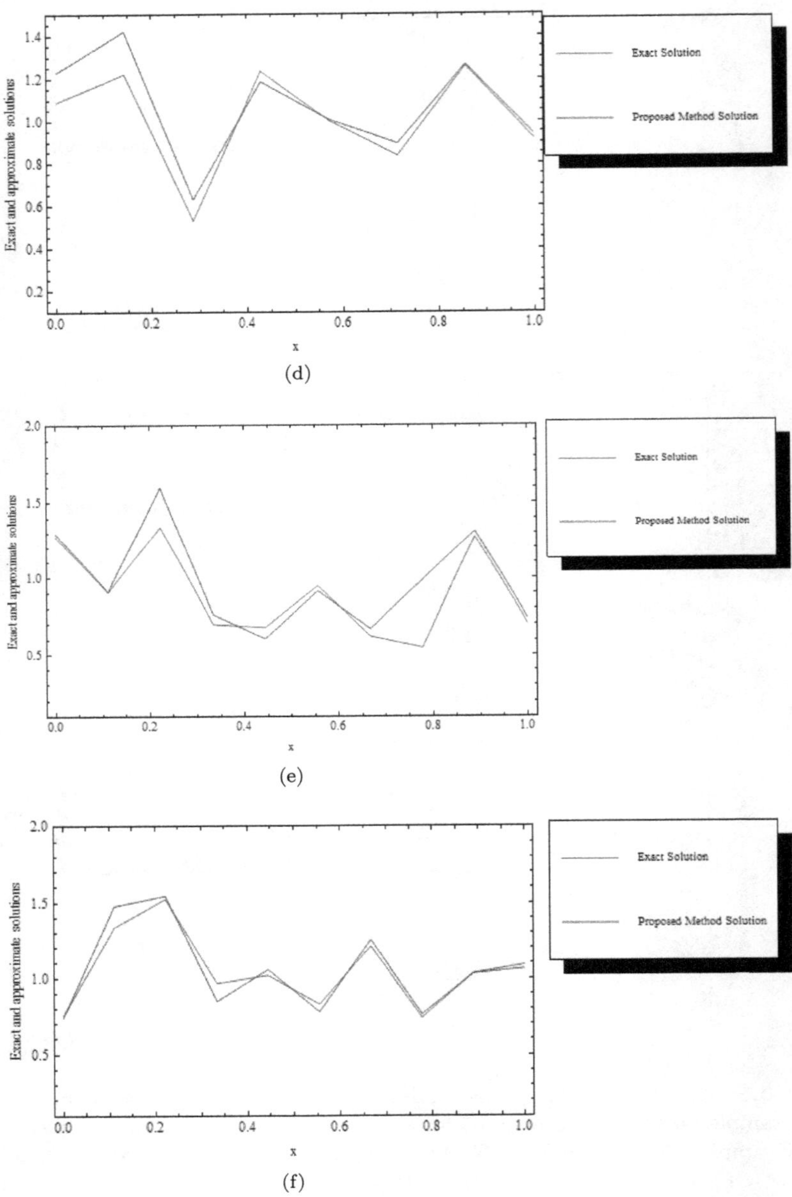

(d)

(e)

(f)

Fig. 5.5. *(Continued)*

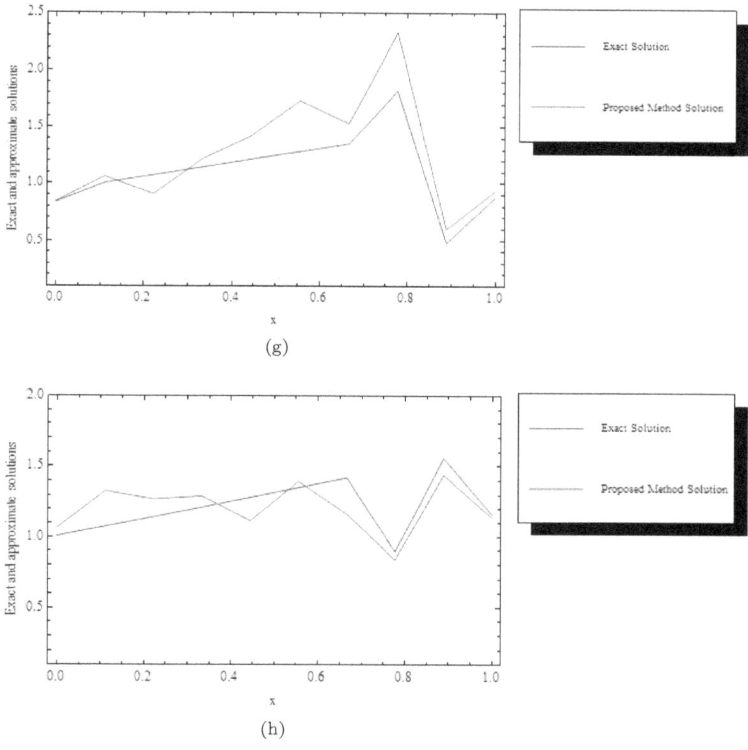

(g)

(h)

Fig. 5.5. (*Continued*)

5.5 Summary

In this chapter, second-kind CWs have been implemented on the fractional stochastic integral equations to obtain the numerical solutions. The BPFs and their relations to second-kind CWs are employed to derive stochastic OM and fractional stochastic operational matrix of second-kind CWs. These operational matrices have been used to discretise the fractional stochastic integral equation. Two illustrated examples have been presented in order to justify the efficiency and applicability of the proposed method. Moreover, the corresponding obtained numerical results have been compared with the exact solutions to establish the accuracy and efficiency of the proposed method.

Also, Bernstein polynomials have been applied to obtain solutions for the nonlinear fractional stochastic integral equations. Fractional stochastic operational matrix based on Bernstein polynomial has been used to discretise the nonlinear fractional stochastic integral equation. Two illustrated examples have been presented in comparison to their respective exact solutions in order to justify the accuracy of the proposed method. Additionally, the proposed method has been compared with modifications of hat functions method which confirm the plausibility of the new technique.

Chapter 6

Numerical Solutions of Stochastic Differential Equations Arising in Physical Phenomena

6.1 Introduction

Stochastic partial differential equations (SPDEs) also arise when deterministic are considered from random preliminary conditions or as tractable approximations to complex deterministic programs. SPDEs are partial differential equations that include random fluctuations, which occur in nature and are missing from deterministic PDE descriptions. The noise term in SPDE is included to capture phenomena not present in the corresponding deterministic equation [16,17].

Many models in physics, chemistry, engineering, etc. are described by SPDEs. Since exact solutions of these equations are rarely known, numerical analysis of SPDEs has been recently the subject of many articles. One of the most important problems that have many applications is stochastic Fisher equation [141,142]. High-dimensional multiscale dynamical systems that are subject to noise can be modelled accurately using SPDEs with a multiscale structure [143]. Generically speaking, noise might also enter the physical system either as temporal fluctuations of internal degrees of freedom or as random variations of some external control parameters; internal randomness often reflects itself in additive noise phrases [144].

The combination of logistics and diffusion is called reaction-diffusion equation which describes a situation in which both dispersion and population growth occur. The reaction-diffusion equation is a differential equation which accounts for diffusion and growth. It is an equation which characterises travelling wave phenomenon. The Fisher equation is one of the reaction-diffusion equations and is widely used in the study of biological invasion [145–148]. It additionally represents a model equation for the evolution of a neutron populace in a nuclear reactor and a prototype model for a spreading flame [146]. The Fisher equation was introduced to explain the spreading of genes and also has applications in different fields of research ranging from ecology to plasma physics [141,142,145–150].

Stochastic differential equations (SDEs) have always been impactful in a lot of areas such as science, economics and finance. A strong framework in advanced probability and stochastic processes leads to a better understanding of SDEs. The FitzHugh–Nagumo model is one of the classical standard models in neuroscience. FitzHugh–Nagumo systems are used to describe a prototype of an excitable system. Hodgkin and Huxley [151] were the first to propose the model in their pioneer work in 1952. Afterwards, FitzHugh–Nagumo proposed a model which will have been considered in this work.

Among the numerical schemes, spectral collocation methods are powerful tools for the approximate solutions of PDEs. The principle of the scheme is that the solution is represented by a finite Chebyshev series with unknown coefficients; which is substituted into the differential equation for obtaining the coefficients so that the equation is satisfied at certain points within the range under consideration. The number of points is chosen so that, along with the initial or boundary conditions, there are enough equations to find the unknown coefficients [152,153].

SPDEs have been getting a great deal of attention from researchers around the world. Singh and Saha Ray [154] applied a semi-implicit Euler–Maruyama scheme for obtaining solutions of Fisher equation. Zheng and Huang [155,156] have also provided a discussion regarding the stability of the stochastic FitzHugh–Nagumo system. Tuckwell and Rodriguez [157] performed simulations on the stochastic FitzHugh–Nagumo system and compared the solutions with the differential equations for both sustained and intermittent deterministic current inputs with superimposed noise.

6.2 Outline of Present Study

In this chapter, numerical solutions of the stochastic Fisher equation (SFE) have been obtained by using a semi-implicit finite difference scheme. The samples for the Wiener process have been obtained from cylindrical Wiener process and Q-Wiener process. Stability and convergence of the proposed finite difference scheme have been discussed scrupulously. The sample paths obtained from cylindrical Wiener process and Q-Wiener process have been also shown graphically.

Also, the numerical solution of stochastic FitzHugh–Nagumo equation has been obtained by Chebyshev spectral collocation. Semi-implicit Euler–Maruyama scheme has been used for temporal variable to discretise the stochastic FitzHugh–Nagumo equation. A detailed stability analysis for stochastic FitzHugh–Nagumo equation has also been discussed. Graphical representations of obtained results of stochastic FitzHugh–Nagumo equation have been discussed to provide a clear idea about the behaviour of the solutions.

6.2.1 *Framework for SPDE driven by $L^2(D)$-valued Q-Wiener process*

6.2.1.1 *Q-Wiener process assumption*

Assumption 6.2.1.1. Let U be a separable Hilbert space with norm $\|.\|_U$ and inner product $\langle \cdot, \cdot \rangle_U$. $Q \in \mathfrak{L}(U)$ is non-negative definite and symmetric, where $\mathfrak{L}(U)$ is the set of bounded linear operators $L : U \to U$. Furthermore, Q has an orthonormal basis $\{x_j : j \in \mathbb{N}\}$ of eigenfunctions with corresponding eigenvalues $q_j \geq 0$ such that $\sum_{j \in \mathbb{N}} q_j < \infty$ (i.e., Q is of trace class).

Let $(\Omega, \mathcal{F}, \mathcal{F}_t, \mathbb{P})$ be a filtered probability space. From the definition of probability space and the Gaussian distribution $N(0, Q)$ on Hilbert space. The Q-Wiener process is defined as follows.

Definition 6.2.1.2. A U-valued stochastic process $\{W(t) : t \geq 0\}$ is a Q-Wiener process if

(a) $W(0) = 0$ a.s.,
(b) $W(t)$ is continuous as a function $\mathbb{R}^+ \to U$, for each $\omega \in \Omega$,
(c) $W(t)$ is $\mathcal{F}_t$-adapted and $W(t) - W(s)$ is independent of $\mathcal{F}_s$ for $s < t$, and
(d) $W(t) - W(s) \sim N(0, (t - s)Q)$ for all $0 \leq s \leq t$.

Theorem 6.2.1.3. *Let Q satisfy Assumption 6.2.1.1. Then, $W(t)$ is a Q-Wiener process if and only if*

$$W(t) = \sum_{j=1}^{\infty} \sqrt{q_j}\chi_j\beta_j(t) \quad a.s., \tag{6.1}$$

where $\beta_j(t)$ are iid $\mathcal{F}_t$-Brownian motions and the series converges in $L^2(\Omega, U)$. Moreover, Eq. (6.1) converges in $L^2(\Omega, C([0,T], U))$ for any $T > 0$.

Proof. It may be referred to Ref. [16]. $\square$

Corollary 6.2.1.4. *From Eq. (6.1),*

$$Cov(\langle W(t), \chi_j \rangle_U, \langle W(t), \chi_k \rangle_U) = t q_j \delta_{jk}. \tag{6.2}$$

Hence, $W(t) \sim N(0, tQ)$.

Consider a bounded domain D and let $H = U = L^2(D)$ and Q satisfy Assumption 6.2.1.1. Then, for a kernel $q \in L^2(D \times D)$, in terms of the Q-Wiener process,

$$Cov(W(t, \boldsymbol{x}), W(t, \boldsymbol{y})) = t\, q(\boldsymbol{x}, \boldsymbol{y}), \quad for \quad (\boldsymbol{x}, \boldsymbol{y}) \in D \times D,$$

and $W(1, \boldsymbol{x})$ is a mean-zero Gaussian random field with covariance $q(\boldsymbol{x}, \boldsymbol{y})$.

6.2.1.2 *Cylindrical Wiener process*

In case of $Q = I$, which is not a trace class on an infinite dimensional space U (as $q_j = 1$ for all j) so that Eq. (6.1) does not converge in $L^2(\Omega, U)$. To extend the definition of a Q-Wiener process, a cylindrical Wiener process is introduced.

Definition 6.2.1.5. The cylindrical Wiener process (also called space-time white noise) is the U-valued stochastic process $W(t)$ defined by

$$W(t) = \sum_{j=1}^{\infty} \chi_j\beta_j(t),$$

where $\{\chi_j\}$ is any orthonormal basis of U and $\beta_j(t)$ are iid $\mathcal{F}_t$-Brownian motions.

6.2.2 *Stochastic Fisher equation*

Consider SFE of the form

$$dU = [U_{xx} + 6U(1 - U)]dt + dW(t), \quad 0 < x < 1, \quad t > 0, \quad (6.3)$$

which is the perturbation of the Fisher equation [158] of the form

$$u_t = [u_{xx} + 6u(1 - u)], \ 0 < x < 1, \ t > 0. \tag{6.4}$$

Here, $W(t)$ is an $L^2(D)$-valued Q-Wiener process.

6.2.3 *Stochastic FitzHugh–Nagumo equation*

The stochastic FitzHugh–Nagumo equation (SFNE) is of the form

$$\begin{aligned}
dq &= (\Delta q - \psi(q) - r)dt + dW_1(t), \\
dr &= (\sigma q - \gamma r)dt + dW_2(t),
\end{aligned} \tag{6.5}$$

where $x \in D \subset \mathbb{R}$, $t \geq s$, $\psi(q) = q(1 - q)(\beta - q)$ and $W_i(t)$, $i = 1, 2$ is the Wiener process.

6.3 Semi-Implicit Finite Difference Method

To discretise in time, the semi-implicit Euler–Maruyama method (EMM) has been applied. Let $\{X_t\}$ be an Itô process on $t \in [t_0, T]$ satisfying the following SDE [45]:

$$\begin{cases} dX_t = a(t, X_t)dt + b(t, X_t)dW_t, \\ X_{t_0} = X_0. \end{cases}$$

For a given time-discretisation

$$t_0 < t_1 < \cdots < t_n = T,$$

an EMM is a continuous time stochastic process $\{Y(t), t_0 \leq t \leq T\}$ satisfying the iterative scheme [98]

$$Y_{n+1} = Y_n + a(t_n, Y_n)\,\Delta t_{n+1} + b(t_n, Y_n)\,\Delta W_{n+1},$$
$$n = 0, 1, 2, \ldots, N - 1,$$

with initial value

$$Y_0 = X_0,$$

where $Y_n = Y(t_n)$, $\Delta t_{n+1} = t_{n+1} - t_n$ and $\Delta W_{n+1} = W(t_{n+1}) - W(t_n)$. Here, each random number ΔW_n is computed as $\Delta W_n = z_n \sqrt{\Delta t_n}$, where z_n is chosen from standard normal distribution $N(0,1)$.

Consider the equidistant discretised times $t_n = t_0 + n\delta$ with $\delta = \Delta_n = \frac{(T-t_0)}{N}$ for some integer N large enough so that $\delta \in (0,1)$.

6.4 Implementation of EMM for Stochastic Fisher Equation

6.4.1 *Semi-implicit EMM*

In this section, the algorithm of EMM has been described.

Step 1. Consider a nonlinear SPDE of the following form:

$$dU = [\varepsilon U_{xx} + f(U)]dt + \sigma dW(t), \quad U(0,x) = U_0(x), \qquad (6.6)$$

with homogenous Dirichlet boundary conditions on (0,a) and with parameters $\varepsilon, \sigma > 0$, a reaction term, $f \colon \mathbb{R} \to \mathbb{R}$, and $W(t)$ is a Q-Wiener process on $L^2(0,a)$.

Step 2. Introducing the grid points $x_j = jh$ for $h = \frac{a}{J}$ and $j = 0, \ldots, J$. Equation (6.6) can be discretised as

$$dU(t, x_j) = \left[\varepsilon \frac{U(t, x_{j+1}) - 2U(t, x_j) + U(t, x_{j-1})}{h^2} + f(U(t, x_j))\right] dt$$
$$+ \sigma dW(t),$$
$$j = 1, 2, \ldots, J - 1.$$

For $j = 1$,

$$dU(t, x_1) = \left[\varepsilon \frac{U(t, x_2) - 2U(t, x_1) + U(t, x_0)}{h^2} + f(U(t, x_1))\right] dt$$
$$+ \sigma dW(t).$$

For $j = 2$,

$$dU(t, x_2) = \left[\varepsilon \frac{U(t, x_3) - 2U(t, x_2) + U(t, x_1)}{h^2} + f(U(t, x_2))\right] dt$$
$$+ \sigma dW(t).$$

For $j = 3$,

$$dU(t, x_2) = \left[\varepsilon \frac{U(t, x_3) - 2U(t, x_2) + U(t, x_1)}{h^2} + f(U(t, x_2))\right] dt$$
$$+ \sigma dW(t).$$

Similarly, for $j = J - 1$,

$$dU(t, x_{J-1}) = \left[\varepsilon \frac{U(t, x_J) - 2U(t, x_{J-1}) + U(t, x_{J-2})}{h^2}\right.$$
$$\left. + f(U(t, x_{J-1}))\right] dt + \sigma dW(t).$$

Step 3. Let $U_J(t)$ be the solution of

$$d\boldsymbol{U}_J = [-\varepsilon \boldsymbol{A}^D \boldsymbol{U}_J + \boldsymbol{f}(\boldsymbol{U}_J)]dt + \sigma d\boldsymbol{W}_J(t) + \varepsilon \boldsymbol{B} dt,$$

resulting from the spatial centred difference approximation, where

$$\boldsymbol{U}_J(t) = [U(t, x_1), U(t, x_2), \ldots, U(t, x_{j-1})]^T,$$
$$\boldsymbol{W}_J(t) = [W(t, x_1), W(t, x_2), \ldots, W(t, x_{J-1})]^T,$$

$$\boldsymbol{A}^D = \frac{1}{h^2} \begin{pmatrix} 2 & -1 & & & \\ -1 & 2 & -1 & & \\ & -1 & 2 & -1 & \\ & & \ddots & \ddots & -1 \\ & & & -1 & 2 \end{pmatrix} \quad \text{and} \quad \boldsymbol{B} = \frac{1}{h^2} \begin{pmatrix} U(t, x_0) \\ \vdots \\ U(t, x_J) \end{pmatrix}.$$

Here, $\boldsymbol{W}_J(t) \sim N(\boldsymbol{0}, t\boldsymbol{C})$ where C is the matrix with entries $q(x_i, x_j)$ for $i, j = 1, \ldots, J - 1$.

Step 4. The semi-implicit EMM with time step $\Delta t > 0$ yields an approximation $\boldsymbol{U}_{J,n}$ to $\boldsymbol{U}_J(t_n)$ at $t_n = n\Delta t$ defined by

$$\boldsymbol{U}_{J,n+1} = (\boldsymbol{I} + \Delta t \varepsilon \boldsymbol{A}^D)^{-1}[\boldsymbol{U}_{J,n} + \boldsymbol{f}(\boldsymbol{U}_{J,n})\Delta t + \sigma \Delta \boldsymbol{W}_n]$$

$$+ \varepsilon(\boldsymbol{I} + \Delta t \varepsilon \boldsymbol{A}^D)^{-1}\boldsymbol{B}, \ n = 0, 1, \ldots, N - 1, \qquad (6.7)$$

with $\boldsymbol{U}_{J,0} = \boldsymbol{U}_J(0)$ and $\Delta \boldsymbol{W}_n = \boldsymbol{W}_J(t_{n+1}) - \boldsymbol{W}_J(t_n)$. Here, circular embedding method is used to generate the increments, i.e., $\Delta \boldsymbol{W}_n$, where $\Delta \boldsymbol{W}_n$, where $\Delta \boldsymbol{W}_n \sim N(\boldsymbol{0}, \Delta t \boldsymbol{C})$.

6.4.1.1 *Stability and convergence*

Theorem 6.4.1.1. *If $\boldsymbol{f}(\boldsymbol{U})\colon \mathbb{R}^{J-1} \to \mathbb{R}^{J-1}$ is Lipschitz continuous function. Then, the necessary and sufficient condition for the difference Eq. (6.7) to be stable is*

$$\|(\boldsymbol{I} + \Delta t \varepsilon \boldsymbol{A}^D)^{-1}\| < \frac{1}{(1 + \Delta t L)}.$$

Proof. Lax–Richtmyer equivalence theorem states that the consistent finite difference method for the well-posed linear initial value problem, the method (finite difference) is convergent if and only if it is stable.

Let $\boldsymbol{e}_{n+1} = \boldsymbol{U}_{J,n+1} - \boldsymbol{u}_{J,n+1}$ be the error at the $(n+1)^{\text{th}}$ time level. Also, let $\boldsymbol{U}_{J,n+1}$ be the computed value and $\boldsymbol{u}_{J,n+1}$ be the exact value.

Let $\boldsymbol{e}_n = [e_1^n, e_2^n, \ldots, e_{J-1}^n]^T$, where $e_i^n = U(t_n, x_i) - u(t_n, x_i)$, $i = 0, 1, \ldots, J - 1$.

Now,

$$\boldsymbol{U}_{J,n+1} = (\boldsymbol{I} + \Delta t \varepsilon \boldsymbol{A}^D)^{-1}[\boldsymbol{U}_{J,n} + \boldsymbol{f}(\boldsymbol{U}_{J,n})\Delta t + \sigma \Delta \boldsymbol{W}_n]$$

$$+ \varepsilon(\boldsymbol{I} + \Delta t \varepsilon \boldsymbol{A}^D)^{-1}\boldsymbol{B}, n = 0, 1, \ldots, N - 1,$$

where $\boldsymbol{A}^D = \dfrac{1}{h^2}\begin{pmatrix} 2 & -1 & & & \\ -1 & 2 & -1 & & \\ & -1 & 2 & -1 & \\ & & \ddots & \ddots & -1 \\ & & & -1 & 2 \end{pmatrix}$ and $\boldsymbol{B} = \dfrac{1}{h^2}\begin{pmatrix} U(t, x_0) \\ \vdots \\ U(t, x_J) \end{pmatrix}.$

Therefore,

$$\|e_{n+1}\| \le \|(I + \Delta t\varepsilon A^D)^{-1}\|\|[e_{J,n} + \Delta t(f(U_{J,n}) - f(u_{J,n}))]\|.$$
(6.8)

Assume that $f(U) \colon \mathbb{R}^{J-1} \to \mathbb{R}^{J-1}$ is Lipschitz continuous so that

$$\|f(U_{J,n}) - f(u_{J,n})\| \le L\|U_{J,n} - u_{J,n}\|,$$
(6.9)

where $L > 0$ is Lipschitz constant.

Using Eq. (6.9), from Eq. (6.8),

$$\|e_{n+1}\| \le \|(I + \Delta t\varepsilon A^D)^{-1}\|(\|e_n\| + \Delta tL\|e_n\|)$$
$$= \|(I + \Delta t\varepsilon A^D)^{-1}\|(1 + \Delta tL)\|e_n\|.$$

Hence, continuing recursively yields

$$\|e_n\| \le (1 + \Delta tL)\|(I + \Delta t\varepsilon A^D)^{-1}\|\|e_{n-1}\|$$
$$\le (1 + \Delta tL)^2\|(I + \Delta t\varepsilon A^D)^{-1}\|^2\|e_{n-2}\|$$
$$\vdots$$
$$\le (1 + \Delta tL)^n\|(I + \Delta t\varepsilon A^D)^{-1}\|^n\|e_0\|.$$

Thus, if $\|(I + \Delta t\varepsilon A^D)^{-1}\| < \frac{1}{(1+\Delta tL)}$ implies $\|e_n\| \to 0$ as $n \to \infty$.

Therefore, according to Lax–Richtmyer stability, $\|(I + \Delta t\varepsilon A^D)^{-1}\| < \frac{1}{1+\Delta tL}$ is the necessary and sufficient condition for the difference equation (6.7) to be stable. Hence, it implies convergence. $\qquad\square$

6.4.2 *Finite difference scheme for stochastic Fisher equation*

Consider the stochastic Fisher equation

$$dU = [U_{xx} + 6U(1 - U)]dt + dW_t, \ 0 < x < l, \ t > 0, \qquad (6.10)$$

with initial and boundary conditions

$$U(x,0) = \frac{1}{(1 + e^x)^2} 0 \le x \le 1,$$

$$U(0,t) = \frac{1}{(1 + e^{-5t})^2} 0 \le t \le T,$$

$$U(1,t) = \frac{1}{(1 + e^{1-5t})^2} 0 \le t \le T.$$

The grid points $x_j = jh$ for $j = 0, \ldots, J$ with $h = \frac{1}{J}$, and time step $\Delta t > 0$ yields an approximation $\boldsymbol{U}_{J,n}$ to $\boldsymbol{U}_J(t_n)$ at $t_n = n\Delta t$.

Equation (6.10) can be approximated by central difference approximation as

$$dU(t, x_i) = \left[\frac{U(t, x_{i+1}) - 2U(t, x_i) + U(t, x_{i-1})}{h^2} \right.$$

$$\left. + 6U(t, x_i)(1 - U(t, x_i)) \right] dt + dW(t),$$

$$i = 1, 2, \ldots, J - 1. \quad (6.11)$$

For $i = 1$,

$$dU(t, x_1) = \left[\frac{U(t, x_2) - 2U(t, x_1) + U(t, x_0)}{h^2} \right.$$

$$\left. + 6U(t, x_1)(1 - U(t, x_1)) \right] dt + dW(t). \quad (6.12)$$

For $i = 2$,

$$dU(t, x_2) = \left[\frac{U(t, x_3) - 2U(t, x_2) + U(t, x_1)}{h^2} \right.$$

$$\left. + 6U(t, x_2)(1 - U(t, x_2)) \right] dt + dW(t). \quad (6.13)$$

For $i = 3$,

$$dU(t, x_3) = \left[\frac{U(t, x_4) - 2U(t, x_3) + U(t, x_2)}{h^2} \right.$$

$$\left. + 6U(t, x_3)(1 - U(t, x_3)) \right] dt + dW(t). \quad (6.14)$$

Similarly, for $i = J - 1$,

$$dU(t, x_{J-1}) = \left[\frac{U(t, x_J) - 2U(t, x_{J-1}) + U(t, x_{J-2})}{h^2} \right.$$

$$\left. + 6U(t, x_{J-1})(1 - U(t, x_{J-1})) \right] dt + dW(t).$$

$$(6.15)$$

Thus,

$$\boldsymbol{U}_J(t) = [U(t,x_1), U(t,x_2), \ldots, U(t,x_{J-1})]^T,$$
$$\boldsymbol{W}_J(t) = [W(t,x_1), W(t,x_2), \ldots, W(t,x_{J-1})]^T.$$

Now, Eq. (6.15) can be discretised with respect to t as

$$\boldsymbol{U}_{J,n+1} = \boldsymbol{U}_{J,n} + \left\lceil -\boldsymbol{A}^D \boldsymbol{U}_{J,n+1} + 6\boldsymbol{U}_{J,n}(1 - \boldsymbol{U}_{J,n}) \right\rceil \Delta t + \Delta \boldsymbol{W}_n$$
$$+ \boldsymbol{B}\Delta t, \ n = 0, 1, \ldots, N-1.$$

Hence,

$$\boldsymbol{U}_{J,n+1} = (\boldsymbol{I} + \Delta t \boldsymbol{A}^D)^{-1}[\boldsymbol{U}_{J,n} + 6\boldsymbol{U}_{J,n}(1 - \boldsymbol{U}_{J,n})\Delta t + \Delta \boldsymbol{W}_n]$$
$$+ (\boldsymbol{I} + \Delta t \boldsymbol{A}^D)^{-1}\boldsymbol{B}, \quad n = 0, 1, \ldots, N-1.$$

6.4.3 *Numerical discussion for stochastic Fisher equation*

Case 1 (cylindrical Wiener process). The covariance $Q = I$. Consider an orthonormal basis $\{\sqrt{2/a}\,\sin(j\pi x/a)\}$ of $L^2(0,a)$, then from Definition 6.2.1.2,

$$W^J(t,x) = \sqrt{2/a}\sum_{j=1}^{J} \sin\left(\frac{j\pi x}{a}\right) \beta_j(t),$$

for i.i.d. Brownian motions $\beta_j(t)$.

Therefore,

$$Cov(W^J(t,x_i), W^J(t,x_k)) = E\lceil W^J(t,x_i)W^J(t,x_k)\rceil$$
$$= \frac{2t}{a}\sum_{j=1}^{J} \sin\left(\frac{j\pi x_i}{a}\right) \sin\left(\frac{j\pi x_k}{a}\right).$$

Thus,

$$\mathbf{W}_J(t) \sim N(\mathbf{0}, (t/h)\boldsymbol{I}).$$

Now, in this numerical simulation of Case 1, consider $h = 0.005$, and $J = 200$, $\Delta t = 0.00005$ and $N = 10$. For $n = 5$, the sample path for $U(x, 0.00025)$ is shown in Fig. 6.1.

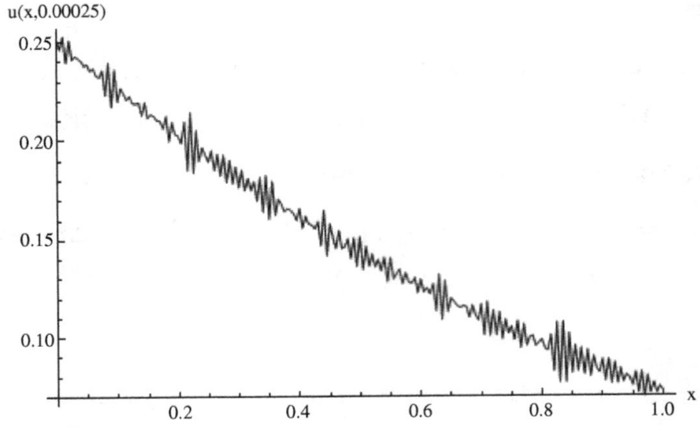

Fig. 6.1. The sample path for $U(x, 0.00025)$ with $h = 0.005$, $J = 200$, $\Delta t = 0.00005$ and $N = 10$.

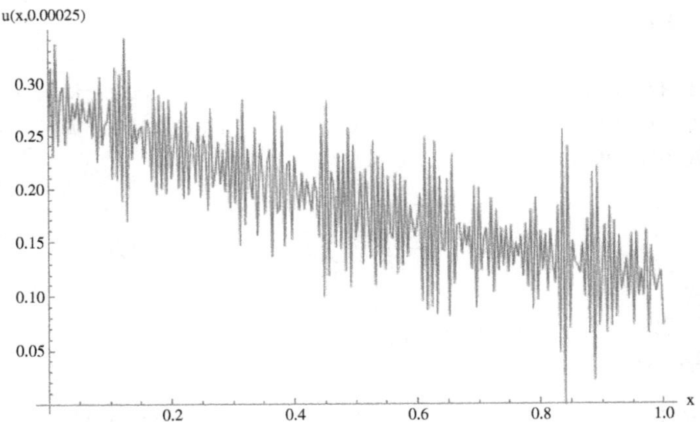

Fig. 6.2. The sample path for $U(x, 0.00025)$ with $h = 0.004$, $J = 250$, $\Delta t = 0.00005$ and $N = 10$.

In this numerical simulation Case 1, consider $h = 0.004$, and $J = 250$, $\Delta t = 0.00005$ and $N = 10$. For $n = 5$, the sample path for $U(x, 0.00025)$ is shown in Fig. 6.2.

Case 2 (Q-Wiener process). Let $W(t)$ be the Q-Wiener process on $L^2(0, a)$ with kernel $q(x, y) = e^{-|x-y|/l}$ for a correlation length $l > 0$.

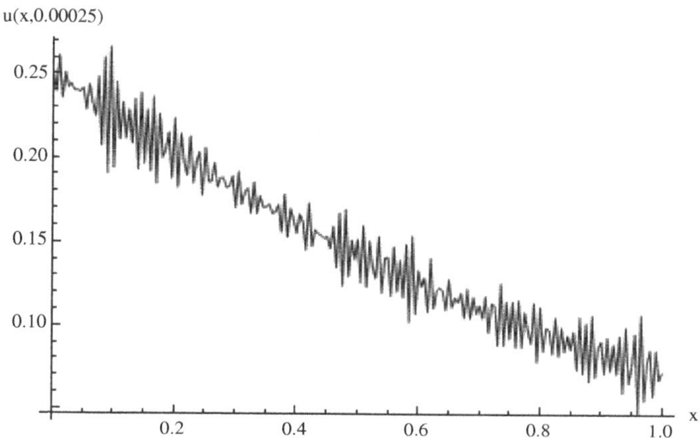

Fig. 6.3. The sample path for $U(x, 0.00025)$ with $h = 0.005$, $J = 200$, $\Delta t = 0.00005$ and $N = 10$.

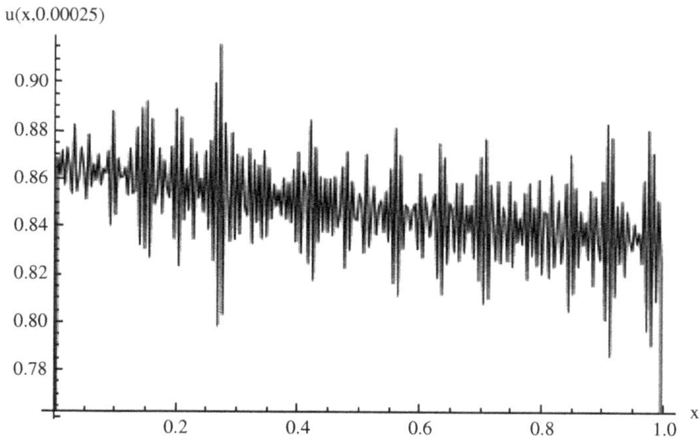

Fig. 6.4. The sample path for $U(x, 0.00025)$ with $h = 0.004$, $J = 250$, $\Delta t = 0.00005$ and $N = 10$.

In the present numerical simulation for Case 2, consider $h = 0.005$, $J = 200$, $\Delta t = 0.00005$ and $N = 10$. For $n = 5$ and correlation length $l = 1$, the sample path for $U(x, 0.00025)$ is shown in Fig. 6.3.

In the present numerical simulation for Case 2, consider $h = 0.004$, $J = 250$, $\Delta t = 0.00005$ and $N = 10$. For $n = 5$ and correlation length $l = 1$, the sample path for $U(x, 0.00025)$ is shown in Fig. 6.4.

6.5 Implementation of Chebyshev Spectral Collocation Method for Stochastic FitzHugh–Nagumo Equation

Consider the stochastic FitzHugh–Nagumo equation

$$q_t = \Delta q - u(1-q)(\beta - q) - r + \frac{dW_1(t)}{dt},$$

$$r_t = \sigma q - \gamma r + \frac{dW_2(t)}{dt}, \tag{6.16}$$

where $a < x < b$, $t > 0$ and $W_i(t)$, $i = 1, 2$, is the Wiener process. The initial conditions are

$$q(x, 0) = 2, \quad x \in [-2.5, 2.5],$$

$$r(x, 0) = 0,$$

with boundary conditions

$$q(a, t) = q(b, t) = 0,$$

$$r(a, t) = r(b, t) = 0.$$

The solutions of $q(x, t)$ and $r(x, t)$ can be approximated as [53]

$$q(x, t) = \sum_{j=0}^{N} {}'' a_j \overset{*}{T}_j(x),$$

$$r(x, t) = \sum_{j=0}^{N} {}'' b_j \overset{*}{T}_j(x). \tag{6.17}$$

The collocation points are

$$x_n = \frac{1}{2}\left((a + b) - (b - a)\cos\left(\frac{\pi n}{N}\right)\right), \quad n = 0, 1, \ldots, N, \tag{6.18}$$

where $\overset{*}{T}_j(x_n) = T_j\left(\frac{2x_n - (b+a)}{b-a}\right)$ is the jth Chebyshev polynomial of the first kind. The double primes summation represents a sum with the first and last terms halved.

The orthogonality relation is given by

$$\sum_{n=0}^{N} {}'' \overset{*}{T}_i(x_n)\overset{*}{T}_j(x_n) = \alpha_i \delta_{ij},$$

with $\alpha_i = \begin{cases} \frac{N}{2}, & i \neq 0, N; \\ N, & i = 0, N. \end{cases}$

So, Eq. (6.17) becomes

$$a_j = \frac{2}{N} \sum_{n=0}^{N} {}'' \overset{*}{T}_j(x_n) q(x_n, t),$$

$$b_j = \frac{2}{N} \sum_{n=0}^{N} {}'' \overset{*}{T}_j(x_n) r(x_n, t).$$

Now, $q_x(x, t)$ can be given by

$$q_x(x_i, t) = \sum_{n=0}^{N} {}'' a_j \overset{*}{T}'_j(x_i)$$

$$= \sum_{n=0}^{N} {}'' \left(\frac{2}{N} \sum_{j=0}^{N} {}'' \overset{*}{T}'_j(x_i) \overset{*}{T}_j(x_n) \right) q(x_n, t)$$

$$= \sum_{n=0}^{N} [A_x]_{in} \, q(x_n, t),$$

where $[A_x]_{in} = \frac{2c_n}{N} \sum_{j=0}^{N} {}'' \overset{*}{T}'_j(x_i) \overset{*}{T}_j(x_n), \quad i, n = 0, 1, \ldots, N.$

$c_0 = c_N = 1/2$ and $c_n = 1$ for $n = 1, 2, \ldots, N-1$.

The first derivative of the Chebyshev functions is formed as follows [159]:

$$\overset{*}{T}'_j(x_i) = 2j\lambda \sum_{n=0, n+j}^{j-1} c_n \overset{*}{T}_n(x_i),$$

where $\lambda = \frac{2}{b-a}$.

So, $q_{xx}(x, t)$ is approximated as

$$q_{xx}(x_i, t) = \sum_{n=0}^{N} [A_x]_{in} q_x(x_n, t)$$

$$= \sum_{j=0}^{N} \left(\sum_{n=0}^{N} [A_x]_{in} q_x(x_n, t) \right) q(x_j, t)$$

$$= \sum_{j=0}^{N} [B_x]_{ij} q(x_j, t),$$

where $B_x = A_x^2$ and the elements of the matrix B_x are

$$[B_x]_{ij} = \sum_{n=0}^{N} [A_x]_{in}[A_x]_{nj}, \quad i,j = 0,1,\ldots,N.$$

Now, the system is

$$dq_i(t) = \left(\sum_{j=0}^{N}[B_x]_{ij}q_j(t) - q_i(t)(1 - q_i(t))(a - q_i(t)) - r_i(t) \right) dt$$
$$+ dW_1(t),$$
$$dr_i(t) = [\sigma q_i(t) - \gamma r_i(t)]dt + dW_2(t),$$

where $q_i(t) = [q_1(t), q_2(t), \ldots, q_{N-1}(t)]^T$, $r_i(t) = [r_1(t), r_2(t), \ldots, r_{N-1}(t)]^T$.

The above system of differential equation can be solved using the semi-implicit Euler–Maruyama scheme [154]

$$q_i(t_{k+1}) = q_i(t_k) + \left(\sum_{j=0}^{N}[B_x]_{ij}q_j(t_k) - q_i(t_k)(1 - q_i(t_k))(\beta - q_i(t_k)) \right.$$
$$\left. - r_i(t_k) \right)\tau + \Delta W_1(t_k),$$
$$r_i(t_{k+1}) = r_i(t_k) + [\sigma q_i(t_k) - \gamma r_i(t_k)]\tau + \Delta W_i(t_k),$$
$$i = 0,1,\ldots,N-1,$$

where $q_i(t_k) = q(x(i), t(k))$, $r_i(t_k) = r(x(i), t(k))$, $\tau = t_{k+1} - t_k$ and $\Delta W_n = z_n\sqrt{\tau}$, where z_n is chosen from standard normal distribution $N(0,1)$.

6.5.1 *Stochastic stability of FitzHugh–Nagumo equation*

Lemma 6.5.1.1. *For any initial value (q_s, r_s) and for some $T > 0$, there exists a unique solution $(q(t, s, x), r(t, s, x)) \in L^2(\Omega, C[0, T], H)$ of Eq. (6.5).*

For $t, s \in \mathbb{R}$,

$$W_{\lambda,s}(t) = \int_s^t \exp(-\lambda(t-s))dW_1(s),$$

$$W_{\delta,s}(t) = \int_s^t \exp(-\delta(t-s))dW_2(s).$$

From Ref. [157], there exists a unique weak solution to Eq. (6.5). It means $Q(t, s, x) = q(t, s, x) - W_{\lambda,s}(t)$ and $R(t, s, x) = r(t, s, x) - W_{\delta,s}(t)$ is the unique solution to the following equations:

$$\begin{aligned} Q_t &= \Delta Q - \psi(Q + W_{\lambda,s}(t)) - (R + W_{\delta,s}(t)), \\ R_t &= \sigma(Q + W_{\lambda,s}(t)) - \gamma(R + W_{\delta,s}(t)), \end{aligned} \tag{6.19}$$

with the initial conditions $Q(s) = q_s = x$ and $R(s) = r_s = y$.

Here, the Hilbert space $L^2(D) \times L^2(D)$ is represented by H. $(\cdot, \cdot)$ represents an ordered pair of H and $\langle \cdot, \cdot \rangle_\Gamma$ stands for the inner product of $\Gamma = L^2(D)$. $|| \cdot ||$ and $|| \cdot ||_H$ are the usual norms on $L^2(D)$ and $H = L^2(D) \times L^2(D)$, respectively.

Define

$$\mu_\Delta = \sup\{\mu > 0; \langle \Delta q, q \rangle_\Gamma \leq -\mu|q|^2, \forall q \in D(\Delta)\},$$

where $D(\Delta)$ implies the domain of Laplace operator Δ with Dirichlet boundary conditions. So, according to Poincare's inequality, such a μ_Δ exists.

Let

$$\mu_\psi = \inf\{\mu \in \mathbb{R}; \langle \psi(q) - \psi(r), q - r \rangle_\Gamma \leq \mu|q - r|^2, \forall q, r \in D(\psi)\},$$

where $D(\psi)$ is the domain of ψ.

Lemma 6.5.1.2. *Consider that $-\mu_\Delta + \mu_h < 0$. Then, there exist constants $C, \omega > 0$ such that*

$$||(Q, R)||_H \leq C(e^{-\omega(t-s)}||(q_s, r_s)||_H$$

$$+ \int_s^t e^{-\omega(t-\tau)}(||\psi(W_{\lambda,s})||_{L^2(D)}$$

$$+ ||W_{\lambda,s}||_{L^2(D)} + ||W_{\delta,s}||_{L^2(D)})(\tau)d\tau).$$

Proof. Consider a new norm $||(Q,R)||_\chi^2 = ||Q||_{L^2(D)}^2 + \frac{1}{\sigma}||R||_{L^2(D)}^2$. Exerting the inner product of Eq. (6.19) with (Q,R) yields

$$\frac{1}{2}\frac{d||Q||_{L^2(D)}^2}{dt} + \frac{1}{2\sigma}\frac{d||R||_{L^2(D)}^2}{dt} = \langle \Delta Q, Q\rangle_\Gamma + \langle \psi(Q+W_{\lambda,s}), Q)_\Gamma\rangle$$

$$- \langle (R+W_{\delta,s}, Q\rangle_\Gamma + \langle (Q+W_{\lambda,s}), R\rangle_\Gamma - \frac{1}{\sigma}\langle \gamma(R+W_{\delta,s}(t)), R\rangle_\Gamma.$$

Under the definition of a new norm, it can be obtained as

$$\frac{1}{2}\frac{d||(Q,R)||_\chi^2}{dt} = \langle \Delta Q, Q\rangle_\Gamma - \gamma\frac{1}{\sigma}\langle R, R\rangle_\Gamma + \langle \psi(Q+W_{\lambda,s}), Q\rangle_\Gamma$$

$$- \frac{\gamma}{\sigma}\langle W_{\delta,s}, R\rangle_\Gamma + \langle W_{\lambda,s}, R\rangle_\Gamma - \langle W_{\delta,s}, Q\rangle_\Gamma.$$

From the definition of μ_Δ,

$$\frac{1}{2}\frac{d||(Q,R)||_\chi^2}{dt} \leq -\mu_\Delta||Q||_{L^2(D)}^2 - \gamma\frac{1}{\sigma}||R||_{L^2(D)}^2$$

$$+ \langle \psi(Q+W_{\lambda,s}) - \psi(W_{\lambda,s}), Q\rangle_\Gamma + \langle \psi(W_{\lambda,s}), Q\rangle_\Gamma$$

$$- \frac{\gamma}{\sigma}\langle W_{\delta,s}, R\rangle_\Gamma + \langle W_{\lambda,s}, R\rangle_\Gamma - \langle W_{\delta,s}, Q\rangle_\Gamma.$$

From the definition of μ_ψ,

$$\frac{1}{2}\frac{d||(Q,R)||_\chi^2}{dt} \leq -\mu_\Delta||Q||_{L^2(D)}^2 - \gamma\frac{1}{\sigma}||R||_{L^2(D)}^2 + \mu_\psi||Q||_{L^2(D)}$$

$$+ \langle \psi(W_{\lambda,s}), Q\rangle_\Gamma - \frac{\gamma}{\sigma}\langle W_{\delta,s}, R\rangle_\Gamma$$

$$+ \langle W_{\lambda,s}, R\rangle_\Gamma - \langle W_{\delta,s}, Q\rangle_\Gamma,$$

whence

$$\frac{1}{2}\frac{d||(Q,R)||_\chi^2}{dt} \leq -\mu_\Delta||Q||_{L^2(D)}^2 - \gamma\frac{1}{\sigma}||R||_{L^2(D)}^2 + \mu_\psi||Q||_{L^2(D)}$$

$$+ ||\psi(W_{\lambda,s})||_{L^2(D)}||Q||_{L^2(D)}$$

$$+ \frac{\gamma}{\sigma}||W_{\delta,s}||_{L^2(D)}||R||_{L^2(D)} + ||W_{\lambda,s}||_{L^2(D)}||R||_{L^2(D)}$$

$$+ ||W_{\delta,s}||_{L^2(D)}||Q||_{L^2(D)}.$$

Now, from the equivalence of norms, there exist constants $C_1, \tilde{\omega} > 0$ such that

$$\frac{1}{2}\frac{d\|(Q,R)\|_\chi^2}{dt} \leq -\tilde{\omega}\|(Q,R)\|_\chi^2 + C_1(\|\psi(W_{\lambda,s})\|_{L^2(D)} + \|W_{\lambda,s}\|_{L^2(D)}$$
$$+ \|W_{\delta,s}\|_{L^2(D)})\|(Q,R)\|_\chi.$$

From Grönwall's inequality, it can be written as

$$\frac{d(e^{2\tilde{\omega}t}\|(Q,R)\|_\chi^2)}{dt} \leq e^{2\tilde{\omega}t}2C_1(\|\psi(W_{\lambda,s})\|_{L^2(D)} + \|W_{\lambda,s}\|_{L^2(D)}$$
$$+ \|W_{\delta,s}\|_{L^2(D)})\|(Q,R)\|_\chi.$$

Further on integrating yields

$$\|(Q,R)\|_\chi \leq e^{-2\tilde{\omega}(t-s)}\|(q_s,r_s)\|_\chi + \int_s^t e^{-2\tilde{\omega}(t-\tau)}2C_1(\|\psi(W_{\lambda,s})\|_{L^2(D)}$$
$$+ \|W_{\lambda,s}\|_{L^2(D)} + \|W_{\delta,s}\|_{L^2(D)})(\tau)d\tau.$$

From the equivalence of norms, there exist constants $C, \omega > 0$; C depends only on smooth bounded domain D, such that

$$\|(Q,R)\|_H \leq C(e^{-\omega(t-s)}\|(q_s,r_s)\|_H + \int_s^t e^{-\omega(t-\tau)}(\|\psi(W_{\lambda,s})\|_{L^2(D)}$$
$$+ \|W_{\lambda,s}\|_{L^2(D)} + \|W_{\delta,s}\|_{L^2(D)})(\tau)d\tau).$$

$$\square$$

Proposition 6.5.1.3. *Consider* $-\mu_\Delta + \mu_h < 0$. *Then, a random variable* η *exists, such that for any* $(x,y) \in H, j > 0$,

$$E\|(q(0,-j,x),r(0,-j,y)) - \eta\|_H \leq Me^{-\tilde{C}j}(\|(x,y)\|_H + 1),$$

for some constant $M > 0$.

Proof. For $j > j_1 > 0$, it can be assumed that

$$\|(q(0,-j,x),r(0,-j,y)) - (q(0,-j_1,x),r(0,-j_1,y))\|_H$$
$$\leq Ce^{-\tilde{C}j_1}\|(q(-j_1,-j,x),r(-j_1,-j,y)) - (x,y)\|_H. \quad (6.20)$$

Setting $(\tilde{q}(t), \tilde{r}(t)) = (q(t, -j, x), r(t, -j, y)) - (q(t, -j_1, x), r(t, -j_1, y))$, we obtain

$$\tilde{q}_t = \Delta\tilde{q} - (\psi(q(t, -j, x)) - \psi(q(t, -j_1, x))) - \tilde{r},$$
$$\tilde{r}_t = \sigma\tilde{q} - \gamma\tilde{r},$$

with initial conditions

$$\tilde{q}(-j_1) = q(-j_1, j, x) - x \text{ and } \tilde{r}(-j_1) = r(-j_1, j, x) - y.$$

Taking inner products of the above system with $\tilde{q}, \tilde{r}$ yields

$$\frac{d\left(||\tilde{q}||^2_{L^2(D)} + \frac{1}{\sigma}||\tilde{r}||^2_{L^2(D)}\right)}{dt} = 2(\Delta\tilde{q}, \tilde{q})_\Gamma - \frac{2\gamma}{\sigma}||\tilde{r}||^2_{L^2(D)}$$
$$-2\langle\psi(q(t, -j, x)) - \psi(q(t, -j_1, x)),$$
$$q(t, -j, x) - q(t, -j_1, x)\rangle_\Gamma.$$

This implies

$$\frac{d||(\tilde{q}, \tilde{r})||^2_\chi}{dt} = 2\langle\Delta\tilde{q}, \tilde{q}\rangle_\Gamma - \frac{2\gamma}{\sigma}||\tilde{r}||^2_{L^2(D)} - 2\langle\psi(q(t, -j, x))$$
$$-\psi(q(t, -j_1, x)), q(t, -j, x) - q(t, -j_1, x)\rangle_\Gamma.$$

By the definition of μ_Δ,

$$\frac{d||(\tilde{q}, \tilde{r})||^2_\chi}{dt} \leq -2\mu_\Delta||\tilde{q}||^2_{L^2(D)} - \frac{2\gamma}{\sigma}||\tilde{r}||^2_{L^2(D)} - 2\langle\psi(q(t, -j, x))$$
$$-\psi(q(t, -j_1, x)), q(t, -j, x) - q(t, -j_1, x)\rangle_\Gamma.$$

Therefore, there exists $\tilde{C} > 0$ such that

$$\frac{d||(\tilde{q}, \tilde{r})||^2_\chi}{dt} \leq -2\tilde{C}||(\tilde{q}, \tilde{r})||^2_\chi.$$

Applying Grönwall's lemma, there exists $\tilde{C}_1 > 0$ such that

$$||(\tilde{q}, \tilde{r})||^2_{L^2(D)} \leq \tilde{C}_1||(\tilde{q}(-j_1), \tilde{r}(-j_1))||^2_{L^2(D)}e^{-2\tilde{C}(j_1+t)}.$$

By equivalence of norms, there exists $C > 0$ such that

$$||(q(t, -j, x) - q(t, -j_1, x), r(t, -j, y) - r(t, -j_1, y))||_H$$
$$\leq C||(q(-j_1, j, x) - x, r(-j_1, -j, y) - y)||_H e^{-2\tilde{C}(j_1+t)}.$$

This implies

$$||(q(t,-j,x),r(t,-j,y)) - (q(t,-j_1,x),r(t,-j_1,y))||_H$$

$$\leq C||(q(-j_1,j,x),r(-j_1,-j,y)) - (x,y)||_H e^{-2\tilde{C}(j_1+t)}.$$

Putting $t = 0$,

$$||(q(0,-j,x),r(0,-j,y)) - (q(0,-j_1,x),r(0,-j_1,y))||_H$$

$$\leq Ce^{-\tilde{C}j_1}||(q(-j_1,j,x),r(-j_1,-j,y)) - (x,y)||_H. \qquad (6.21)$$

Hence, the claim (6.20) is proved.
From Lemma 6.5.1.2,

$$||(Q(-j_1,-j,x),R(-j_1,-j,y))||_H \leq C(e^{-\omega(j-j_1)}||(q_s,r_s)||_H$$

$$+ \int_{-j}^{-j_1} e^{-\omega(-j_1-\tau)}(||\psi(W_{\lambda,-j})||_{L^2(D)}$$

$$+||W_{\delta,-j}||_{L^2(D)} + ||W_{\lambda,-j}||_{L^2(D)})(\tau)d\tau).$$

Thus,

$$||(q(-j_1,-j,x),r(-j_1,-j,y))||_H \leq C(e^{-\omega(j-j_1)}||(q_s,r_s)||_H$$

$$+ \int_{-j}^{-j_1} e^{-\omega(-j_1-\tau)}(||\psi(W_{\lambda,-j})||_{L^2(D)} + ||W_{\delta,-j}||_{L^2(D)}$$

$$+||W_{\lambda,-j}||_{L^2(D)})(\tau)d\tau + ||(W_{\delta,-j},W_{\lambda,-j})||_H).$$

Taking expectation and subsequently applying Cauchy–Schwarz inequality, from Eq. (6.21), the following has been obtained:

$$E(||(q(0,-j,x),r(0,-j,y)) - (q(0,-j_1,x),r(0,-j_1,y))||_H)$$

$$\leq Ce^{-\tilde{C}j_1}E(||(q(-j_1,-j,x),r(-j_1,-j,y)) - (x,y)||_H)$$

$$\leq Ce^{-\tilde{C}j_1}\left(Ce^{-\omega(j-j_1)}||(x,y)||_H + ||(x,y)||_H\right.$$

$$+E\sup_{-j\leq t\leq -j_1}\left(\frac{1}{\omega}||\psi(W_{\lambda,-j})(t)||_{L^2(D)}\right.$$

$$\left.\left.+\frac{1+\omega}{\omega}(||W_{\lambda,-j}(t)||_{L^2(D)}+||W_{\delta,-j}(t)||_{L^2(D)})\right)\right)$$

$$\leq Ce^{-\tilde{C}j_1}\left((Ce^{-\omega(-j_1+j)}+1)||(x,y)||_H\right.$$

$$+E\sup_{-j\leq t\leq -j_1}\left(\frac{1}{\omega}||\psi(W_{\lambda,-j})(t)||_{L^2(D)}\right.$$

$$\left.\left.+\frac{1+\omega}{\omega}(||W_{\lambda,-j}(t)||_{L^2(D)}+||W_{\delta,-j}(t)||_{L^2(D)})\right)\right).$$

$$(6.22)$$

Setting $K = \sup_{-j\leq t\leq -j_1}\left(\frac{1}{\omega}||\psi(W_{\lambda,-j})(t)||_{L^2(D)} + \frac{1+\omega}{\omega}(||W_{\lambda,-j}(t)||_{L^2(D)} + ||W_{\delta,-j}(t)||_{L^2(D)})\right)$, Eq. (6.22) yields

$$E||(q(0,-j,x),r(0,-j,y)) - (q(0,-j_1,x),r(0,-j_1,y))||_H$$

$$\leq Ce^{-\tilde{C}j_1}((Ce^{-\omega(-j_1+j)}+1)||(x,y)||_H + K).$$

Now, there exists some constant $M > 0$, then

$$E||(q(0,-j,x),r(0,-j,y)) - (q(0,-j_1,x),r(0,-j_1,y))||_H$$

$$\leq Me^{-\tilde{C}j_1}(||(x,y)||_H + 1).$$

The above estimate manifests that $(q(0,-j,x),r(0,-j,y))$ converges in $L^1(\tilde{\omega},\mathbb{F},\mathrm{P};H)$ as $j \to +\infty$ to some random variable η and therefore

$$E||(q(0,-j,x),r(0,-j,y)) - \eta||_H \leq Me^{-\tilde{C}j}(||(x,y)||_H + 1). \qquad \square$$

Consider the deterministic FitzHugh–Nagumo system as

$$q_t = \Delta q - \psi(q) - r,$$

$$r_t = \sigma q - \gamma r, \qquad\qquad (6.23)$$

where $x \in D \subset \mathbb{R}$ and $t \geq s \in \mathbb{R}$.

Denote $(q^*(t,s,x),r^*(t,s,y))$ as the weak solution to Eq. (6.23).

Furthermore, for small $\theta > 0$, consider the stochastic FitzHugh–Nagumo system as

$$q_t = \Delta q - \psi(q) - r + \frac{dW_1^\theta(t)}{dt},$$

$$r_t = \sigma q - \gamma r + \frac{dW_2^\theta(t)}{dt}, \qquad\qquad (6.24)$$

where $x \in D \subset \mathbb{R}$, $t \geq s \in \mathbb{R}$.

According to Lemma 6.5.1.1, there exists a unique weak solution $(q^\theta(t, s, x), r^\theta(t, s, y))$ to Eq. (6.24) for different $\theta > 0$.

Now, there is a special case of Lemma 6.5.1.2 which is presented as Lemma 6.5.1.4.

Lemma 6.5.1.4. *Assume that* $-\mu_\Delta + \mu_h < 0$. *Then, there exist constants* $C, \omega > 0$ *so that*

$$\|(q^*(t, s, x), r^*(t, s, y))\|_H \leq Ce^{-\omega(t-s)}\|(x, y)\|_H.$$

Theorem 6.5.1.5. *Assume that* $-\mu_\Delta + \mu_h < 0$. *Then, the FitzHugh–Nagumo systems are stochastic stable.*

Proof. For a fixed $(x, y) \in H$, it can be defined as

$$(\tilde{q}(t), \tilde{r}(t)) = (q^\theta(t, -j, x), r^\theta(t, -j, y)) - (q^*(t, -j, x), r^*(t, -j, y)),$$

for $t, \tau > 0$. Then, the following is obtained:

$$\tilde{q}_t = \Delta\tilde{q} - (\psi(q^\theta(t, -j, x)) - \psi(q^*(t, -j, x))) - \tilde{r} + \frac{dW_1^\theta(t)}{dt},$$

$$\tilde{r}_t = \sigma\tilde{q} - \gamma\tilde{r} + \frac{dW_2^\theta(t)}{dt},$$

with the initial conditions

$$\tilde{q}(-j) = 0 \text{ and } \tilde{r}(-j) = 0.$$

Now, define

$$W_{\lambda,-j}^\theta(t) = \int_{-j}^t \exp(-\lambda(t - s))dW_1(s),$$

$$W_{\delta,-j}^\theta(t) = \int_{-j}^t \exp(-\delta(t - s))dW_2(s). \tag{6.25}$$

Similar to Lemma 6.5.1.1, it can be easily proven that $\tilde{Q}(t, -j, x) = \tilde{q}(t, -j, x) - W_{\lambda,-j}^\theta(t)$ and $\tilde{R}(t, -j, y) = \tilde{r}(t, -j, y) - W_{\delta,-j}^\theta(t)$ is the unique weak solution of the following:

$$\tilde{Q}_t = \Delta\tilde{Q} - (\psi(q^\theta(t, -j, x)) - \psi(q^*(t, -j, x))) - (\tilde{R} + W_{\delta,-j}^\theta(t)),$$

$$\tilde{R}_t = \sigma(\tilde{Q} + W_{\lambda,-j}^\theta(t)) - \gamma(\tilde{R} + W_{\delta,-j}^\theta(t)), \tag{6.26}$$

with the initial conditions

$$\tilde{Q}(-j) = \tilde{q}_s = x \quad \text{and} \quad \tilde{R}(-j) = \tilde{r}_s = y.$$

Taking the inner product of Eq. (6.26) with $(\tilde{Q}, \tilde{R})$ and according to Lemma 6.5.1.2, it can be shown that

$$||(\tilde{Q}(t, -j, x), \tilde{R}(t, -j, y))||_H \leq C(e^{-\omega(t+j)}||(\tilde{q}_s, \tilde{r}_s)||_H$$
$$+ \int_{-j}^{t} e^{-\omega(t-\tau)}(||\psi(W_{\lambda,-j}^{\theta})||_{L^2(D)}$$
$$+ ||W_{\lambda,-j}^{\theta}||_{L^2(D)} + ||W_{\delta,-j}^{\theta}||_{L^2(D)})(\tau)d\tau.$$

Then, it follows that for $t = 0$,

$$||(\tilde{q}(0, -j, x), \tilde{r}(0, -j, y))||_H \leq C(e^{-\omega j}||(\tilde{q}_s, \tilde{r}_s)||_H$$
$$+ \int_{-j}^{0} e^{\omega\tau}(||\psi(W_{\lambda,-j})||_{L^2(D)} + ||W_{\lambda,-j}^{\theta}||_{L^2(D)}$$
$$+ ||W_{\delta,-j}^{\theta}||_{L^2(D)})(\tau)d\tau + ||W_{\delta,-j}^{\theta}(0)||_H + ||W_{\lambda,-j}^{\theta}(0)||_H.$$

Taking $j \to +\infty$, it reduces to

$$||\eta^{\theta} - (0,0)||_H \leq C \int_{-\infty}^{0} e^{\omega\tau}(||\psi(W_{\lambda,-\infty})||_{L^2(D)} + ||W_{\lambda,-\infty}^{\theta}||_{L^2(D)}$$
$$+ ||W_{\delta,-\infty}^{\theta}||_{L^2(D)})(\tau)d\tau$$
$$+ ||W_{\delta,-\infty}^{\theta}(0)||_H + ||W_{\lambda,-\infty}^{\theta}(0)||_H$$
$$\leq \frac{C}{\omega} \sup_{t \leq 0} ||\psi(W_{\lambda,-\infty}^{\theta})(t)||_{L^2(D)}$$
$$+ \frac{C+\omega}{\omega} \sup_{t \leq 0} ||W_{\lambda,-\infty}^{\theta}(t)||_{L^2(D)}$$
$$+ \frac{C+\omega}{\omega} \sup_{t \leq 0} ||W_{\delta,-\infty}^{\theta}(t)||_{L^2(D)}.$$

Now, from Eq. (6.25), there exits some constant $\hat{C} > 0$, such that

$$\left\|W_{\lambda,-\infty}^{\theta}(t)\right\|_{L^2(D)} \leq \hat{C}\theta \left\|\int_{-\infty}^{t} \exp(-\lambda(t-s))dW_1(s)\right\|_{L^2(D)}, \text{ and}$$

$$\left\|W_{\delta,-\infty}^{\theta}(t)\right\|_{L^2(D)} \leq \hat{C}\theta \left\|\int_{-\infty}^{t} \exp(-\delta(t-s))dW_2(s)\right\|_{L^2(D)}.$$

So, it can be written as

$$\lim_{\theta \to 0} E\|\eta^{\theta} - (0,0)\|_H = 0.$$

Hence, it is proved. $\qquad\qquad\qquad\qquad\qquad\qquad\qquad\qquad\qquad\square$

6.5.2   *Numerical discussion for stochastic FitzHugh–Nagumo equation*

The numerical results for the stochastic FitzHugh–Nagumo equation have been discussed in this section. Here, the value of the parameters are given by $\alpha = 0.13$, $\varepsilon = 0.0035$, $\beta = 0.75$ and $\gamma = 0.0035 \times 2.70$. The numerical results on computational domain $(x,t) \in [-40, 40] \times [0, 20]$ have been investigated. The numerical solution is obtained by taking from a mesh with $h = 0.8$ and $\tau = 0.05$. The behaviour of the solutions for $q(x,t)$ and $r(x,t)$has been discussed in Figs. 6.5(a) and (b), 6.6(a) and (b), respectively.

6.6   Summary

In this chapter, a semi-implicit FDM has been successfully applied here for finding the solution for the stochastic Fisher equation. Finite difference scheme has been applied to discretise the stochastic Fisher equation. The sample paths obtained from cylindrical Wiener process have been shown in Figs. 6.1 and 6.2. Also, the sample paths obtained from Q-Wiener process have been shown in Figs. 6.3 and 6.4.

The stochastic FitzHugh–Nagumo equation has also been solved by Chebyshev spectral collocation method and semi-implicit EMM. In Section 6.5.1, the stability of the stochastic FitzHugh–Nagumo equation has been discussed. The algorithm for Chebyshev spectral collocation method has also been discussed very thoroughly and its

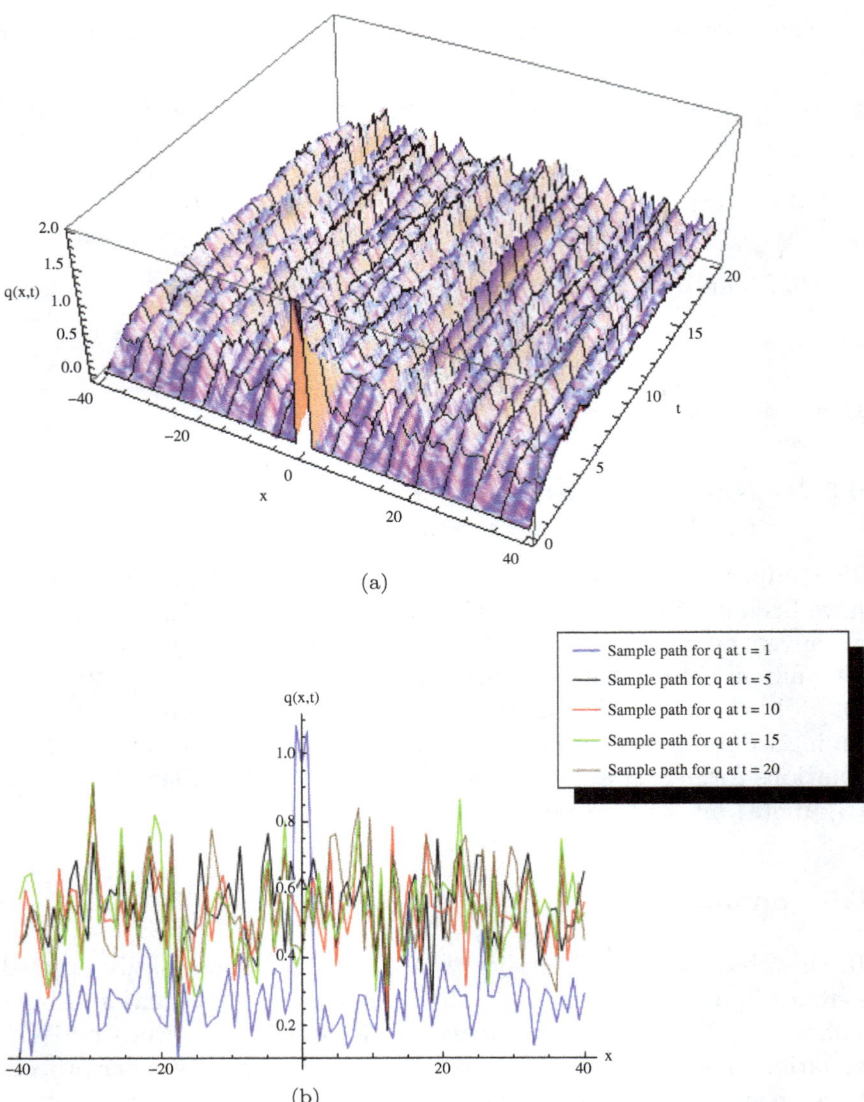

Fig. 6.5. (a) Approximate realisation for $q(x, t)$ over the domain $(x, t) \in [-40, 40] \times [0, 20]$. (b) Sample paths of $q(x, t)$ for different times.

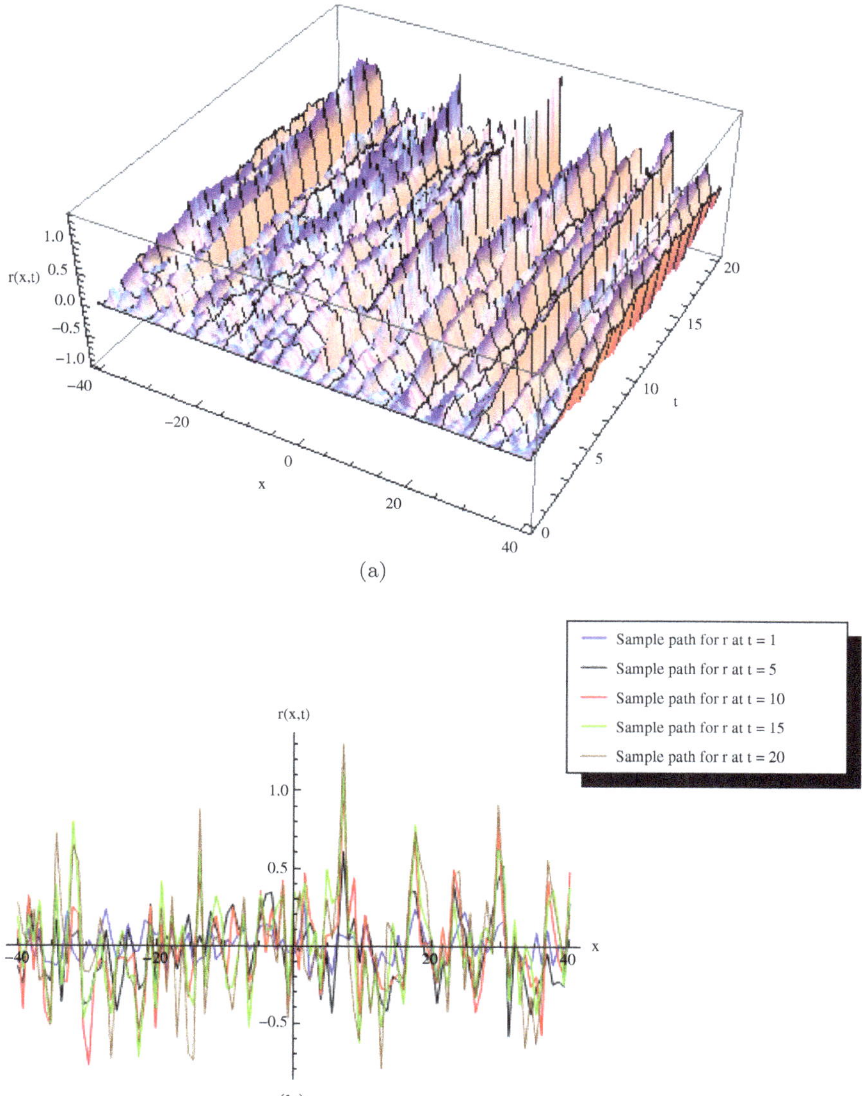

Fig. 6.6. (a) Approximate realisation for $r(x,t)$ over the domain $(x,t) \in [-40, 40] \times [0, 20]$. (b) Sample paths of $r(x,t)$ for different times.

efficiency has also been tested in the numerical section with graphical explanation of the behaviour of the sample paths. The graphs show the mean sample path for different time instances and separate two-dimensional plots for sample paths at different time points have also been shown for a more clear idea about the dynamical behaviour of the governing equation.

Chapter 7

Numerical Solutions of Stochastic Point Kinetics Equations

7.1 Introduction

The neutron diffusion and point kinetics equations are very important models in nuclear engineering. The neutron diffusion concept allows us to establish the comportment of average neutron dynamics. The equations are formulated as a set of ODEs that can exhibit a rather stiff solution [160,161] and this solution helps in understanding the dynamic behaviour of the power level of a nuclear reactor. Recently, various researchers have been working towards forming prototypes for studying the dynamic behaviour of a nuclear reactor using point kinetics equations.

Many authors have specified that stochastic point kinetics equations (SPKEs) continue to be an important set of equations. The temperature feedback in the point kinetics equations provides a very vital approximation which establishes the transient behaviour of reactor power and other system variables of the reactor core model that are very tightly coupled [160,162,163].

Nuclear reactor deals with transient neutron flux changes which occur during the startup and shutdown of a reactor or due to accidental disturbances in the reactor steady-state operation [164,165]. However, in a more realistic situation, the point kinetics equations are stochastic in nature [166,167], and in this work, it has been used to study the random variations of the neutron density and delayed neutron precursor concentrations with respect to time.

In order to improve the efficiency of the numerical methods for solving SDEs which are stiff in both the deterministic and stochastic components, some attempts have been made to propose split-step methods. In this chapter, a discussion on the derivative-free Milstein method and split-step forward Euler–Maruyama method for solving stiff SDEs has been provided. It is to be important to note that the proposed efficient derivative-free Milstein scheme uses a special approximation of the derivative which turns out to be very efficient. The computational cost for the derivative-free Milstein method is lesser than the Milstein method [168]. The stable results obtained show that the split-step forward Euler–Maruyama method is suitable for solving stiff SDEs. It is also noteworthy that the split-step forward Euler–Maruyama method is more stable than the original method on the premise that the cost is invariable.

In recent years, various researchers have provided significant models for studying the behaviour of a nuclear reactor using PKEs. Hayes and Allen [167] implemented stochastic piecewise constant approximation method (stochastic PCA) to a system of stiff SDEs. Saha Ray [169] established 1.5 strong Taylor and Euler–Maruyama methods as sensible computational alternatives to Stochastic PCA in solving the point kinetics equations. Nahla and Edress [170] solved the SPKEs with step, ramp and sinusoidal reactivity respectively by analytical exponential model. Patra and Saha Ray [171] obtained solutions for the point kinetics equations by explicit finite difference method and Haar wavelet operational method [172]. Espinosa-Paredes, Polo-Labarrios, Espinosa-Martínez and Valle-Gallegos [173] obtained the solution of the fractional neutron point kinetics model.

7.2 Outline of Present Study

In this chapter, a comparison between two numerical approximation methods, i.e., Euler–Maruyama and 1.5 strong Taylor methods (EMM and STM), has been established in this article. The SPKEs consist of a system of Itô SDEs and this system is solved over each time-step size using Euler–Maruyama and 1.5 strong Taylor methods. The obtained results establish the accuracy of both the methods in solving the SPKEs in the presence of Newtonian temperature feedback.

Similarly, a comparative study between two numerical approximation methods, viz., split-step forward EMM and derivative-free Milstein method (DFMM), has been established. These SPKEs have been solved for step and ramp external reactivities using split-step forward Euler–Maruyama method and derivative-free Milstein method. The obtained numerical results show that the split-step approximations are straightforward and effective methods in studying the behaviour of neutron density of the SPKEs which also have been represented graphically.

7.2.1 *Stochastic nonlinear point reactor kinetics equations*

The deterministic nonlinear PKE is useful in the context of diffusion phenomena because of the highly heterogeneous configuration of nuclear reactors. The nonlinear PKE in the presence of Newtonian temperature feedback effects and multigroup of delayed neutrons is a system of stiff nonlinear ordinary differential equations and takes the following form [174–178]:

$$\frac{dN(t)}{dt} = \left(\frac{\rho(t) - \beta}{l}\right) N(t) + \sum_{i=1}^{m} \lambda_i C_i(t), \tag{7.1}$$

$$\frac{dC_i(t)}{dt} = \frac{\beta_i}{l} N(t) - \lambda_i C_i(t), \quad i = 1, 2, 3, \ldots, m, \tag{7.2}$$

$$\rho(t) = \rho_{ex}(t) - \alpha[T(t) - T_0], \tag{7.3}$$

$$\frac{dT(t)}{dt} = K_c N(t), \tag{7.4}$$

where $N(t)$ is the neutron population, t is the time, $\rho(t)$ is the total reactivity dependent on the temperature, $\rho_{ex}(t)$ represents external reactivity, $C_i(t)$ is the precursor concentration of i-group of delayed neutrons, $\beta = \sum_{i=1}^{m} \beta_i$ is the total fraction of delayed neutrons, β_i is the fraction of i-group of delayed neutrons, λ_i is the decay constant of i-group of delayed neutrons, l is the prompt neutron generation time, m is the total number of delayed neutrons groups, $T(t)$ is the temperature of the reactor, T_0 is the initial temperature of the reactor,

α is the temperature coefficient of reactivity and K_c is the reciprocal of the thermal capacity of the reactor.

Integrating Eq. (7.4) with respect to time t and substituting the result into Eq. (7.3), it can be written as [161,177–179]

$$\rho(t) = \rho_{ex}(t) - \rho_f(t), \quad \rho_f(t) = \sigma \int_0^t N(\tau)d\tau, \qquad (7.5)$$

where $\sigma = \alpha K_c$ is the nonlinear coefficient part. $\rho_f(t)$ represents the Newtonian temperature feedback effects and is dependent on the neutron density $N(t)$.

The deterministic point kinetics equations (7.1) and (7.2) are separated into the following terms:

(a) $\left(\frac{\rho(t)}{l} N(t)\right) \equiv$ These neutrons are termed as prompt neutrons. The prompt neutrons are emitted by the direct fission products, immediately after the fission process [165].

(b) $\left(\frac{\beta_i}{l} N(t) - \lambda_i C_i(t)\right) \equiv$ These neutrons are termed as delayed neutrons. Nuclei of some of the fission products may beta-decay into daughter nuclei which then immediately emit a neutron known as precursors of delayed neutrons [165].

Equations (7.1), (7.2) and (7.5) are the couple of stiff nonlinear ODEs, where $\rho_f(t)$ depends upon neutron density $N(t)$. This parameter represents the Newtonian temperature feedback effects.

Consider that the deterministic point kinetics equations (7.1) and (7.2) are separated into two terms: prompt $(\frac{\rho(t)}{l} N(t))$ and delayed $(\frac{\beta_i}{l} N(t) - \lambda_i C_i(t))$ neutrons. Using the centre limit theorem, the Itô SDE for the nonlinear point reactor kinetics equations yields [179,184]

$$\frac{d\mathbf{\Psi}(t)}{dt} = A(t)\mathbf{\Psi}(t) + \mathbf{B}^{\frac{1}{2}}(t)\frac{d\mathbf{W}(t)}{dt}, \qquad (7.6)$$

where $\mathbf{\Psi}(t) = \begin{pmatrix} N(t) \\ C_1(t) \\ C_2(t) \\ \vdots \\ C_m(t) \end{pmatrix}$, $\mathbf{W}(t) = \begin{pmatrix} W_0(t) \\ W_1(t) \\ W_2(t) \\ \vdots \\ W_m(t) \end{pmatrix}$ and $W_0(t), W_1(t),$ $W_2(t), \ldots, W_m(t)$ are standard Wiener processes, or scalar standard

Brownian motion, over $[0, T]$ is the set of the random variables that depend continuously on time t and satisfies the following properties [169]:

(a) $W(0) = 0$ with probability 1.
(b) For $0 \leq s \leq t \leq T$, the random variable given by the increment $W(t) - W(s)$ is normally distributed with mean 0 and variance $t - s$; equivalently, $W(t) - W(s) \sim \sqrt{t - s}N(0, 1)$, where $N(0, 1)$ denotes a normally distributed random variable with zero mean and unit variance.
(c) For $0 \leq s < t < u < v \leq T$, the increments $W(t) - W(s)$ and $W(v) - W(u)$ are independent.

Here, $\boldsymbol{A}(t)$ is called the coefficient matrix and is

$$\boldsymbol{A}(t) = \begin{pmatrix} \dfrac{\rho(t) - \beta}{l} & \lambda_1 & \lambda_2 & \ldots & \lambda_m \\ \dfrac{\beta_1}{l} & -\lambda_1 & 0 & \ldots & 0 \\ \dfrac{\beta_2}{l} & 0 & -\lambda_2 & \ldots & 0 \\ \vdots & \vdots & \vdots & \ddots & \vdots \\ \dfrac{\beta_m}{l} & 0 & 0 & \ldots & -\lambda_m \end{pmatrix}. \tag{7.7}$$

$\boldsymbol{B}(t)$ is the covariance matrix which is evaluated in Ref. [168] as

$$\boldsymbol{B}(t) = \begin{pmatrix} \mu_0(t) & -\mu_1(t) & -\mu_2(t) & \ldots & -\mu_m(t) \\ -\mu_1(t) & \mu_1(t) & 0 & \ldots & 0 \\ -\mu_2(t) & 0 & \mu_2(t) & \ldots & 0 \\ \vdots & \vdots & \vdots & \ddots & \vdots \\ -\mu_m(t) & 0 & 0 & \ldots & \mu_m(t) \end{pmatrix}, \tag{7.8}$$

where $\mu_0(t) = (\frac{\rho(t) + \beta}{l})N(t) - \sum_{i=1}^{m} \lambda_i C_i(t)$, $\mu_i(t) = \frac{\beta_i}{l}N(t) - \lambda_i C_i(t)$, $i = 1, 2, 3, \ldots, m$.

7.3 The Order 1.5 STM

A stochastic Taylor expansion for Itô SDEs was first described in Wagner and Platen [180].

Let $\{X_t\}$ be an Itô process on $t \in [t_0, T]$ satisfying the following SDE [169]:

$$\begin{cases} dX_t = a(t, X_t)dt + b(t, X_t)dW_t, \\ X_{t_0} = X_0. \end{cases} \tag{7.9}$$

For a given time-discretisation $t_0 < t_1 < \cdots < t_n = T$, the order 1.5 strong Itô–Taylor method [98,181] for numerical solution of Eq. (7.9) is

$$w_{n+1} = w_n + a\Delta t_n + b\Delta W_n + \frac{1}{2}bb_x(\Delta W_n^2 - \Delta t_n) + a_x b\Delta Z_n$$

$$+ \frac{1}{2}\left(aa_x + \frac{1}{2}b^2 a_{xx}\right)\Delta t_n^2$$

$$+ \left(ab_x + \frac{1}{2}b^2 b_{xx}\right)(\Delta W_n \Delta t_n - \Delta Z_n)$$

$$+ \frac{1}{2}b(bb_{xx} + b_x^2)\left(\frac{1}{3}\Delta W_n^2 - \Delta t_n\right)\Delta W_n, \tag{7.10}$$

for $n = 0, 1, 2, \ldots, N - 1$ with initial value $w_0 = X_0$.

Here, the random variable ΔZ_n is normally distributed with mean 0 and variance $E(\Delta Z_n^2) = \frac{1}{3}\Delta t_n^3$ and correlated with ΔW_n by covariance and $E(\Delta Z_n \Delta W_n) = \frac{1}{2}\Delta t_n^2$. The partial derivatives are denoted by subscripts. Here, ΔZ_n can be generated as

$$\Delta Z_n = \frac{1}{2}\Delta t_n\left(\Delta W_n + \frac{\Delta V_n}{\sqrt{3}}\right), \tag{7.11}$$

where $\Delta V_n \sim \sqrt{\Delta t_n}N(0, 1)$. Here, the approximation $w_n = w(t_n)$ is the continuous time stochastic process $w = \{w(t), t_0 \leq t < T\}$, the time step-size $\Delta t_n = t_n - t_{n-1}$ and $\Delta W_n = W(t_n) - W(t_{n-1})$.

7.4 Euler–Maruyama Method

An EMM is a continuous time stochastic process $\{Y(t), t_0 \leq t < T\}$ satisfying the iterative scheme [98,179] for Eq. (7.9)

$$Y_{m+1} = Y_n + a(t_n, Y_n)\Delta t_{n+1} + b(t_n, Y_n)\Delta W_{n+1}, n = 0, 1, 2, \ldots, N-1,$$

with initial value

$$Y_0 = X_0,$$

where $Y_n = Y(t_n), \Delta t_{n+1} = t_{n+1} - t_n$ and $\Delta W_{n=1} = W(t_{n+1}) - W(t_n)$. Here, each random number ΔW_n is computed as $\Delta W_n = z_n\sqrt{\Delta t_n}$, where z_n is chosen from standard normal distribution $N(0, 1)$. The equidistant discretised times $t_n = t_0 + n\delta$ with $\delta = \Delta_n = \frac{(T-t_0)}{N}$ have been considered for some integer N large enough so that $\delta \in (0, 1)$.

7.5 Split-Step Forward Euler–Maruyama Method

To discretise in time, the split-step forward Euler–Maruyama method has been applied. Let $\{X_t\}$ be an Itô process on $t \in [t_0, T]$ satisfying the following SDE [98,169]:

$$\begin{cases} dX_t = a(t, x_t)dt + b(t, X_t)dW_t, \\ X_{t_0} = X_0. \end{cases} \tag{7.12}$$

For a given time-discretisation

$$t_0 < t_1 < \cdots < t_n = T,$$

split-step forward EMM is a continuous time stochastic process $\{Y(t), t_0 \leq t \leq T\}$ satisfying the iterative scheme [182]

$$\bar{Y}_n = Y_n + \Delta t_{n+1}a(t_n, Y_n), \tag{7.13}$$

$$Y_{n+1} = \bar{Y}_n + b(t_n, \bar{Y}_n)\Delta W_{n+1}, \quad n = 0, 1, 2, \ldots, M-1, \tag{7.14}$$

with initial value

$$Y_0 = X_0,$$

where $Y_n = Y(t_n), \Delta t_{n+1} = t_{n+1} - t_n$ and $\Delta W_{n+1} = W(t_{n+1}) - W(t_n)$. Here, each random number ΔW_n is computed as $\Delta W_n = z_n\sqrt{\Delta t_n}$, where z_n is chosen from standard normal distribution $N(0,1)$.

The equidistant discretised times $t_n = t_0 = n\delta$ with $\delta = \frac{(T-t_0)}{M}$ have been considered for some integer M large enough so that $\delta \in (0,1)$.

7.6   DFMM

The following method approximates this derivative in accordance to the Runge–Kutta approach [183].

The derivative-free Milstein method for numerical solution of Eq. (7.12) is presented as follows:

$$\bar{Y}_n = Y_n + \Delta t_{n+1}a(t_n, Y_n) + \sqrt{\Delta t_{n+1}}b(t_n, Y_n), \qquad (7.15)$$

$$Y_{n+1} = Y_n + \Delta t_{n+1}a(t_n, Y_n) + a(t_n, Y_n)\Delta W_{n+1}$$

$$+ \frac{1}{2\sqrt{\Delta t_{n+1}}}[b(t_n, \bar{Y}_n) - b(t_n, Y_n)][(\Delta W_{n+1})^2$$

$$-\Delta t_{n+1}], \qquad (7.16)$$

for $n = 0, 1, 2, \ldots, N - 1$ with initial value $Y_0 = X_0$, where $Y_n = Y(t_n), \Delta t_{n+1} = t_{n+1} - t_n$ and $\Delta W_{n+1} = W(t_{n+1}) - W(t_n)$. Here, each random number ΔW_n is computed as $\Delta W_n = z_n\sqrt{\Delta t_n}$, where z_n is chosen from $N(0,1)$.

7.7   Numerical Solutions of SPKE in Presence of Newtonian Temperature Feedback Effects

In this section, the time domain $[0, T]$ has been assumed to be divided into very small time intervals, i.e., $t_k = kh$ and $k = 0, 1, 2, \ldots, M$. Therefore, in the interval $[t_k, t_{k+1}]$, the total reactivity $\rho(t)$ has been approximated accordingly to the definition of Riemann integral

$$\rho(t_k) \approx \rho_{ex}(t_k) - h\sigma \sum_{j=0}^{k} N(t_j). \tag{7.17}$$

7.7.1 *Implementation of 1.5 STM for SPKE in presence of Newtonian temperature feedback effect*

In this section, strong order 1.5 STM has been applied to Eq. (7.6) in time interval $[t_k, t_{k+1}]$ yielding

$$\boldsymbol{\Psi}(t_{k+1}) = \boldsymbol{\Psi}(t_k) + \boldsymbol{A}(t_k)\boldsymbol{\Psi}(t_k)h + \boldsymbol{B}^{\frac{1}{2}}(t_k)\Delta\boldsymbol{W}(t_k)$$
$$+\boldsymbol{A}(t_k)\boldsymbol{B}^{\frac{1}{2}}(t_k)\Delta\boldsymbol{Z}_k + \frac{1}{2}(\boldsymbol{A}(t_k)\boldsymbol{\Psi}(t_k))\boldsymbol{A}(t_k)h^2, \tag{7.18}$$

where $\Delta\boldsymbol{Z}_k = \frac{1}{2}h(\Delta\boldsymbol{W}(t_k) + \frac{\Delta\boldsymbol{V}(t_k)}{\sqrt{3}})$ and $\Delta\boldsymbol{V}(t_k) = \sqrt{h}N(0,1)$ with initial condition

$$\boldsymbol{\Psi}(0) = \begin{pmatrix} N_0 \\ \dfrac{\beta_1 N_0}{l\lambda_1} \\ \dfrac{\beta_2 N_0}{l\lambda_2} \\ \vdots \\ \dfrac{\beta_I N_0}{l\lambda_m} \end{pmatrix}. \tag{7.19}$$

7.7.2 *Euler–Maruyama method for stochastic point kinetic model in presence of Newtonian temperature feedback effect*

In this section, Euler–Maruyama method has been applied to Eq. (7.6) in time interval $[t_k, t_{k+1}]$ yielding

$$\boldsymbol{\Psi}(t_{k+1}) = \boldsymbol{\Psi}(t_k) + \boldsymbol{A}(t_k)\boldsymbol{\Psi}(t_k)h + \boldsymbol{B}^{\frac{1}{2}}(t_k)\Delta\boldsymbol{W}(t_k), \tag{7.20}$$

where $\Delta\boldsymbol{W}(t_k) = \sqrt{h}\boldsymbol{S}_k$ with initial condition

$$\boldsymbol{\Psi}(0) = \begin{pmatrix} N_0 \\ \dfrac{\beta_1 N_0}{l\lambda_1} \\ \dfrac{\beta_2 N_0}{l\lambda_2} \\ \vdots \\ \dfrac{\beta_I N_0}{l\lambda_m} \end{pmatrix}, \tag{7.21}$$

$$\boldsymbol{S}_k = \begin{pmatrix} S_0 \\ S_1 \\ S_2 \\ \vdots \\ S_m \end{pmatrix}, \tag{7.22}$$

where $S_0, S_1, S_2, \ldots, S_m$ are random variables chosen from $N(0,1)$ with mean equal to 0 and variance equal to 1.

7.7.3   *Split-step forward Euler–Maruyama method for stochastic point kinetic model in presence of Newtonian temperature feedback effect*

In this section, split-step forward Euler–Maruyama has been applied to Eq. (7.6) in time interval $[t_k, t_{k+1}]$

$$\bar{\boldsymbol{\Psi}}(t_k) = \boldsymbol{\Psi}(t_k) + \boldsymbol{A}(t_k)\boldsymbol{\Psi}(t_k)h, \tag{7.23}$$

$$\boldsymbol{\Psi}(t_{k+1}) = \bar{\boldsymbol{\Psi}}(t_k) + \bar{\boldsymbol{B}}^{\frac{1}{2}}(t_k)\Delta\boldsymbol{W}(t_k), k = 0, 1, 2, \ldots, M - 1, \tag{7.24}$$

where $\Delta\boldsymbol{W}(t_k) = \sqrt{h}\boldsymbol{S}_k$ with initial condition

$$\boldsymbol{\Psi}(0) = \begin{pmatrix} N_0 \\ \dfrac{\beta_1 N_0}{l\lambda_1} \\ \dfrac{\beta_2 N_0}{l\lambda_2} \\ \vdots \\ \dfrac{\beta_m N_0}{l\lambda_m} \end{pmatrix}, \tag{7.25}$$

$$\boldsymbol{S}_k = \begin{pmatrix} S_{0_k} \\ S_{1_k} \\ S_{2_k} \\ \vdots \\ S_{m_k} \end{pmatrix}, \tag{7.26}$$

where $S_{0_k}, S_{1_k}, S_{2_k}, \ldots, S_{m_k}$, $k = 0, 1, 2, \ldots, M-1$, are chosen from $N(0,1)$ with zero mean and unit variance.

$$\boldsymbol{\Psi}(t_k) = \begin{pmatrix} N(t_k) \\ C_1(t_k) \\ C_2(t_k) \\ \vdots \\ C_m(t_k) \end{pmatrix}, \tag{7.27}$$

$$\bar{\boldsymbol{\Psi}}(t_k) = \begin{pmatrix} \bar{N}(t_k) \\ \bar{C}_1(t_k) \\ \bar{C}_2(t_k) \\ \vdots \\ \bar{C}_m(t_k) \end{pmatrix}, \tag{7.28}$$

$$\bar{\boldsymbol{B}}(t_k) = \begin{pmatrix} \bar{\mu}_0(t_k) & -\bar{\mu}_1(t_k) & -\bar{\mu}_2(t_k) & \cdots & -\bar{\mu}_m(t_k) \\ -\bar{\mu}_1(t_k) & \bar{\mu}_1(t_k) & 0 & \cdots & 0 \\ -\bar{\mu}_2(t_k) & 0 & \bar{\mu}_2(t_k) & \cdots & 0 \\ \vdots & \vdots & \vdots & \ddots & \vdots \\ -\bar{\mu}_m(t_k) & 0 & 0 & \cdots & \bar{\mu}_m(t_k) \end{pmatrix}, \tag{7.29}$$

where $\bar{\mu}_0(t_k) = \left(\frac{\bar{\rho}(t_k)+\beta}{l} \right) \bar{N}(t_k) - \sum_{i=1}^m \lambda_i \bar{C}_i(t_k)$, $\bar{\mu}_i(t) = \frac{\beta_i}{l} \bar{N}(t) - \lambda_i \bar{C}_i(t)$, $i = 1, 2, 3, \ldots, m$.

7.7.4 *Derivative-free Milstein method for stochastic point kinetic model in presence of Newtonian temperature feedback effect*

In this section, derivative-free Milstein method has been applied to Eq. (7.6) in time interval $[t_k, t_{k+1}]$

$$\bar{\boldsymbol{\Psi}}(t_k) = \boldsymbol{\Psi}(t_k) + \boldsymbol{A}(t_k)\boldsymbol{\Psi}(t_k)h + \boldsymbol{B}^{\frac{1}{2}}(t_k)\sqrt{h},$$

$$k = 0, 1, 2, \ldots, M-1, \tag{7.30}$$

$$\boldsymbol{\Psi}(t_{k+1}) = \boldsymbol{\Psi}(t_k) + \boldsymbol{A}(t_k)\boldsymbol{\Psi}(t_k)h + \boldsymbol{B}^{\frac{1}{2}}(t_k)\Delta\boldsymbol{W}(t_k)$$

$$+\frac{1}{2\sqrt{h}}[\bar{\boldsymbol{B}}^{\frac{1}{2}}(t_k) - \boldsymbol{B}^{\frac{1}{2}}(t_k)][(\Delta\boldsymbol{W}(t_k))^2 - h],$$

$$k = 0, 1, 2, \ldots, M - 1, \tag{7.31}$$

where $\Delta\boldsymbol{W}(t_k) = \sqrt{h}\boldsymbol{S}_k$ with initial condition

$$\boldsymbol{\Psi}(0) = \begin{pmatrix} N_0 \\ \dfrac{\beta_1 N_0}{l\lambda_1} \\ \dfrac{\beta_2 N_0}{l\lambda_2} \\ \vdots \\ \dfrac{\beta_I N_0}{l\lambda_m} \end{pmatrix}, \tag{7.32}$$

$$\boldsymbol{S}_k = \begin{pmatrix} S_{0_k} \\ S_{1_k} \\ S_{2_k} \\ \vdots \\ S_{m_k} \end{pmatrix}, \tag{7.33}$$

where $S_{0_k}, S_{1_k}, S_{2_k}, \ldots, S_{m_k}$, $k = 0, 1, 2, \ldots, M - 1$, are chosen from $N(0, 1)$ with zero mean and unit variance.

7.8 Numerical Solutions of SPKE in Presence of Newtonian Temperature Feedback Effects Using Euler–Maruyama and Strong Order 1.5 Taylor Methods

7.8.1 *Step external reactivity*

The numerical solutions have been obtained in this section for the stochastic point kinetic model of U^{235} nuclear reactor [161] with the following parameters: $\lambda_1 = 0.0124(s^{-1}), \lambda_2 = 0.0305(s^{-1}),$ $\lambda_3 = 0.111(s^{-1}), \lambda_4 = 0.301(s^{-1}), \lambda_5 = 1.13(s^{-1}), \lambda_6 = 3.0(s^{-1}),$ $\beta_1 = 0.00021, \beta_2 = 0.00141, \beta_3 = 0.00127, \beta_4 = 0.00255,$

$\beta_5 = 0.00074, \beta_6 = 0.00027, \beta = 0.00645, l = 5.0 \times 10^{-5}(s)$, the temperature coefficient of reactivity is $\alpha = 5.0 \times 10^{-5}(K^{-1})$ and the reciprocal of thermal capacity of the reactor is $k_c = 0.05(K/MW_s)$. The initial conditions are $N(0) = 1$ (neutron) and $C_i(0) = \frac{\beta_i N(0)}{l\lambda_i}$.

For different step external reactivity, i.e., $\rho_{ex} = 0.5\beta, \rho_{ex} = 0.75\beta$ and $\rho_{ex} = \beta$, the peaks of the mean neutron population with regard to their time are presented in Table 7.1 using 500 trials for $\rho_{ex} = 0.5\beta, \rho_{ex} = 0.75\beta$ and $\rho_{ex} = \beta$. The peak of the mean neutron population obtained by using above discussed numerical methods for SPKE has been tabulated to establish the efficiency of both the schemes.

In Figs. 7.1–7.3, the mean neutron population and the sample paths for two individual neutron populations have been plotted for step external reactivity.

The dynamical behaviour of mean neutron population along with two sample paths for step reactivity are shown in Fig. 7.1 with $\rho_{ex} = 0.5\beta$.

Table 7.1. Peak of the mean neutron population with regard to its time for stochastic nonlinear model at step external reactivity.

	EMM		Taylor 1.5 strong order	
ρ_{ex}	Peak	Time (s)	Peak	Time (s)
0.5β	46.2303	28.42	55.6596	22.41
0.75β	164.129	8.98	185.44	7.1725
β	766.174	1.0725	1192.48	0.725

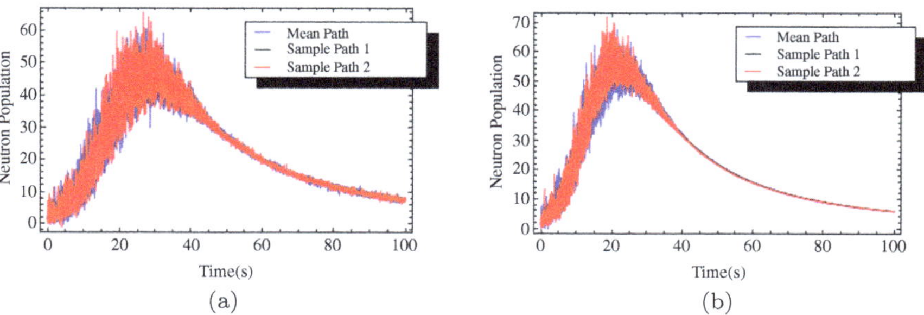

Fig. 7.1. (a) $\rho_{ex} = 0.5\beta$ (EMM). (b) $\rho_{ex} = 0.5\beta$ (Taylor 1.5 strong order).

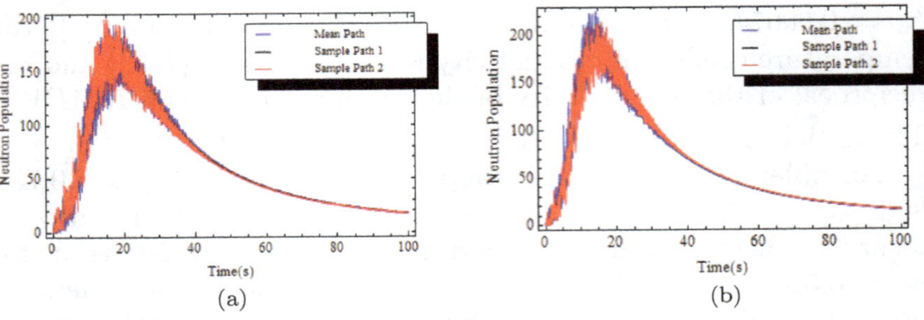

Fig. 7.2.   (a) $\rho_{ex} = 0.75\beta$ (EMM). (b) $\rho_{ex} = 0.75\beta$ (Taylor 1.5 strong order).

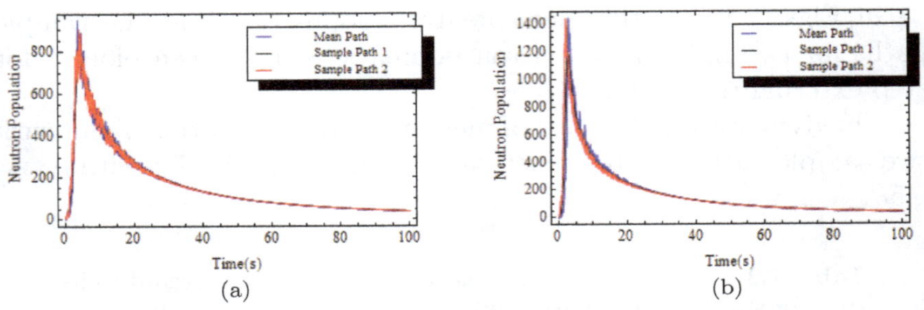

Fig. 7.3.   (a) $\rho_{ex} = \beta$ (EMM). (b) $\rho_{ex} = \beta$ (Taylor 1.5 strong order).

The dynamical behaviour of mean neutron population along with two sample paths for step reactivity are shown in Fig. 7.2 with $\rho_{ex} = 0.75\beta$.

The dynamical behaviour of mean neutron population along with two sample paths for step reactivity is shown in Fig. 7.3 with $\rho_{ex} = \beta$.

7.8.2   *Ramp external reactivity*

The numerical solutions have been obtained for the nonlinear SPKE of U^{235} nuclear reactor with the same parametric values as in Section 7.8.1. Here, the external reactivity is a function of time such as $\rho_{ex} = 0.1t$ and $\rho_{ex} = 0.01t$, the nonlinear coefficient σ takes 10^{-11} or 10^{-13}. The initial conditions are $N(0) = 1$ (neutron) and $C_i(0) = \frac{\beta_i N(0)}{l\lambda_i}$.

For different ramp external reactivities, i.e., $\rho_{ex} = 0.1t$ and $\rho_{ex} = 0.01t$, the peak of the mean neutron population with regard to its time is presented in Table 7.2 using 500 trials. The peak of the

Table 7.2. Peak of the mean neutron population with regard to its time for stochastic nonlinear model at ramp external reactivity $\rho(t) = at - \sigma \int_0^t N(\tau)d\tau$.

a	σ	EMM		Taylor 1.5 strong order		Analytical exponential technique [174]		DNP $(B=0)$ [174]	
		Peak	Time (s)	Peak	Time (s)	Peak	Time (s)	Peak	Time (s)
0.01	10^{-11}	$1.72654E+10$	1.118	$1.74004E+10$	1.101	$1.673436E+10$	0.854	$1.904837E+10$	0.846
	10^{-13}	$2.09001E+12$	1.169	$2.13183E+12$	1.151	$2.082531E+12$	0.877	$2.389026E+12$	0.868
0.1	10^{-11}	$1.82682E+11$	0.235	$1.86337E+11$	0.227	$1.790577E+11$	0.142	$2.315740E+11$	0.134
	10^{-13}	$2.25166E+13$	0.251	$2.31433E+13$	0.241	$2.143778E+13$	0.150	$2.839821E+13$	0.141

mean neutron population obtained by using above discussed numerical methods for SPKE has been tabulated to establish the efficiency of both the schemes.

In Figs. 7.4–7.7, the mean neutron population and the sample paths for two individual neutron populations have been plotted for ramp external reactivity.

The dynamical behaviour of mean neutron population along with two sample paths for ramp reactivity are shown in Fig. 7.4 with $\rho_{ex} = 0.01t, \sigma = 10^{-11}$.

The dynamical behaviour of mean neutron population along with two sample paths for ramp reactivity are shown in Fig. 7.5 with $\rho_{ex} = 0.01t, \sigma = 10^{13}$.

The dynamical behaviour of mean neutron population along with two sample paths for ramp reactivity are shown in Fig. 7.6 with $\rho_{ex} = 0.1t, \sigma = 10^{-11}$.

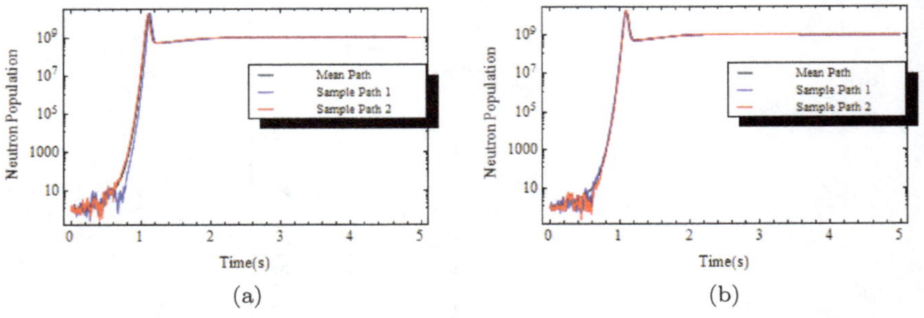

(a)   (b)

Fig. 7.4. (a) $\rho_{ex} = 0.01t, \sigma = 10^{-11}$ (EMM). (b) $\rho_{ex} = 0.01t, \sigma = 10^{-11}$ (Taylor 1.5 strong order).

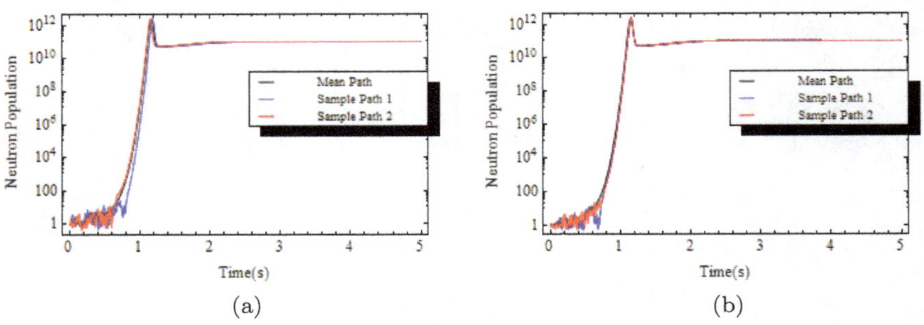

(a)   (b)

Fig. 7.5. (a) $\rho_{ex} = 0.01t, \sigma = 10^{-13}$ (EMM). (b) $\rho_{ex} = 0.01t, \sigma = 10^{-13}$ (Taylor 1.5 strong order).

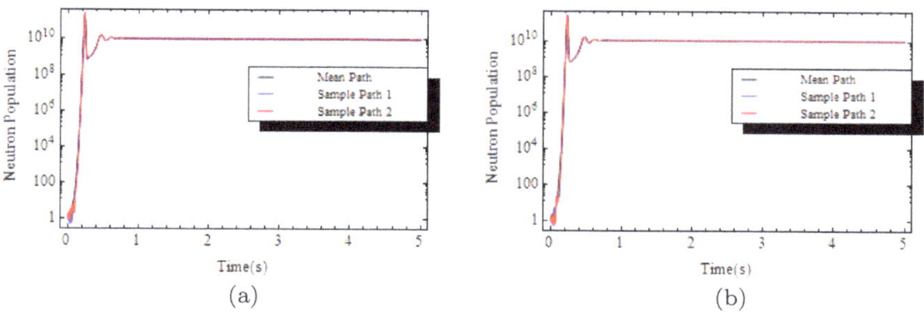

(a)   (b)

Fig. 7.6. (a) $\rho_{ex} = 0.1t, \sigma = 10^{-11}$ (EMM). (b) $\rho_{ex} = 0.1t, \sigma = 10^{-11}$ (Taylor 1.5 strong order).

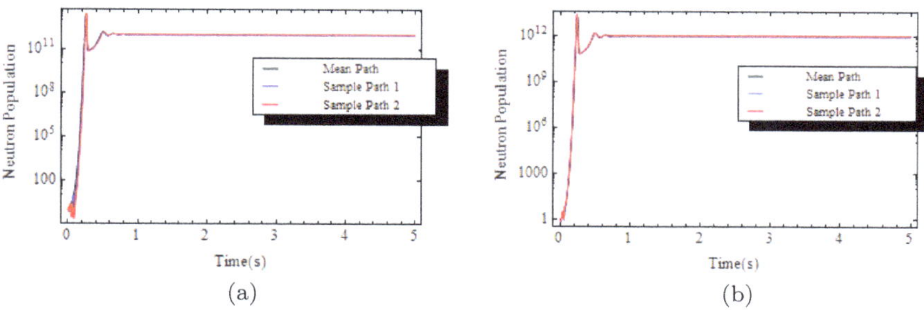

(a)   (b)

Fig. 7.7. (a) $\rho_{ex} = 0.1t, \sigma = 10^{-13}$ (EMM). (b) $\rho_{ex} = 0.1t, \sigma = 10^{-13}$ (Taylor 1.5 strong order).

The dynamical behaviour of mean neutron population along with two sample paths for ramp reactivity are shown in Fig. 7.7 with $\rho_{ex} = 0.1t, \sigma = 10^{-13}$.

7.9 Numerical Solutions of SPKE in Presence of Newtonian Temperature Feedback Effects Using Split-Step Forward Euler–Maruyama and Derivative-Free Milstein Methods

Split-step Euler–Maruyama and derivative-free Milstein methods have been applied to obtain solutions of SNPKE in the presence of Newtonian temperature feedback and i-group of delayed neutrons (here $i = 6$) for a U^{235} nuclear reactor, in case for step and ramp external reactivities.

7.9.1 *Step external reactivity*

In this section, the split-step numerical scheme is applied to solve the nonlinear stochastic point kinetic model of U^{235} nuclear reactor [161,174] with the following parameters: $\lambda_1 = 0.0124(s^{-1})$, $\lambda_2 = 0.0305(s^{-1})$, $\lambda_3 = 0.111(s^{-1})$, $\lambda_4 = 0.301(s^{-1})$, $\lambda_5 = 1.13(s^{-1})$, $\lambda_6 = 3.0(s^{-1})$, $\beta_1 = 0.00021$, $\beta_2 = 0.00141$, $\beta_3 = 0.00127$, $\beta_4 = 0.00255$, $\beta_5 = 0.00074$, $\beta_6 = 0.00027$, $\beta = 0.00645$, $l = 5.0 \times 10^{-5}(s)$, the temperature coefficient of reactivity is $\alpha = 5.0 \times 10^{-5}(K^{-1})$ and the reciprocal of the thermal capacity of the reactor is $K_c = 0.05(K/MW_s)$. The initial conditions are $N(0) = 1$ (neutron/cm^3) and $C_i(0) = \frac{\beta_i N(0)}{l\lambda_i}$.

The peaks of the mean neutron population obtained using 500 trials for three cases of step external reactivity $\rho_{ex} = 0.5\beta$, $\rho_{ex} = 0.75\beta$ and $\rho_{ex} = \beta$ are cited in Table 7.3. The peaks of the mean neutron population for the nonlinear SPKE by split-step forward EMM and DFMM have been compared to establish the efficiency of both the schemes.

In Figs. 7.8–7.10, the mean neutron population and the sample paths for two individual neutron populations have been plotted for $\rho_{ex} = 0.5\beta$, $\rho_{ex} = 0.75\beta$ and $\rho_{ex} = \beta$, respectively.

The dynamical behaviour of mean neutron population along with two sample paths for step reactivity are shown in Fig. 7.8 with $\rho_{ex} = 0.5\beta$.

The dynamical behaviour of mean neutron population along with two sample paths for step reactivity are shown in Fig. 7.9 with $\rho_{ex} = 0.75\beta$.

Table 7.3. Peak of the mean neutron population with regard to its time for stochastic nonlinear model at step external reactivity.

ρ_{ex}	Split-step forward EMM		Derivative-free Milstein method	
	Peak	Time (s)	Peak	Time (s)
0.5β	46.4939	28.34	46.2606	27.84
0.75β	163.707	8.795	164.22	8.95
β	760.589	1.065	769.238	1.0575

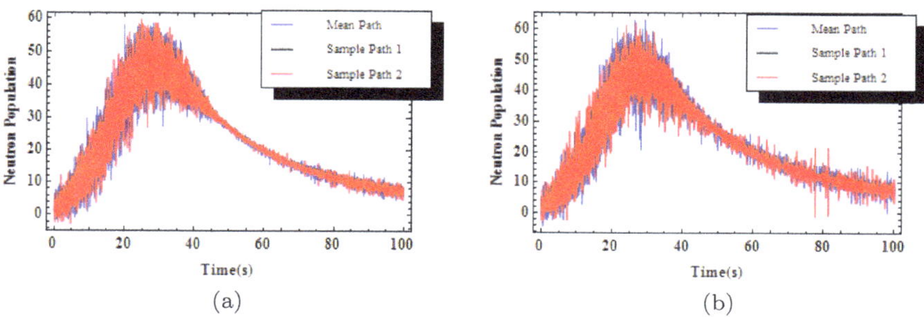

Fig. 7.8. (a) $\rho_{ex} = 0.5t$, (split-step EMM). (b) $\rho_{ex} = 0.5\beta$ (derivative-free Milstein).

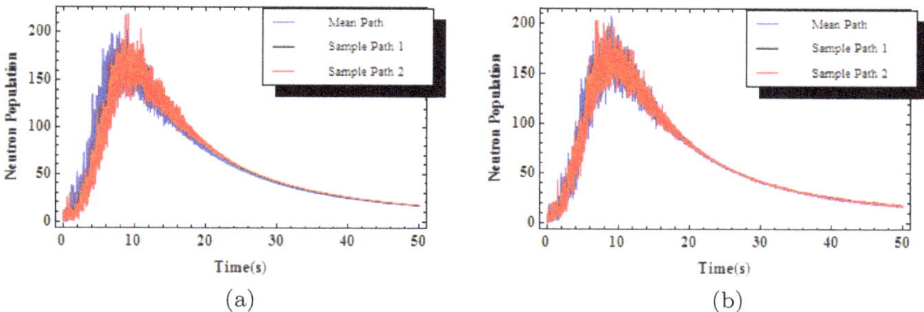

Fig. 7.9. (a) $\rho_{ex} = 0.75\beta$ (split-step EMM). (b) $\rho_{ex} = 0.75\beta$ (derivative-free Milstein).

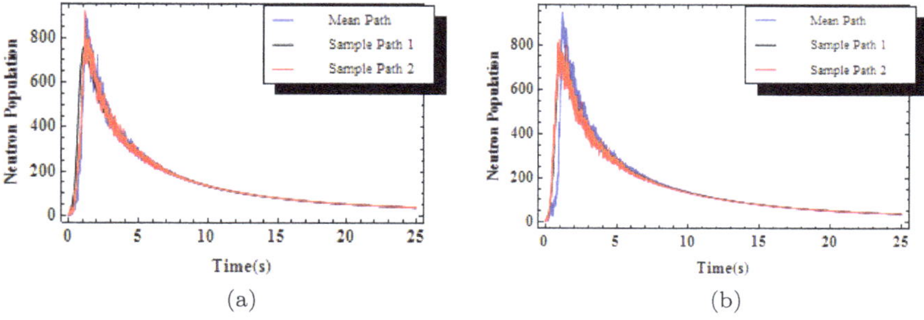

Fig. 7.10. (a) $\rho_{ex} = \beta$ (split-step EMM). (b) $\rho_{ex} = \beta$ (derivative-free Milstein).

The dynamical behaviour of mean neutron population along with two sample paths for step reactivity are shown in Fig. 7.10 with $\rho_{ex} = \beta$.

7.9.2   *Ramp external reactivity*

In this section, the split-step numerical scheme is applied to SPKE of U^{235} nuclear reactor with the same parametric values as step external reactivity in Section 7.9.1. Here, the external reactivity is a function of time such as $\rho_{ex} = 0.1t$ and $\rho_{ex} = 0.01t$; the nonlinear coefficient σ takes 10^{-11} or 10^{-13}. The initial conditions are $N(0) = 1$ (neutron/cm^3) and $C_i(0) = \frac{\beta_i N(0)}{l\lambda_i}$.

The peak of the mean neutron population obtained for time interval size $h = 0.001(s)$ and 500 trials for two cases of ramp external reactivity, i.e., $\rho_{ex} = 0.1t$, $\rho_{ex} = 0.01t$, $\rho_{ex} = 0.001t$ and $\rho_{ex} = 0.003t$ are presented in Table 7.4. The peaks of the mean neutron population for the nonlinear SPKE by split-step forward EMM and DFMM have been compared to establish the efficiency of both the schemes.

In Figs. 7.11–7.14, the mean neutron population and the sample paths for two individual neutron populations have been plotted for $\rho_{ex} = 0.1t$ and $\rho_{ex} = 0.01t$, respectively.

Table 7.4.   Peak of the mean neutron population with regard to its time for stochastic nonlinear model at ramp external reactivity $\rho(t) = at - \sigma \int_0^t N(\tau)d\tau$.

		Split-step forward EMM		Derivative-free Milstein method		Analytical exponential technique [174]	
a	σ	Peak	Time (s)	Peak	Time (s)	Peak	Time (s)
0.01	10^{-11}	$1.68604E+10$	1.118	$1.69492E+10$	1.118	$1.673436E+10$	0.854
	10^{-13}	$2.12034E+12$	1.169	$2.12802E+12$	1.168	$2.082531E+12$	0.877
0.1	10^{-11}	$1.88849E+11$	0.235	$1.89642E+11$	0.235	$1.790577E+11$	0.142
	10^{-13}	$2.24448E+13$	0.251	$2.26026E+13$	0.251	$2.143778E+13$	0.150

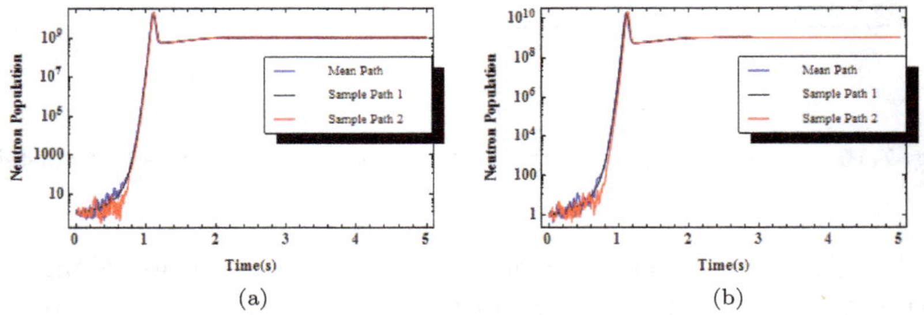

Fig. 7.11.   (a) $\rho_{ex} = 0.01t$, $\sigma = 10^{-11}$ (split-step EMM). (b) $\rho_{ex} = 0.01t$, $\sigma = 10^{-11}$ (derivative-free Milstein).

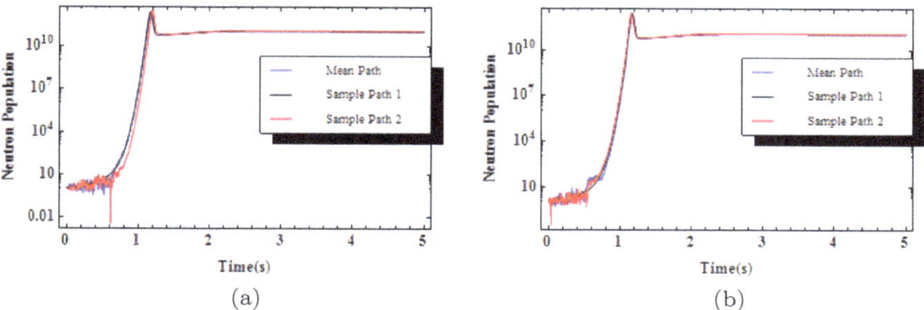

Fig. 7.12. (a) $\rho_{ex} = 0.01t$, $\sigma = 10^{-13}$ (split-step EMM). (b) $\rho_{ex} = 0.01t$, $\sigma = 10^{-13}$ (derivative-free Milstein).

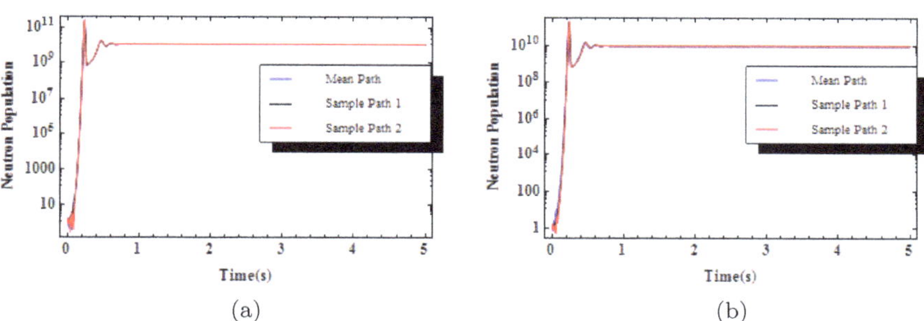

Fig. 7.13. (a) $\rho_{ex} = 0.1t$, $\sigma = 10^{-11}$ (split-step EMM). (b) $\rho_{ex} = 0.1t$, $\sigma = 10^{-11}$ (derivative-free Milstein).

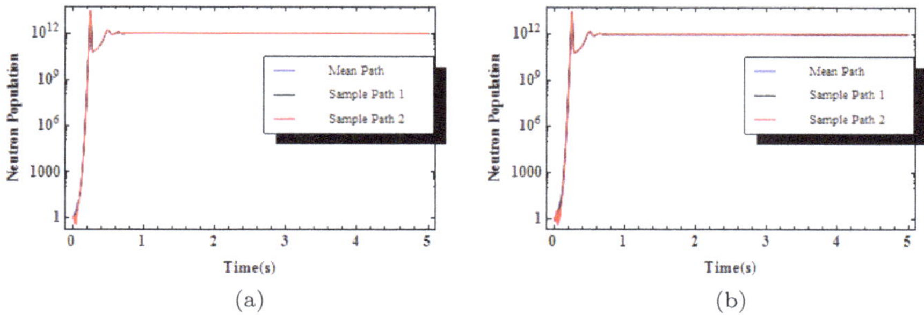

Fig. 7.14. (a) $\rho_{ex} = 0.1t$, $\sigma = 10^{-13}$ (split-step EMM). (b) $\rho_{ex} = 0.1t$, $\sigma = 10^{-13}$ (derivative-free Milstein).

The dynamical behaviour of mean neutron population along with two sample paths for ramp reactivity are shown in Fig. 7.11 with $\rho_{ex} = 0.01t$, $\sigma = 10^{-11}$.

Table 7.5. Peak of the mean neutron population with regard to its time for stochastic nonlinear model at ramp external reactivity $\rho(t) = at - \sigma \int_0^t N(\tau)d\tau$ for different time interval sizes, i.e., $h = 0.001(s)$ and $h = 0.0005(s)$, for split-step forward EMM.

a	σ	Split-step forward EMM ($h = 0.001(s)$)		Split-step forward EMM ($h = 0.0005(s)$)	
		Peak	Time (s)	Peak	Time (s)
0.001	10^{-11}	14.1361	4.991	14.073	5
	10^{-13}	14.0805	5.001	16.285	4.995
0.003	10^{-11}	$4.719E+9$	2.923	$4.74397E+9$	2.9175
	10^{-13}	$6.0092E+11$	3.020	$6.05314E+11$	3.15

Table 7.6. Peak of the mean neutron population with regard to its time for stochastic nonlinear model at ramp external reactivity $\rho(t) = at - \sigma \int_0^t N(\tau)d\tau$ for different time interval sizes, i.e., $h = 0.001(s)$ and $h = 0.0005(s)$, for derivative-free Milstein method.

a	σ	Derivative-free Milstein method ($h = 0.001(s)$)		Derivative-free Milstein method ($h = 0.0005(s)$)	
		Peak	Time (s)	Peak	Time (s)
0.001	10^{-11}	14.399	5.001	14.5955	5.001
	10^{-13}	14.0373	5.001	14.2704	5
0.003	10^{-11}	$4.74073E+9$	2.921	$4.75339E+9$	2.9190
	10^{-13}	$6.0057E+11$	0.021	$6.14374E+11$	3.0170

The dynamical behaviour of mean neutron population along with two sample paths for ramp reactivity are shown in Fig. 7.12 with $\rho_{ex} = 0.01t$, $\sigma = 10^{-13}$.

The dynamical behaviour of mean neutron population along with two sample paths for ramp reactivity are shown in Fig. 7.13 with $\rho_{ex} = 0.1t$, $\sigma = 10^{-11}$.

The dynamical behaviour of mean neutron population along with two sample paths for ramp reactivity are shown in Fig. 7.14 with $\rho_{ex} = 0.1t$, $\sigma = 10^{-13}$.

The comparison between the peaks of the mean neutron population obtained for different time interval sizes $h = 0.001(s)$ and $h = 0.0005(s)$ for 500 trials for two cases of ramp external reactivities, i.e., $\rho_{ex} = 0.001t$ and $\rho_{ex} = 0.003t$, is presented in Tables 7.5 and 7.6.

In Figs. 7.15–7.18, the mean neutron population and the sample paths for two individual neutron populations have been plotted for $\rho_{ex} = 0.01t$ and $\rho_{ex} = 0.1t$ respectively over the interval $[0, 1]$.

The dynamical behaviour of mean neutron population along with two sample paths for ramp reactivity are shown in Fig. 7.15 with $\rho_{ex} = 0.01t$, $\sigma = 10^{-11}$ over the interval $[0, 1]$.

The dynamical behaviour of mean neutron population along with two sample paths for ramp reactivity are shown in Fig. 7.16 with $\rho_{ex} = 0.01t$, $\sigma = 10^{-13}$ over the interval $[0, 1]$.

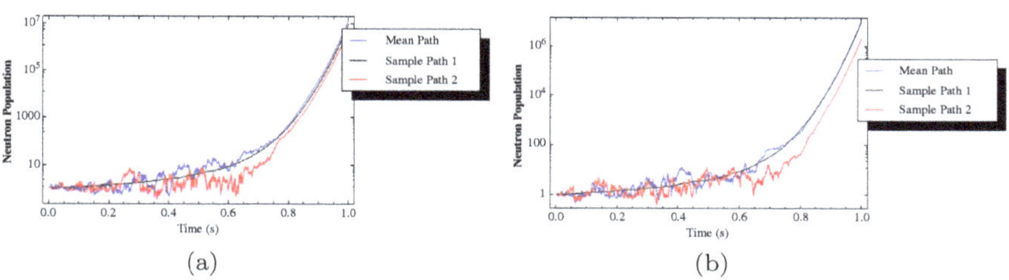

Fig. 7.15. (a) $\rho_{ex} = 0.01t$, $\sigma = 10^{-11}$ (split-step EMM). (b) $\rho_{ex} = 0.01t$, $\sigma = 10^{-11}$ (derivative-free Milstein).

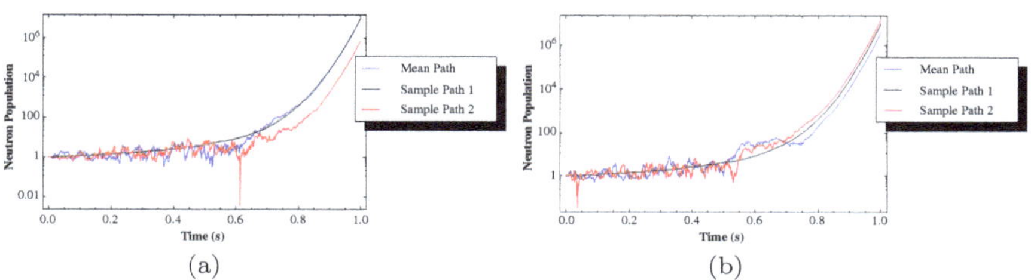

Fig. 7.16. (a) $\rho_{ex} = 0.01t$, $\sigma = 10^{-13}$ (split-step EMM). (b) $\rho_{ex} = 0.01t$, $\sigma = 10^{-13}$ (derivative-free Milstein).

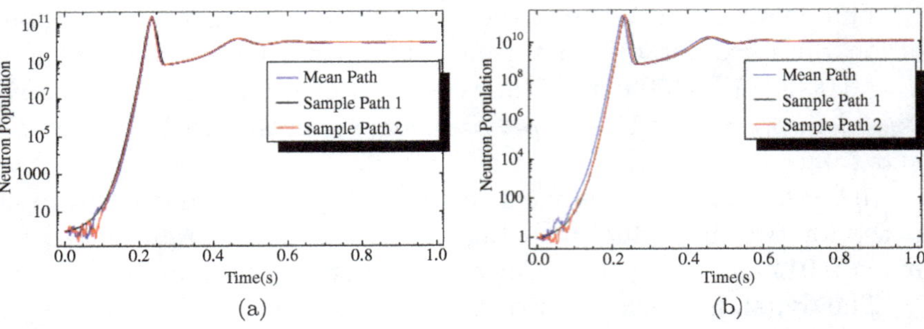

(a)   (b)

Fig. 7.17. (a) $\rho_{ex} = 0.1t$, $\sigma = 10^{-11}$ (split-step EMM). (b) $\rho_{ex} = 0.1t$, $\sigma = 10^{-11}$ (derivative-free Milstein).

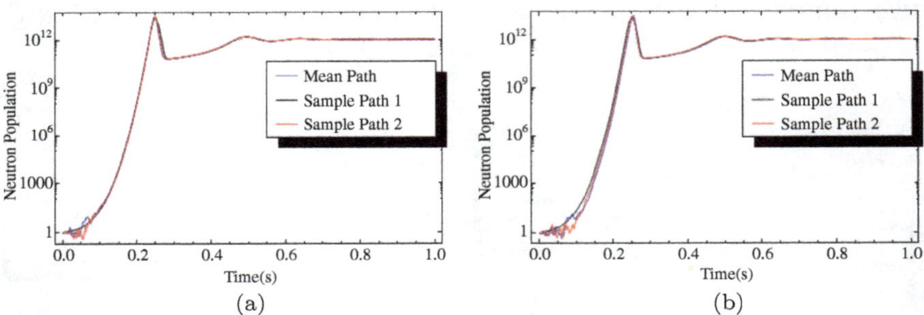

(a)   (b)

Fig. 7.18. (a) $\rho_{ex} = 0.1t$, $\sigma = 10^{-13}$ (split-step EMM). (b) $\rho_{ex} = 0.1t$, $\sigma = 10^{-13}$ (derivative-free Milstein).

The dynamical behaviour of mean neutron population along with two sample paths for ramp reactivity are shown in Fig. 7.17 with $\rho_{ex} = 0.1t$, $\sigma = 10^{-11}$ over the interval $[0, 1]$.

The dynamical behaviour of mean neutron population along with two sample paths for ramp reactivity are shown in Fig. 7.18 with $\rho_{ex} = 0.1t$, $\sigma = 10^{-13}$ over the interval $[0, 1]$.

7.10 Summary

In this chapter, the SPKEs have been solved by Euler–Murayama, 1.5 strong Taylor, split-step forward Euler–Murayama and derivative-free Milstein methods. The numerical results have been presented in the tables and are also graphically demonstrated in order to justify the accuracy of the proposed schemes, as the split-step schemes offer efficient results with low computational cost as compared to the original Milstein and forward EMM methods.

In case of step external reactivity, the mean neutron population increases until it achieves its maximum value, then it starts decreasing with time due to Newtonian temperature feedback effects. Also, the fluctuations of the sample paths decrease when the mean neutron population decreases after achieving its maximum value. For ramp external reactivity, the mean neutron population increases until it achieves its peak value due to the increasing external reactivity ρ_{ex} and then it tries to reach the equilibrium state. Also, at low power levels such as at startup, random fluctuations in the neutron density and neutron precursor concentrations can be significant, but it becomes almost stable with the increase in time.

Chapter 8

Numerical Solutions of Fractional Stochastic Point Kinetics Equation

8.1 Introduction

Stochastic point kinetics equations (SPKEs) have been a very important model in nuclear engineering. The SPKEs are a system of coupled nonlinear SDEs. The SPKEs model a system of Itô SDE, specifically, neutron population and delayed neutron precursors. The physical dynamical system has been established to be a population process, and techniques have been employed in Hayes and Allen [167] to transform the deterministic PKEs into a system of SDE. The fractional diffusion model is normally applied for large variations of neutron cross-sections which preclude the use of the classical neutron diffusion equations [183,185–187]. Various methods have been developed for improvement in controlling the processes taking place in the nuclear reactor.

In recent developments dynamic, multiphysics phenomena face a lot of challenges regarding accurate numerical schemes, which results in severe computational requirements. One approach to reducing the severe computational requirements of standard low-order simulations is to employ higher-order formulations. In the hierarchy of high-order methods, compact schemes represent an attractive choice for reducing dispersion and anisotropy errors.

Nowak *et al.* [188] presented results concerning numerical solutions to a fractional neutron point kinetics model for a nuclear reactor. Numerical solutions of stochastic point kinetics equations by implementing stochastic principal component analysis (PCA) have been obtained by Hayes and Allen [167] which provided a very succinct idea about the randomness of neutron density and precursor concentrations. Saha Ray [169] showed that the Euler–Maruyama method and strong order 1.5 Taylor numerical schemes are well-founded estimators in comparison to stochastic PCA. Saha Ray and Patra [189] applied the Grunwald–Letnikov definition of fractional derivative for solving SPKEs. Nahla and Edress [170] showed the efficiency of the analytical exponential method (AEM) for obtaining solutions of SPKEs.

8.2 Outline of Present Study

In this chapter, the SPKEs are a system of Itô SDEs, and the solution for this system has been obtained by higher-order approximations for ramp and step external reactivities. In this chapter, a fractional model for the fractional SPKEs has been studied and analyzed. Implementation of higher-order approximation scheme has been done for the fractional stochastic nonlinear point reactor kinetics equations and fractional stochastic nonlinear PKEs with temperature feedback reactivity. The higher-order approximation scheme is efficient and the efficiency of the proposed has been discussed in the results section. The sample mean neutron population for each reactivity has been plotted with two sample paths for each reactivity to show the mean population behaviour. A brief summary has been provided at the end to conclude this chapter.

8.2.1 *Fractional stochastic nonlinear point reactor kinetics equations*

The fractional Itô SDE for the NPKEs with temperature feedback effects [171,175–178] obtained from centre limit theorem can be written as [190]

$$
{}_{0}^{C}D_{t}^{\alpha}\boldsymbol{\Psi}(t) = \boldsymbol{A}(t)\boldsymbol{\Psi}(t) + \boldsymbol{Q} + \boldsymbol{B}^{\frac{1}{2}}(t)\frac{d\boldsymbol{W}(t)}{dt}, \tag{8.1}
$$

where α is the fractional order of the derivative and $0 < \alpha \leq 1$,

$$\Psi(t) = \begin{pmatrix} N(t) \\ C_1(t) \\ C_2(t) \\ \vdots \\ C_m(t) \end{pmatrix}, \quad W(t) \begin{pmatrix} W_0(t) \\ W_1(t) \\ W_2(t) \\ \vdots \\ W_m(t) \end{pmatrix} \quad \text{and} \quad Q = \begin{pmatrix} q \\ 0 \\ 0 \\ \vdots \\ 0 \end{pmatrix},$$

m is the total number of delayed neutrons groups and $W_0(t)$, $W_1(t)$, $W_2(t), \ldots, W_m(t)$ are standard Wiener processes as defined in Ref. [167] with $N(t)$ being the neutron population and $C_i(t)$ being the precursor concentration of i-group of delayed neutrons.

The coefficient matrix is represented as $A(t)$ and is of the following

$$\text{form: } = \begin{pmatrix} W_0(t) \\ W_1(t) \\ W_2(t) \\ \vdots \\ W_m(t) \end{pmatrix}$$

$$A(t) = \begin{pmatrix} \frac{\rho-\beta}{l} & \lambda_1 & \lambda_2 & \cdots & \lambda_m \\ \frac{\beta_1}{l} & -\lambda_1 & 0 & \cdots & 0 \\ \frac{\beta_2}{l} & 0 & -\lambda_2 & \cdots & 0 \\ \vdots & \vdots & \vdots & \ddots & \vdots \\ \frac{\beta_m}{l} & 0 & 0 & \cdots & -\lambda_m \end{pmatrix}, \tag{8.2}$$

where ρ is the total reactivity, $\beta = \sum_{i=1}^{m} \beta_i$ is the total fraction of delayed neutrons, β_i is the fraction, λ_i is the decay constant of i-group of delayed neutrons and l is the prompt neutron generation time.

The covariance matrix $B(t)$ is evaluated in Ref. [190] as

$$B(t) = \begin{pmatrix} \mu_0(t) & -\mu_1(t) & -\mu_2(t) & \cdots & -\mu_m(t) \\ -\mu_1(t) & \mu_1(t) & 0 & \cdots & 0 \\ -\mu_2(t) & 0 & \mu_2(t) & \cdots & 0 \\ \vdots & \vdots & \vdots & \ddots & \vdots \\ -\mu_m(t) & 0 & 0 & \cdots & \mu_m(t) \end{pmatrix}, \tag{8.3}$$

where $\mu_0(t) = \left(\frac{\rho+\beta}{l}\right) N(t) - \sum_{i=1}^{m} \lambda_i C_i(t)$, $\mu_i(t) = \frac{\beta_i}{l} N(t) - \lambda_i C_i(t)$, $i = 1, 2, 3, \ldots, m$.

8.3　Implementation of Higher-Order Approximation Method for Fractional Stochastic Point Kinetic Model

8.3.1　*The higher-order approximation scheme*

Numerical algorithms with Caputo derivative have recently received attention and attracted increasing interests. The Caputo derivative operator for $\alpha \in (0,1)$ is defined as

$$
{}_{0}^{C}D_{t}^{\alpha}f(t) = \frac{1}{\Gamma(1-\alpha)} \int_{0}^{t} (t-s)^{-\alpha} f'(s)ds,
$$

in which $\Gamma(\cdot)$ is the Euler gamma function.

In this section, an $(3-\alpha)^{\text{th}}$ order scheme for Caputo ${}_{0}^{C}D_{t}^{\alpha}\Psi(t)$ with $\alpha \in (0,1)$ has been discussed for obtaining the solutions for the fractional stochastic nonlinear point reactor kinetics equations (SNPKEs). Let $0 = t_0 < t_1 < \cdots < t_M = T$, and the equidistant discretized times $t_n = t_0 + n\tau$ with $\tau = \Delta_n = \frac{(T-t_0)}{M}$ has been considered for some integer M large enough so that $\tau \in (0,1)$. Now, using the Taylor expansion to $\Psi'(s)$, $\Psi(t_{i-1})$, $\Psi(t_{i+1})$ at the point $t = t_i$ $(0 \le i < n)$, one gets

$$
\Psi'(s) = \Psi'(t_i) + \Psi''(t_i)(s - t_i) + \frac{\Psi'''(t_i)}{2!}(s - t_i)^2 + O((s - t_i)^3),
$$

$$
s \in (t_i, t_{i+1}),
$$

$$
\Psi'(t_i) = \frac{\Psi(t_{i+1}) - \Psi(t_{i-1})}{2\tau} - \frac{\Psi'''(t_i)}{3!}\tau^2 + O(\tau^2), \text{ and}
$$

$$
\Psi''(t_i) = \frac{\Psi(t_{i+1}) - 2\Psi(t_i) + \Psi(t_{i-1})}{\tau^2} - \frac{\Psi^{(4)}(t_i)}{12}\tau^2 + O(\tau^4).
$$

Hence, the following has been obtained:

$$
\Psi'(s) = \frac{\Psi(t_{i+1}) - \Psi(t_{i-1})}{2\tau} + \frac{\Psi(t_{i+1}) - 2\Psi(t_i) + \Psi(t_{i-1})}{\tau^2}(s - t_i)
$$

$$
- \frac{\Psi'''(t_i)}{3!}\tau^2 + \frac{\Psi'''(t_i)}{2!}(s - t_i)^2 + O((s - t_i)^3),
$$

$$
0 < s - t_i < \tau.
$$

Therefore, the Caputo derivative can be discretized as

$$
{}_{0}^{C}D_{i}^{\alpha}\Psi(t)|_{t=t_n} = \frac{1}{\Gamma(1-\alpha)} \int_{0}^{t_n} (t_n - s)^{-\alpha}\Psi'(s)ds
$$

$$
= \frac{1}{\Gamma(1-\alpha)} \sum_{i=0}^{n-1} \int_{t_i}^{t_{i+1}} (t_n - s)^{-\alpha}\Psi'(s)ds
$$

$$
= \frac{1}{\Gamma(1-\alpha)} \sum_{i=0}^{n-1} \int_{t_i}^{t_{i+1}} (t_n - s)^{-\alpha}
$$

$$
\times \left[\frac{\Psi(t_{i+1}) - \Psi(t_{i-1})}{2\tau} + \frac{\Psi(t_{i+1}) - 2\Psi(t_i) + \Psi(t_{i-1})}{\tau^2} \right.
$$

$$
\left. \times(s - t_i) - \frac{\Psi'''(t_i)}{3!} + \frac{\Psi'''(t_i)}{2!}(s - t_i)^2 \right]ds + O(\tau^3)
$$

$$
= \frac{\tau^{-\alpha}}{\Gamma(3-\alpha)} \sum_{i=0}^{n-1} [w_{1,n-1}(\Psi_{i+1} - \Psi_{i-1})
$$

$$
+ w_{2,n-i}(\Psi_{i+1} - 2\Psi_i + \Psi_{i-1})] + r^n,
$$

where $i = 0, 1, \ldots, n-1$, $n = 1, 2, \ldots, M$ and r^n is the truncation error.

Therefore, the Caputo derivative has the following numerical approximation [191]:

$$
{}_{0}^{C}D_{t_n}^{\alpha}\Psi(t_n) = \frac{\tau^{-\alpha}}{\Gamma(3-\alpha)} \sum_{i=0}^{n-1} [w_{1,n-i}(\Psi_{i+1} - \Psi_{i-1})
$$

$$
+ w_{2,n-i}(\Psi_{i+1} - 2\Psi_i + \Psi_{i-1})] + O(\tau^{3-\alpha}), \qquad (8.4)
$$

where $0 < \alpha < 1$, $w_{1,n-i} = \frac{2-\alpha}{2}[(n-i)^{1-\alpha} - (n-i-1)^{1-\alpha}]$, $w_{2,n-i} = (n-i)^{2-\alpha} - (n-i-1)^{2-\alpha} - (2-\alpha)(n-i-1)^{1-\alpha}$ and r^n is the truncation error in the following form:

$$
r^n = \frac{1}{\Gamma(1-\alpha)} \sum_{i=0}^{n-1} \int_{t_i}^{t_{i+1}} (t_n - s)^{-\alpha}
$$

$$
\times[-C_\Psi\tau^2 + 3C_\Psi(s - t_i)^2]ds + O(\tau^3), \qquad (8.5)
$$

where $C_\Psi = \frac{\Psi'''(t_i)}{3!}$ is a constant.

The right-hand side of Eq. (8.5) can be expressed as follows:

$$\frac{1}{\Gamma(1-\alpha)} \sum_{i=0}^{n-1} \int_{t_i}^{t_{i+1}} (t_n - s)^{-\alpha}[-C_{\Psi}\tau^2 + 3C_{\Psi}(s - t_i)^2]ds + O(\tau^3)$$

$$= \frac{C_{\Psi}}{\Gamma(1-\alpha)} \sum_{i=0}^{n-1} (I_1 + .3I_2),$$

where

$$I_1 = -\int_{t_n}^{t_{n+1}} (t_n - s)^{-\alpha}\tau^2 ds = \frac{\tau^{3-\alpha}}{1-\alpha}[(n-i-1)^{1-\alpha} - (n-i)^{1-\alpha}],$$

$$I_2 = \int_{t_n}^{t_{n+1}} (t_n - s)^{-\alpha}(s - t_i)^2 ds$$

$$= \frac{\tau^{3-\alpha}}{1-\alpha}(n-i-1)^{1-\alpha} - \frac{2\tau^{3-\alpha}}{(1-\alpha)(2-\alpha)}(n-i-1)^{2-\alpha}$$

$$- \frac{2\tau^{3-\alpha}}{(1-\alpha)(2-\alpha)(3-\alpha)}[(n-i-1)^{3-\alpha} - (n-i)^{3-\alpha}].$$

Therefore,

$$\frac{C_{\Psi}}{\Gamma(1-\alpha)} \sum_{i=0}^{n-1} (I_1 + .3I_2) = \frac{C_{\Psi}\tau^{3-\alpha}}{\Gamma(2-\alpha)} \left\{ -n^{1-\alpha} - 3[(n-1)^{1-\alpha} \right.$$

$$+ \cdots + 2^{1-\alpha} + 1] - \frac{6}{2-\alpha}$$

$$\times [(n-1)^{3-\alpha} + \cdots + 2^{3-\alpha} + 1]$$

$$\left. + \frac{6}{(2-\alpha)(3-\alpha)}n^{3-\alpha} \right\}.$$

Let

$$S(n) = -3\sum_{i=1}^{n-1} i^{1-\alpha} - \frac{6}{2-\alpha}\sum_{i=1}^{n-1} i^{2-\alpha} + \frac{6}{(2-\alpha)(3-\alpha)}n^{3-\alpha} - n^{1-\alpha}$$

$$= \sum_{i=1}^{n-1} a_i, \quad n \geq 1.$$

If $n = 1$, define $a_0 = s(1) = \frac{6}{(2-\alpha)(3-\alpha)} - 1$. Then, $a_i(i \leq 1)$ can be defined as follows:

$$a_i = S(i+1) - S(i) = \frac{6}{(2-\alpha)(3-\alpha)}[(i+1)^{3-\alpha} - i^{3-\alpha}] - \frac{6}{2-\alpha}i^{2-\alpha}$$
$$- (i+1)^{1-\alpha} - 2i^{1-\alpha}.$$

It can be proven that $|S(n)|$ is bounded for $n \geq 1$ [191,193,194]. This proves that the series $\sum_{i=0}^{\infty} a_i$ converges.

On further simplification, it yields

$$
{}_0^C D_{t_n}^{\alpha} \Psi(t_n) = \frac{\tau^{-\alpha}}{\Gamma(3-\alpha)} \Big[\Psi_n(w_{1,1} + w_{2,1}) + \sum_{i=0}^{n-2} w_{1,n-i} \Psi_{i+1}
$$
$$
- \sum_{i=0}^{n-1} w_{1,n-i} \Psi_{i-1} + \sum_{i=0}^{n-2} w_{2,n-i} \Psi_{i+1}
$$
$$
+ \sum_{i=0}^{n-1} w_{2,n-i}(-2\Psi_i + \Psi_{i-1}) \Big] + O(\tau^{3-\alpha}). \qquad (8.6)
$$

In Eq. (8.6), if $i = 0$, then $\Psi_{i-1} = \Psi_{-1}$ which lies outside of $[0, T]$. Various options have been provided to approach Ψ_{-1}. In numerical calculation, the neighbouring function values have been used to approximate Ψ_{-1}, that is, $\Psi_{-1} = \Psi(0) - \tau\Psi'(0) + \frac{\tau^2}{2}\Psi''(0) + O(\tau^3)$.

1. When $\Psi'(0) = \Psi''(0) = 0$, then $\Psi_{-1} = \Psi_0 + O(\tau^3)$, the convergence order is $O(\tau^{3-\alpha})$.
2. When $\Psi'(0) = 0$, $\Psi''(0) \neq 0$, then $\Psi_{-1} = \Psi_0 + \frac{\tau^2}{2}\Psi''(0) + O(\tau^3)$, the convergence order of Eq. (8.4) is $O(\tau^2)$.
3. When $\Psi'(0) \neq 0$, then the convergence order is $O(\tau)$.

8.3.2 *Solution of SPKEs by higher-order approximation method*

In this section, higher-order approximation to Caputo derivative has been applied to Eq. (8.1)

$$\frac{\tau^{-\alpha}}{\Gamma(3-\alpha)}\left[\mathbf{\Psi}_n(w_{1,1}+w_{2,1})+\sum_{i=0}^{n-2}w_{1,n-i}\mathbf{\Psi}_{i+1}-\sum_{i=0}^{n-1}w_{1,n-i}\mathbf{\Psi}_{i-1}\right.$$

$$\left.+\sum_{i=0}^{n-2}w_{2,n-i}\mathbf{\Psi}_{i+1}+\sum_{i=0}^{n-1}w_{2,n-i}(-2\mathbf{\Psi}_i+\mathbf{\Psi}_{i-1})\right]$$

$$=\mathbf{A}(t)\mathbf{\Psi}(t)+\mathbf{Q}+\mathbf{B}^{\frac{1}{2}}(t)\frac{d\mathbf{W}(t)}{dt}.\tag{8.7}$$

Thus, the above expression can be simplified into an explicit numerical scheme as follows:

$$\mathbf{\Psi}=\frac{1}{(w_{1,1}+w_{2,1})}\left(\Gamma(3-\alpha)\tau^{\alpha}\left(\mathbf{A}(t_{n-1})\mathbf{\Psi}_{n-1}+\mathbf{Q}\right.\right.$$

$$\left.+\mathbf{B}^{\frac{1}{2}}(t_{n-1})\Delta\mathbf{W}(t_n)\right)$$

$$-\left(\sum_{i=0}^{n-2}w_{1,n-i}\mathbf{\Psi}_{i+1}-\sum_{i=0}^{n-1}w_{1,n-i}\mathbf{\Psi}_{i-1}\right.$$

$$\left.\left.\left.+\sum_{i=0}^{n-2}w_{2,n-i}\mathbf{\Psi}_{i+1}+\sum_{i=0}^{n-1}w_{2,n-i}(-2\mathbf{\Psi}_i+\mathbf{\Psi}_{i-1}))\right)\right)\right),\tag{8.8}$$

where $\Delta\mathbf{W}(t_n)=\sqrt{h}\mathbf{S}_n$ and $n=1,2,\ldots,M$ with initial condition

$$\mathbf{\Psi}(0)=\begin{pmatrix}N_0\\\frac{\beta_1 N_0}{l\lambda_1}\\\frac{\beta_2 N_0}{l\lambda_2}\\\vdots\\\frac{\beta_I N_0}{l\lambda_m}\end{pmatrix},\tag{8.9}$$

$$\mathbf{S}_n=\begin{pmatrix}S_{0_n}\\S_{1_n}\\S_{2_n}\\\vdots\\S_{m_n}\end{pmatrix},\tag{8.10}$$

where $S_{0_n},S_{1_n},S_{2_n},\ldots,S_{m_n}$ are random variables chosen from $N(0,1)$ with mean equal to 0 and variance equal to 1.

Theorem 8.3.2.1. *The local truncation error of the scheme is* $O(\tau^{3-\alpha})$.

Proof. The local truncation error of the higher-order scheme for Eq. (8.8)

$$
\begin{aligned}
R_j^k ={}& \frac{\tau^{-\alpha}}{\Gamma(3-\alpha)} \sum_{i=0}^{n-1} [w_{1,n-i}(\boldsymbol{\Psi}_{i+1} - \boldsymbol{\Psi}_{i-1}) \\
& + w_{2,n-i}(\boldsymbol{\Psi}_{i+1} - 2\boldsymbol{\Psi}_i + \boldsymbol{\Psi}_{i-1})] - \boldsymbol{A}(t_{n-1})\boldsymbol{\Psi}_{n-1} \\
& - \boldsymbol{Q} - \boldsymbol{B}^{\frac{1}{2}}(t_{n-1})\Delta \boldsymbol{W}(t_n) \\
={}& \frac{\tau^{-\alpha}}{\Gamma(3-\alpha)} \sum_{i=0}^{n-1} [w_{1,n-1i}(\boldsymbol{\Psi}_{i+1} - \boldsymbol{\Psi}_{i-1}) \\
& + w_{2,n-i}(\boldsymbol{\Psi}_{i+1} - 2\boldsymbol{\Psi}_i + \boldsymbol{\Psi}_{i-1})] - {}_0^C D_{t_n}^\alpha \boldsymbol{\Psi}(t_n) \\
& - \boldsymbol{A}(t_{n-1})(\boldsymbol{\Psi}_{n-1} - \boldsymbol{\Psi}_{n-1}) \\
={}& O(\tau^{3-\alpha}).
\end{aligned}
$$
$\qquad\square$

8.4   Numerical Solutions of Fractional Stochastic Point Kinetic Model

In this section, the solutions of SPKE (here $i = 6$) have been obtained by a higher-order approximation scheme.

8.4.1   *Step external reactivity*

The numerical solutions have been obtained in this section for the fractional stochastic point kinetic model [168] with the following parameters: $\lambda_1 = 0.0127(s^{-1})$, $\lambda_2 = 0.0317(s^{-1})$, $\lambda_3 = 0.115(s^{-1})$, $\lambda_4 = 0.311(s^{-1})$, $\lambda_5 = 1.4(s^{-1})$, $\lambda_6 = 3.87(s^{-1})$, $\beta_1 = 0.000266$, $\beta_2 = 0.001491$, $\beta_3 = 0.001316$, $\beta_4 = 0.002849$, $\beta_5 = 0.000896$, $\beta_6 = 0.000182$, $\beta = 0.007$, $l = 2.0 \times 10^{-5}(s)$, $v = 2.5$ and $q = 0$ with $N(0) = N_0 = 100$ (neutron) and $C_i(0) = \frac{\beta_i N(0)}{l\lambda_i}$.

For different step reactivities, i.e., $\rho = 0.003$ and $\rho = 0.007$, the mean peaks of $N(t)$ at step size $h = 0.001(s)$ with respect to its time for fractional orders $\alpha = 0.96$, 0.98 and 0.99 are presented in Table 8.1 using 500 trials. Also, the results have been compared

Table 8.1. Mean peak of $N(t)$ for different step reactivities.

	$\alpha = 0.96$		$\alpha = 0.98$		$\alpha = 0.99$		EMM $\alpha = 1$ [169]	Taylor 1.5 strong order $\alpha = 1$ [169]	AEM $\alpha = 1$ [185]	ESM $\alpha = 1$ [170]
ρ	Peak	Time (s)	Peak	Time (s)	Peak	Time (s)	Peak	Peak	Peak	Peak
0.003	180.796	0.1	180.033	0.095	180.819	0.083	208.6	199.408	186.30	179.93
0.007	128.655	0.001	128.248	0.001	124.113	0.001	139.568	139.569	134.54	134.96

with that obtained by other methods, namely Euler–Maruyama [169], Taylor 1.5 strong order [45], analytical exponential model (AEM) [185] and efficient stochastic model (ESM) [170] with graphical representations in Figs. 8.1(a)–(c). The solutions obtained by using the above-discussed fractional scheme for SPKE have been tabulated to establish the efficiency of the higher-order approximation method.

8.4.2 *Ramp external reactivity*

The numerical solutions have been obtained for the fractional SNPKE with the same parametric values as those in Section 8.4.1. Here, the reactivity can be represented as $\rho = 0.1\beta t$.

For ramp external reactivity, i.e., $\rho = 0.1\beta t$, the mean peak of $N(t)$ with respect to its time has been presented in Table 8.2 for fractional order $\alpha = 0.96$, 0.98 and 0.99 using step size $h = 0.001(s)$ and for 500 trials. Also, the results have been compared with that obtained by other methods, namely AEM [189] and ESM [180] with graphical representation in Figs. 8.2(a)–(c). The solutions obtained by using the above-discussed fractional scheme for SPKEs have been tabulated to establish the efficiency of the higher-order approximation method.

8.4.3 *Temperature feedback reactivity*

In this section, solutions of fractional stochastic point kinetic model with i-group of delayed neutrons (here $i = 6$) in the presence of Newtonian temperature feedback have been obtained by the higher-order approximation method.

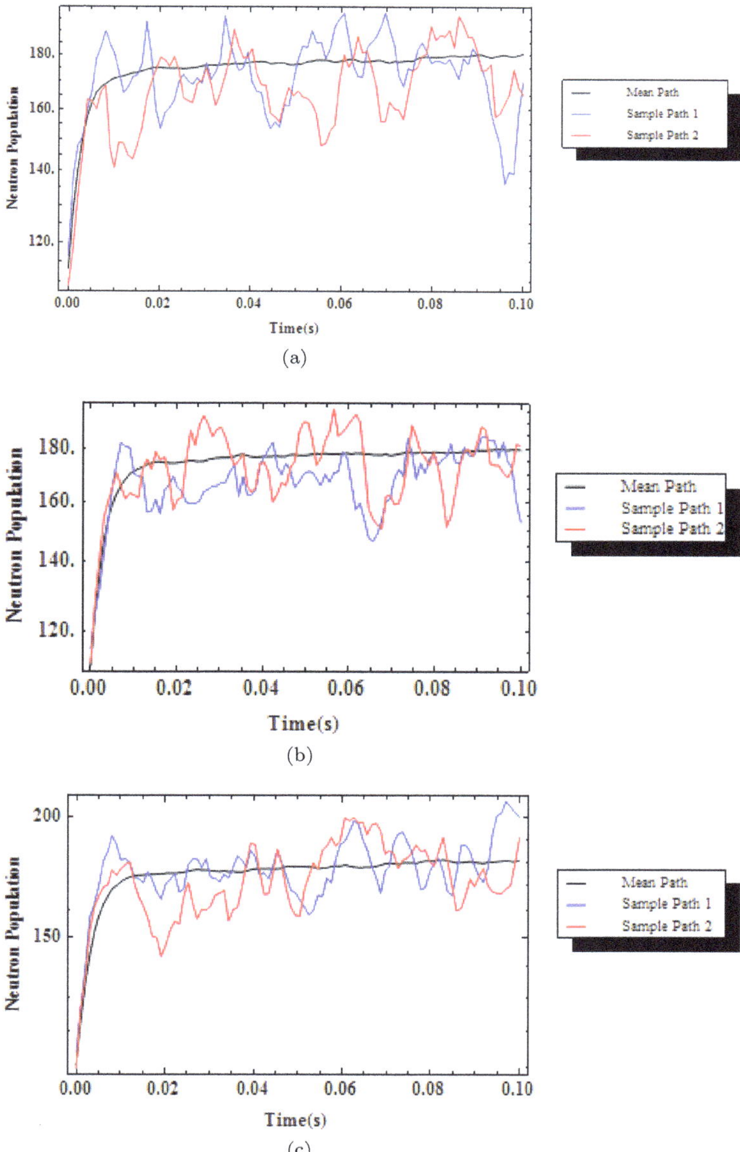

Fig. 8.1.   Mean $N(t)$ and two arbitrary sample paths for (a) step reactivity $\rho = 0.003$ and $\alpha = 0.96$, (b) step reactivity $\rho = 0.003$ and $\alpha = 0.98$, (c) step reactivity $\rho = 0.003$ and $\alpha = 0.99$.

Table 8.2.   Mean peak of $N(t)$ for ramp reactivity $\rho = 0.1\beta t$ and different values of fractional order α.

		$\alpha = 0.96$		$\alpha = 0.98$	$\alpha = 0.99$		AEM $\alpha = 1$ [185]	ESM $\alpha = 1$ [170]
a	σ	Peak	Time (s)	Peak	Peak	Peak	Time (s)	Peak	Peak
$0.1\beta t$	10^{-11}	113.563	0.998	113.275	186.30	113.045	0.1	113.267707	113.116433

The total reactivity of the reactor in the presence of temperature feedback reactivity [174] is of the following form:

$$\rho(t) = \rho_{ex}(t) - \rho_f(t), \quad \rho_f(t) = \sigma \int_0^t N(\tau)d\tau,$$

where $\sigma = \alpha K_c$, $\rho_{ex}(t)$ represents external reactivity, $T(t)$ is the temperature and T_0 is the initial temperature, α is the temperature coefficient and K_c is the reciprocal of the thermal capacity.

In the interval $[t_k, t_{k+1}]$, the total reactivity can be expressed as follows [178]:

$$\rho(t_k) \approx \rho_{ex}(t_k) - h\sigma \sum_{j=0}^{k} N(t_j).$$

8.4.3.1   *Step external reactivity*

The numerical solutions have been obtained in this section for the SPKE of U^{235} nuclear reactor [173,174] with the following parameters: $\lambda_1 = 0.0124(s^{-1})$, $\lambda_2 = 0.0305(s^{-1})$, $\lambda_3 = 0.111(s^{-1})$, $\lambda_4 = 0.301(s^{-1})$, $\lambda_5 = 1.13(s^{-1})$, $\lambda_6 = 3.0(s^{-1})$, $\beta_1 = 0.00021$, $\beta_2 = 0.00141$, $\beta_3 = 0.00127$, $\beta_4 = 0.00255$, $\beta_5 = 0.00074$, $\beta_6 = 0.00027$, $\beta = 0.00645$, $l = 5.0 \times 10^{-5}(s)$, $\alpha = 5.0 \times 10^{-5}(K^{-1})$ and $K_c = 0.05(K/MW_s)$ with $N(0) = N_0 = 1$ (neutron) and $C_i(0) = \frac{\beta_i N(0)}{l\lambda_i}$.

For different step external reactivities, i.e., $\rho_{ex} = 0.5\beta$, $\rho_{ex} = 0.75\beta$ and $\rho_{ex} = \beta$, the mean peaks of $N(t)$ are presented in Table 8.3 for fractional order $\alpha = 0.96$, 0.98 and 0.99 using 500 trials for different step external reactivities 0.5β, 0.75β and β which have been compared with previously obtained results by split-step forward EMM

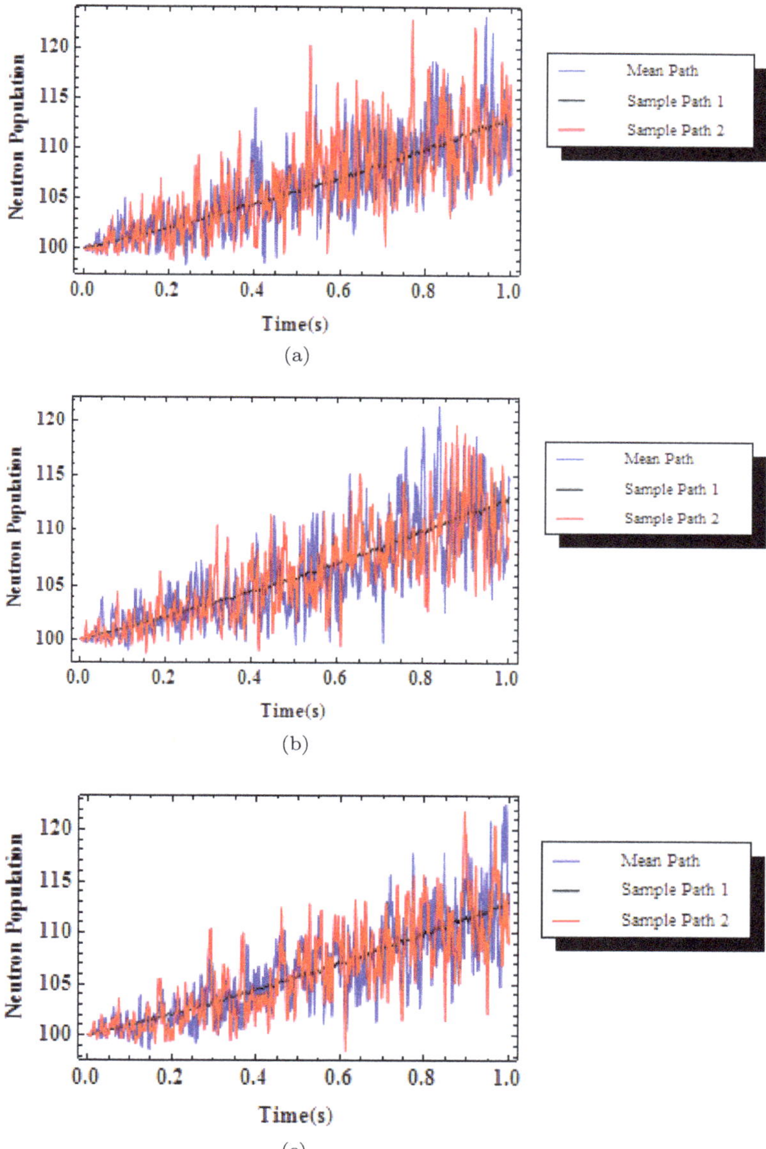

Fig. 8.2.   Mean $N(t)$ and two arbitrary sample paths for (a) ramp reactivity $\rho = 0.1\beta t$ and $\alpha = 0.96$, (b) ramp reactivity $\rho = 0.1\beta t$ and $\alpha = 0.98$, (c) ramp reactivity $\rho = 0.1\beta t$ and $\alpha = 0.99$.

Table 8.3. Mean peak of $N(t)$ for $\rho_{ex} = 0.5\beta$, $\rho_{ex} = 0.75\beta$ and $\rho_{ex} = \beta$.

ρ_{ex}	$\alpha = 0.96$		$\alpha = 0.95$		$\alpha = 0.99$	
	Peak	Time (s)	Peak	Time (s)	Peak	Time (s)
0.5β	42.6182	28.65	44.789	30.29	45.9708	29.25
0.75β	159.21	8.875	160.124	8.895	162.99	9.305
β	801.166	0.985	795.268	1.03	772.893	1.0625

Table 8.4. Comparison between mean peak of $N(t)$ for $\rho_{ex} = 0.5\beta$, $\rho_{ex} = 0.75\beta$ and $\rho_{ex} = \beta$ for $\alpha = 1$ and $\alpha = 0.98$.

ρ_{ex}	SSFEMM [188] ($\alpha = 1$)		DFMM [188] ($\alpha = 1$)		$\alpha = 0.98$	
	Peak	Time (s)	Peak	Time (s)	Peak	Time (s)
0.5β	46.4939	28.34	46.2606	27.84	44.789	30.29
0.75β	163.707	8.795	164.22	8.95	160.124	8.895
β	760.589	1.065	769.238	1.0575	795.268	1.03

(SSFEMM) and derivative-free Milstein method (DFMM) [178] in Table 8.4 with graphical representation in Figs. 8.3(a)–(c), 8.4(a)–(c) and 8.5(a)–(c). The solutions obtained by using the above-discussed fractional scheme for SPKE have been tabulated to establish the efficiency of the higher-order approximation scheme.

For $\alpha = 0.96$, 0.98 and 0.99, the mean $N(t)$ with two arbitrary sample paths have been shown for $\rho_{ex} = 0.5\beta$, $\rho_{ex} = 0.75\beta$ and $\rho_{ex} = \beta$.

8.4.3.2 *Ramp external reactivity*

The numerical solutions have been obtained for the fractional SNPKE of U^{235} nuclear reactor with similar values as those in Section 8.4.3.1. Here, the external reactivity is represented as $\rho_{ex} = 0.1t$ and $\rho_{ex} = 0.01t$, and the nonlinear coefficient σ takes 10^{-11} or 10^{-13}.

For different ramp external reactivities, i.e., $\rho_{ex} = 0.1t$ and $0.01t$, mean peaks of $N(t)$ with respect to its time have been presented in Table 8.5 for fractional order $\alpha = 0.96$, 0.98 and 0.99 using step size

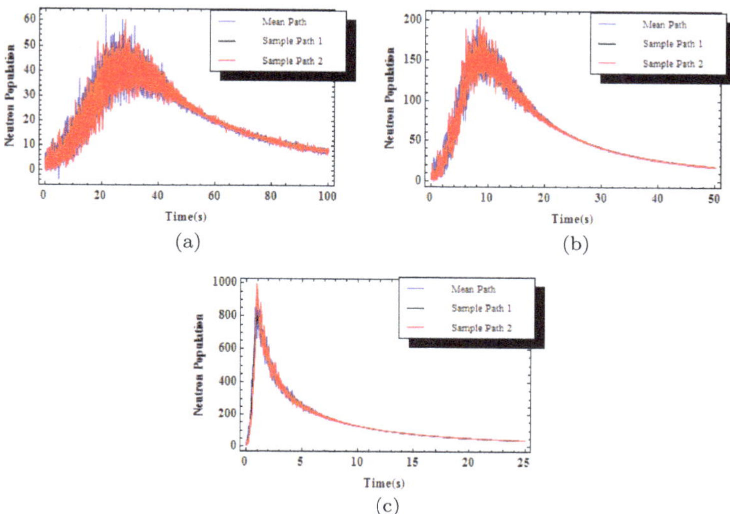

Fig. 8.3. Mean $N(t)$ and two arbitrary sample paths for (a) step reactivity $\rho_{ex} = 0.5\beta$ and $\alpha = 0.96$, (b) step reactivity $\rho_{ex} = 0.75\beta$ and $\alpha = 0.96$, (c) step reactivity $\rho_{ex} = \beta$ and $\alpha = 0.96$.

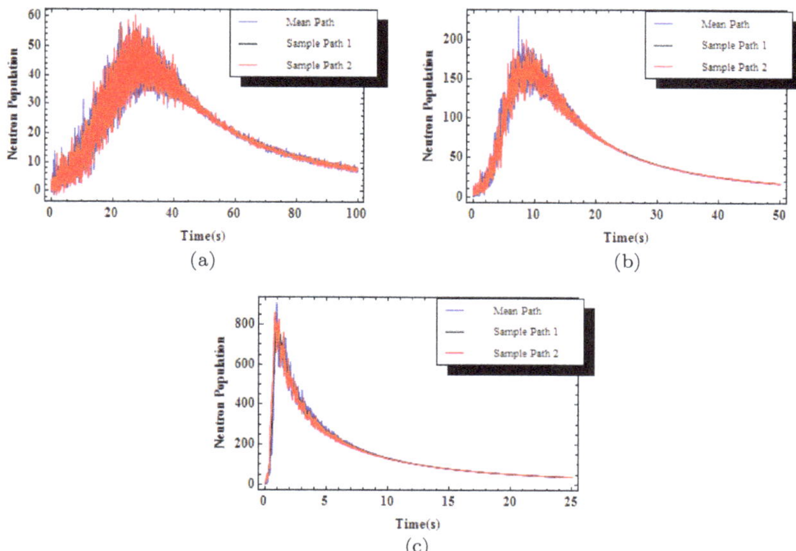

Fig. 8.4. Mean $N(t)$ and two arbitrary sample paths for (a) step reactivity $\rho_{ex} = 0.5\beta$ and $\alpha = 0.98$, (b) step reactivity $\rho_{ex} = 0.75\beta$ and $\alpha = 0.98$, (c) step reactivity $\rho_{ex} = \beta$ and $\alpha = 0.98$.

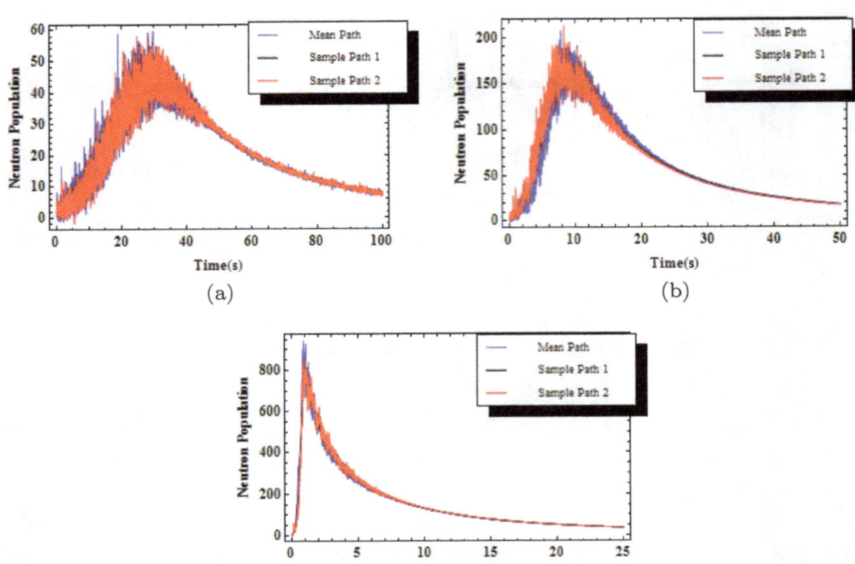

Fig. 8.5. Mean $N(t)$ and two arbitrary sample paths for (a) step reactivity $\rho_{ex} = 0.5\beta$ and $\alpha = 0.99$, (b) step reactivity $\rho_{ex} = 0.75\beta$ and $\alpha = 0.99$, (c) step reactivity $\rho_{ex} = \beta$ and $\alpha = 0.99$.

Table 8.5. Peak of the mean $N(t)$ for $\rho_{ex} = 0.1t$ and $0.01t$, $\rho(t) = at - \sigma \int_0^t N(\tau)d\tau$.

		$\alpha = 0.96$		$\alpha = 0.98$		$\alpha = 0.99$	
a	σ	Peak	Time (s)	Peak	Time (s)	Peak	Time (s)
0.01	10^{-11}	1.69869E + 10	1.084	1.76031E + 10	1.103	1.73211E + 10	1.112
	10^{-13}	2.0663E + 12	1.132	2.18262E + 12	1.151	2.13422E + 2	1.161
0.1	10^{-11}	1.78236E + 11	0.223	1.90873E + 11	0.232	2.01965E + 11	0.235
	10^{-13}	2.34211E + 13	0.238	2.37627E + 13	0.248	2.41795E + 13	0.252

Table 8.6. Comparison between peak of the mean $N(t)$ for $\rho_{ex} = 0.1t$ and $0.01t$, $\rho(t) = at = \sigma \int_0^t N(\tau)d\tau$ for $\alpha = 1$ and $\alpha = 0.98$.

a	σ	SSFEMM [178] ($\alpha = 1$)		DFMM [178] ($\alpha = 1$)		AEM [170] ($\alpha = 1$)		$\alpha = 0.98$	
		Peak	Time (s)	Peak	Time (s)	Peak	Time (s)	Peak	Time (s)
0.01	10^{-11}	1.68604E + 10	1.118	1.69492E + 10	1.118	1.673436E + 10	0.854	1.76031E + 10	1.103
	10^{-13}	2.12034E + 12	1.169	2.12802E + 12	1.168	2.082531E + 12	0.877	2.18262E + 12	1.151
0.1	10^{-11}	1.88849E + 11	0.235	1.89642E + 11	0.235	1.790577E + 11	0.142	1.90873E + 11	0.232
	10^{-13}	2.2448E + 13	0.251	2.26026E + 13	0.251	2.143778E + 13	0.150	2.37627E + 13	0.248

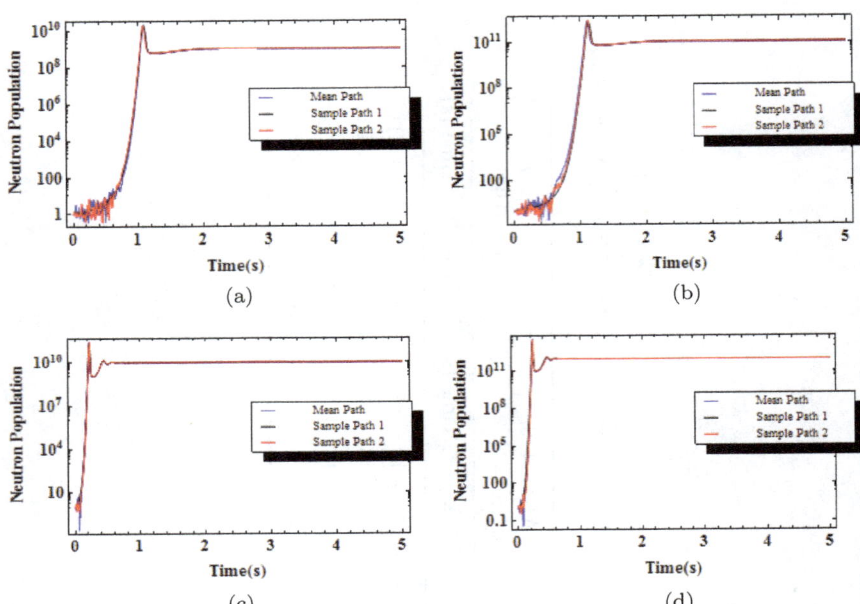

Fig. 8.6.   Mean $N(t)$ and two arbitrary sample paths for (a) ramp reactivity $\rho_{ex} = 0.01t$, $\sigma = 10^{-11}$ and $\alpha = 0.96$, (b) ramp reactivity $\rho_{ex} = 0.01t$, $\sigma = 10^{-13}$ and $\alpha = 0.96$, (c) ramp reactivity $\rho_{ex} = 0.1t$, $\sigma = 10^{-11}$ and $\alpha = 0.96$, (d) ramp reactivity $\rho_{ex} = 0.1t$, $\sigma = 10^{-13}$ and $\alpha = 0.96$.

$h = 0.001(s)$ and for 500 trials and have been compared with previously obtained results [170,178] in Table 8.6 with graphical representation in Figs. 8.6(a)–(c), 8.7(a)–(c) and 8.8(a)–(c). The solutions obtained by using the above-discussed fractional scheme for SPKE have been tabulated to establish the efficiency of the higher-order approximation method.

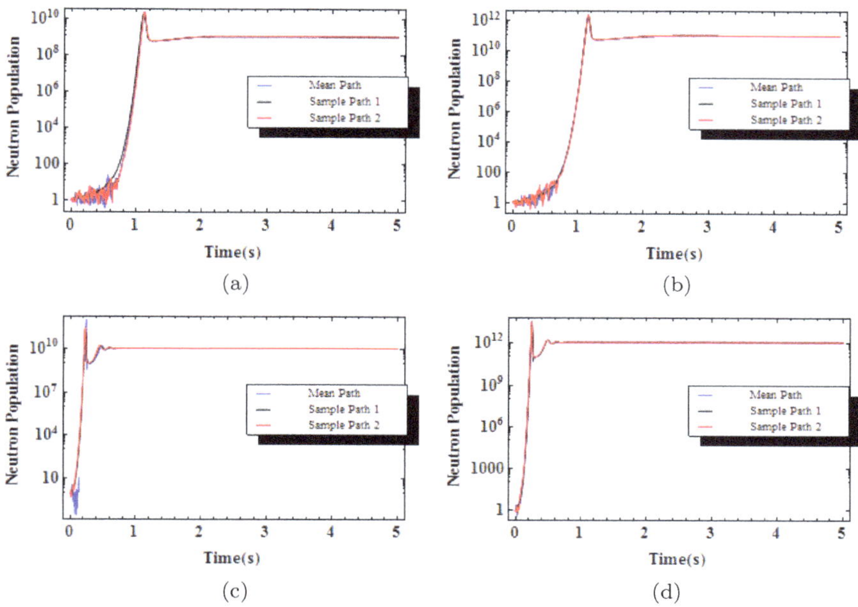

Fig. 8.7. Mean $N(t)$ and two arbitrary sample paths for (a) ramp reactivity $\rho_{ex} = 0.01t$, $\sigma = 10^{-11}$ and $\alpha = 0.98$, (b) ramp reactivity $\rho_{ex} = 0.01t$, $\sigma = 10^{-13}$ and $\alpha = 0.98$, (c) ramp reactivity $\rho_{ex} = 0.1t$, $\sigma = 10^{-11}$ and $\alpha = 0.98$, (d) ramp reactivity $\rho_{ex} = 0.1t$, $\sigma = 10^{-13}$ and $\alpha = 0.98$.

8.4.3.3   *Sinusoidal reactivity*

In this section, the reactivity is in the form of sinusoidal change, i.e., $\rho = \rho_0 \sin\left(\frac{\pi t}{T}\right)$. The numerical solution for this reactivity has been obtained with the following parameters: $\rho_0 = 0.005333$, $\beta_1 = \beta = 0.0079$, $\lambda_1 = 0.077$, $\Lambda = 0.001$, $q = 0$, $N(0) = N_0 = 1$ and time period $T = 100(s)$. The mean peaks of $N(t)$ with respect to its time are presented in Table 8.7 for fractional order $\alpha = 0.96$, 0.98 and 0.99 with graphical representations in Figs. 8.9(a)–(c).

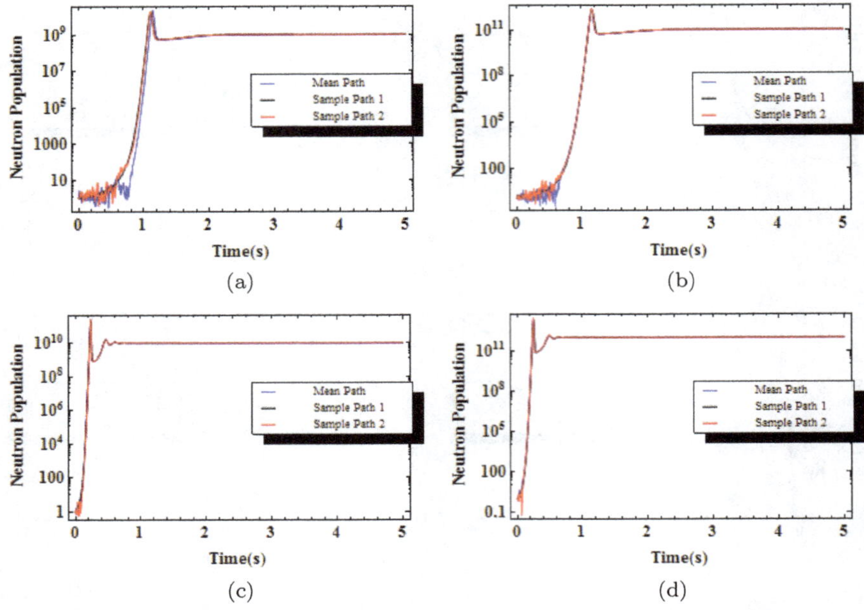

Fig. 8.8. Mean $N(t)$ and two arbitrary sample paths for (a) ramp reactivity $\rho_{ex} = 0.01t$, $\sigma = 10^{-11}$ and $\alpha = 0.99$, (b) ramp reactivity $\rho_{ex} = 0.01t$, $\sigma = 10^{-13}$ and $\alpha = 0.99$, (c) ramp reactivity $\rho_{ex} = 0.1t$, $\sigma = 10^{-11}$ and $\alpha = 0.99$, (d) ramp reactivity $\rho_{ex} = 0.1t$, $\sigma = 10^{-13}$ and $\alpha = 0.99$.

Table 8.7. Peak of the mean $N(t)$ for sinusoidal reactivity $\rho = 0.005333 \sin\left(\frac{\pi t}{T}\right)$ for different values of fractional order α.

ρ	$\alpha = 0.96$		$\alpha = 0.98$		$\alpha = 0.99$	
	Peak	Time (s)	Peak	Time (s)	Peak	Time (s)
$0.005333 \sin\left(\frac{\pi t}{T}\right)$	38.0005	38.28	45.8029	38.18	49.1345	38.99

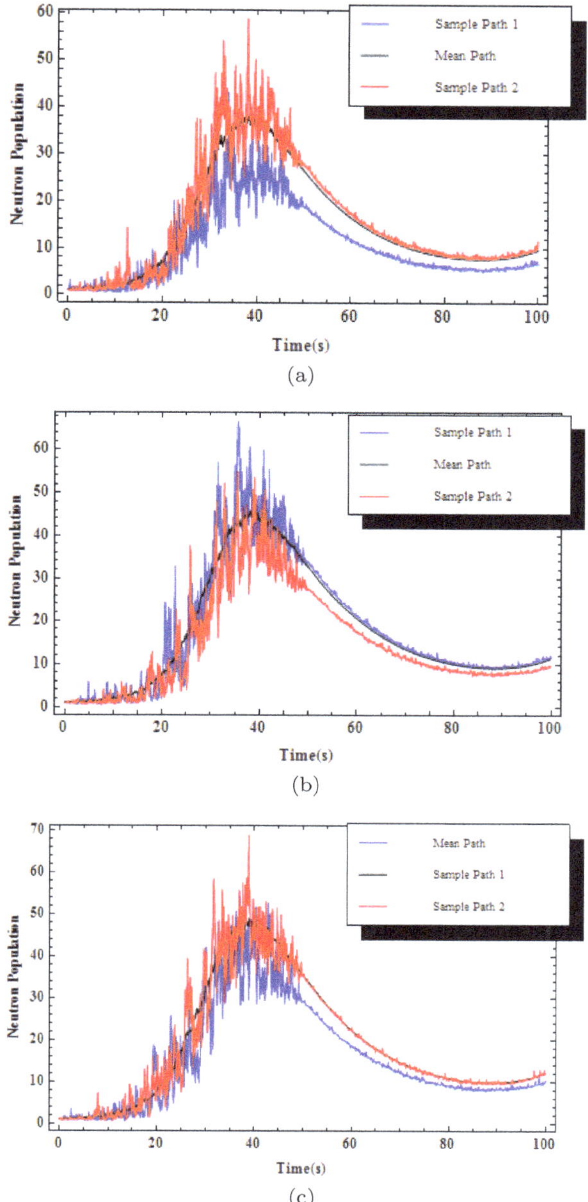

Fig. 8.9. Mean $N(t)$ and two arbitrary sample paths for sinusoidal reactivity for $\rho = 0.005333 \sin\left(\frac{\pi t}{T}\right)$ with (a) fractional order $\alpha = 0.96$, (b) fractional order $\alpha = 0.98$ and (c) fractional order $\alpha = 0.99$.

8.5 Summary

In this chapter, the fractional SPKEs have been solved by higher-order approximation scheme. The fractional SPKEs have been solved with different fractional orders, i.e., α. The obtained numerical solutions for mean $N(t)$ have been presented in the tables and are also graphically demonstrated in order to justify the efficiency of the proposed higher-order approximation method. The results as compared to some previous works, such as Refs. [52,180,188], show that the results obtained by the implemented fractional model are in good agreement with the previous results which further establishes the proficient nature of the proposed scheme. The graphical representation for different reactivities shows the behaviour of the mean neutron population. The random fluctuations at low power levels and the going into equilibrium state after reaching its peak value provide a very succinct idea about the behaviour of $N(t)$ for different reactivities.

Chapter 9

Conclusion and Future Directions

Stochastic differential equations play a crucial role in modelling the dynamics of various real-life phenomena happening around in the real world. Solving these equations using stochastic calculus helps to analyse the behaviour of these phenomena and develop various applications beneficiary to the real world.

In this study, various analytical and numerical methods have been applied to various stochastic differential and integral equations in order to demonstrate the accuracy and efficiency of these implemented schemes. Analysing the approximate and exact solutions, it can be concluded that the analytical and numerical methods provide worthy approximate and exact solutions for stochastic differential and integral equations.

Analytical and numerical methods are powerful techniques for obtaining exact and approximate solutions of stochastic differential and integral equations. Using these methods and obtaining solutions helps us to bring attention of various researchers and students around the world with any interest in this field. Stochastic calculus is a very new and less explored area in the field of applied mathematics. The mission is that the work discussed somewhat helps and encourages readers to explore and appreciate more about this beautiful subject and its applications to real world.

In the following section, conclusions are drawn with respect to various methods and the application problems mentioned in previous chapters.

9.1 Conclusion

- In Chapter 1, the preliminaries of stochastic calculus have been presented. Moreover, the motivation, objectives and the organisation of the dissertation have been discussed in this chapter.
- In Chapter 2, various analytical methods such as Kudryashov method, improved sub-equation method, Jacobi elliptic function (JEF) expansion method and extended auxiliary equation method have been discussed to obtain exact solutions of stochastic differential equations such as Wick-type stochastic Zakharov–Kuznetsov (ZK) equation, Wick-type stochastic Kudryashov–Sinelshchikov equation, Wick-type stochastic modified Boussinesq equations, Wick-type stochastic Kersten–Krasil'shchik coupled KdV-mKdV equations and Wick-type stochastic nonlinear Schrödinger equation.
- In Chapter 3, wavelet methodologies such as hybrid Legendre Block-Pulse functions and second-kind Chebyshev wavelets have been used to obtain numerical solutions of stochastic integral equations. Stochastic Volterra–Fredholm integral equation and stochastic mixed Volterra–Fredholm integral equation have been discussed in this chapter. Hybrid Legendre Block-Pulse functions and second-kind Chebyshev wavelets respectively have been applied to solve the stochastic integral equations mentioned above. The numerical results have been tabulated with comparison to other numerical schemes to show the efficiency and accuracy of the proposed scheme. The graphical representation of the results gives an idea about the behaviour of the obtained sample paths. Error analysis has also been discussed for both the equations.
- In Chapter 4, numerical solutions of stochastic integral equations have been obtained. The equation under discussion in this chapter is the multidimensional stochastic Itô–Volterra integral equation. Hybrid Legendre Block-Pulse functions and second-kind Chebyshev wavelets have been applied respectively to solve the stochastic integral equation as mentioned above. The obtained numerical solutions have been compared with the exact solutions of the equations discussed in the examples. The comparison gives an insight about the accuracy and efficiency of the proposed scheme.
- In Chapter 5, numerical solutions of fractional stochastic integral equations, i.e., fractional stochastic Itô–Volterra integral equation

and nonlinear fractional stochastic Itô–Volterra integral equation, have been obtained. In the equation, $B^H(t)$ has been considered as fractional Brownian with Hurst index $H \in (0, 1)$ on complete probability space $(\Omega, \mathcal{F}, \mathbb{P})$. Second-kind Chebyshev wavelet and Bernstein polynomial approximation have been applied successfully to solve the linear and nonlinear fractional stochastic Itô–Volterra integral equation respectively for different Hurst indices. The graphical representation of the sample path and the results show the effectiveness of the proposed scheme.

- In Chapter 6, a numerical discussion for stochastic Fisher equation has been catered using semi-implicit Euler–Maruyama scheme. Stability and convergence have been also discussed for the stochastic difference equation. Numerical discussion of the Fisher equation has been done for cylindrical Wiener process and Q-Wiener process. The sample paths for Wiener process and Q-Wiener process have been plotted for different step sizes providing a very useful insight into the behaviour of the obtained solutions. Stochastic FitzHugh–Nagumo equation has been solved by Chebyshev spectral collocation method and semi-implicit Euler–Maruyama scheme. A detailed stability analysis of the stochastic FitzHugh–Nagumo equation has been provided. The algorithm for Chebyshev spectral collocation method has also been discussed very thoroughly, and its efficiency has been tested in the numerical section with graphical explanation of the behaviour of the solutions. The graphs show the mean sample path at each time point, and separate two-dimensional plots for sample paths at different time points have also been shown for a more clear idea about the dynamical variation of the governing equation.

- In Chapter 7, the stochastic point reactor kinetics equations have been solved by Euler–Murayama, 1.5 strong Taylor, split-step forward Euler–Murayama and derivative-free Milstein methods. The numerical results have been presented in the tables and are also graphically demonstrated in order to justify the efficiency of the proposed schemes. In case of step external reactivity, the mean neutron population increases until it achieves its maximum value and then it starts decreasing with time due to Newtonian temperature feedback effects. Also, the fluctuations of the sample paths decrease when the mean neutron population decreases after achieving its maximum value. For ramp external reactivity, the mean neutron population increases until it achieves its peak value due to the increasing external reactivity and then it tries to reach the equilibrium state. Also, at low power levels such as at startup, random fluctuations in the neutron density and neutron precursor concentrations can be significant, but it becomes almost stable with increase in time.

- In Chapter 8, the fractional SPKEs have been solved by higher-order approximation scheme. The fractional SPKEs have been solved with different fractional orders, i.e., $\alpha = 0.96, 0.98$ and 0.99. The obtained numerical solutions for mean neutron population have been presented in the tables and are also graphically demonstrated in order to justify the efficiency of the proposed higher-order approximation method. The results as compared to some previous works such as Refs. [52,179,187] show that the results obtained by the implemented fractional model have good agreement with the previous results which further establishes the proficient nature of the proposed scheme. The graphical representation for different reactivities shows the behaviour of the mean neutron population. The random fluctuations at low power levels and the going into equilibrium state after reaching its peak value provide us with a very succinct idea about the behaviour of the nuclear reactor for different reactivities.

9.2   Future Directions

There are several ways in which the research can be pursued to enhance the dissemination of the proposed work, which have been enlisted in the following:

- In future, analytical solutions for more important stochastic nonlinear partial differential equation such as the stochastic Korteweg–de Vries equation derived from the reductive perturbation technique [194], stochastic coupled nonlinear partial differential equations such as stochastic diffusive predator–prey system [195], stochastic Klein–Gordon Schrödinger equation via Jacobi elliptic function expansion method, modified tanh-coth method [196], etc. shall be obtained.

- In future, various types of stochastic integral equations, stochastic differential equations and stochastic integro-differential equations will be solved using CAS wavelet [120], Legendre wavelet [112,113,124], Legendre multi-wavelet [114], Haar wavelet [115], etc. Also, stochastic fractional integral equations shall be explored more for different Hurst parameters for better understanding of the dynamics of the equation.
- Many more real-life stochastic differential equations with applications in physics, biology, engineering, etc. shall be solved using newer techniques. Also, fractional order stochastic differential equations shall be studied with newer and better fundamentals.

References

[1] Yu. G. Kondratiev, "Spaces of entire functions of an infinite number of variables, connected with the rigging of a Fock space", *Selecta Mathematica Sovietica*, vol. **10**, no. 2, pp. 165–180, 1991.

[2] Yu. G. Kondratiev, "Nuclear spaces of entire functions in problems of infinite-dimensional analysis", *Soviet Mathematics – Doklady*, vol. **22**, pp. 588–592, 1980.

[3] Yu. G. Kondratiev, "Wick powers of Gaussian generalized random processes", *Methods of Functional Analysis in Problems of Mathematical Physics*, Institute of Mathematics, Academy of Sciences of the Ukrainian SSR, Kiev, pp. 129–158, 1978 (Russian).

[4] Yu. G. Kondratiev and Yu. S. Samoilenko, "Integral representation of generalized positive definite kernels of an infinite number of variables", *Soviet Mathematics – Doklady*, vol. **17**, pp. 517–521, 1976.

[5] Yu. G. Kondratiev and Yu. S. Samoilenko, "Generalized derivatives of probability measures on R^∞", *Methods of Functional Analysis in Problems of Mathematical Physics*, Institute of Mathematics, Academy of Sciences of the Ukrainian SSR, Kiev, pp. 159–176, 1978 (Russian).

[6] Yu. G. Kondratiev and Yu. S. Samoilenko, "The spaces of trial and generalized functions of infinitely many variables", *Reports on Mathematical Physics*, vol. **14**, no. 3, pp. 323–348, 1978.

[7] T. Hida, "Analysis of Brownian functionals", *Carleton Math. Lecture Notes*, vol. **13**, Carleton University, Ottawa, 1975.

[8] T. Hida, *Brownian Motion*, Springer-Verlag, New York, 1980.

[9] Yu. M. Berezansky and V. A. Tesko, "Spaces of test and generalized functions related to generalized translation operators", *Ukrainian Mathematical Journal*, vol. **55**, no. 12, pp. 1907–1979, 2003.

[10] N. A. Kachanovsky, "On Kondratiev spaces of test functions in the non-Gaussian infinite-dimensional analysis", *Methods of Functional Analysis and Topology*, vol. **19**, no. 4, pp. 301–309, 2013.

[11] Yu. M. Berezansky and Yu. G. Kondratiev, *Spectral Methods in Infinite-Dimensional Analysis*, Providence, RI, Dordrecht-Boston-London, vols. 1, 2, 1995. (Russian edition: Naukova Dumka, Kiev, 1988.)

[12] Yu. M. Berezansky, Z. G. Sheftel and G. F. Us, *Functional Analysis*, Birkhäuser Verlag, Basel, 1996.

[13] S. Dineen, "Complex analysis in locally convex spaces", *North-Holland Mathematics Studies*, vol. **57**, Elsevier North-Holland, Amsterdam, 1981.

[14] A. Løkka, B. Øksendal and F. Proske, "Stochastic partial differential equations driven by Lévy space-time white noise", *The Annals of Applied Probability*, vol. **14**, no. 3, pp. 1506–1528, 2004.

[15] H. Holden, B. Øksendal, J. Ubøe and T. Zhang, *Stochastic Partial Differential Equations*, Springer-Verlag, New York, 2010.

[16] G. J. Lord, C. E. Powell and T. Shardlow, *An Introduction to Computational Stochastic PDEs*, Cambridge University Press, New York, USA, 2014.

[17] T. Sauer, Numerical solution of stochastic differential equations in finance, *Handbook of Computational Finance*, Springer, New York, 2012.

[18] B. B. Mandelbrot and J. W. Van Ness, "Fractional Brownian motions, fractional noises and applications", *SIAM Review*, vol. **10**, no. 4, pp. 422–437, 1968.

[19] K. B. Oldham and J. Spanier, *The Fractional Calculus*, Academic Press, New York, London, 1974.

[20] C. P. Tsokos and W. J. Padgett, *Random Integral Equations with Applications to Life Sciences and Engineering*, Academic Press, New York, 1974.

[21] G. A. Gottwalld, "The Zakharov–Kuznetsov equation as a two-dimensional model for nonlinear Rossby wave", *arXiv:nlin/0312009v1*, 12 pages, 2003.

[22] N. Glatt-Holtz, R. Temam and C. Wang, "Martingale and pathwise solutions to the stochastic Zakharov-Kuznetsov equation with multiplicative noise", *Discrete and Continuous Dynamical Systems Series B*, vol. **19**, no. 4, 39 pages, 2014.

[23] D. Furihata and T. Matsuo, *Discrete Variational Derivative Method*, Chapman & Hall/CRC, 2011.

[24] V. E. Zakharov and E. A. Kuznetsov, "On three-dimensional solitons", *Soviet Physics*, vol. **39**, no. 1974, pp. 285–288, 1974.

[25] M. H. Islam, K. Khan, A. M. Akbar and A. M. Salam, "Exact traveling wave solutions of modified KdV–Zakharov–Kuznetsov equation and viscous Burgers equation", *Springer Plus*, vol. **3**, no. 105, 9 pages, 2014.

[26] H. A. Ghanny, "Analytical approach to exact solutions for the Wick-type stochastic space-time fractional KdV equation", *Chinese Physics Letters*, vol. **31**, no. 6, 10 pages, 2014.

[27] M. S. Mohammed and H. A. Ghanny, "White noise functional solutions for Wick-type stochastic fractional KdV-Burgers-Kuramoto equations", *Chinese Journal of Physics*, vol. **50**, no. 4, pp. 619–627, 2012.

[28] H. A. Ghanny and A. Hyder, "Abundant solutions of Wick-type stochastic fractional 2D KdV equations", *Chinese Physics B*, vol. **23**, no. 6, p. 060503, 2014.

[29] Y. Xie, "Exact solutions for stochastic mKdV equations", *Chaos, Solitons and Fractals*, vol. **19**, pp. 509–513, 2004.

[30] Y. Xie, "Exact solutions for stochastic KdV equations", *Physics Letters A*, vol. **310**, no. 2–3, pp. 161–167, 2003.

[31] Z. Hammouch and T. Mekkaoui, "Approximate analytical solution to a time-fractional Zakharov-Kuznetsov equation", *International Journal of Physical Research*, vol. **1**, no. 2, pp. 28–33, 2013.

[32] H. A. Ghanny, "Exact solutions for stochastic fractional Zakharov-Kuznetsov equations", *Chinese Journal of Physics*, vol. **51**, no. 5, pp. 875–881, 2013.

[33] C. Wang, *Initial and Boundary Value Problems for the Deterministic and Stochastic Zakharov-Kuznetsov Equation in a Bounded Domain*, Bloomington, Indiana University, 2015.

[34] Y. He, S. Li and L. Yao, "Exact solutions of the Kudryashov-Sinelshchikov equation using the multiple G'/G-expansion method", *Mathematical Problems in Engineering*, vol. **2013**, no. 2013, article ID 708049, 7 pages, 2013.

[35] M. Randrüüt, "On the Kudryashov–Sinelshchikov equation for waves in bubbly liquids", *Physics Letters A*, vol. **375**, no. 2011, pp. 3687–3692, 2011.

[36] D. G. Natsis, "Solitary wave solutions of the one-dimensional Boussinesq equations", *Numerical Algorithms*, vol. **44**, no. 3, pp. 281–289, 2007.

[37] H. S. Zhang, H. W. Zhou, G. W. Hong and J. M. Yang, "A fully nonlinear Boussinesq model for water wave propagation", *Proceedings of 32nd Conference on Coastal Engineering (ICCE)*, 2016, Washington, DC.

[38] C. M. Liu, "Preliminary study on higher-order modified Boussinesq equations for internal waves in a two-fluid system", *In 2013 MTS/IEEE OCEANS-Bergen*, IEEE, pp. 1–4, 2013.

[39] M. A. Manna and V. Merle, "Modified Kortweg–de Vries hierarchies in multiple–time variables and the solutions of modified Boussinesq equations", *Proceedings of the Royal Society A*, vol. **454**, no. 1973, pp. 1445–1456, 1998.

[40] E. Aksoy, A. C. Çevikel and A. Bekir, "Soliton solutions of $(2+1)$-dimensional time-fractional Zoomeron equation", *Optik*, vol. **127**, no. 17, pp. 6933–6942, 2016.

[41] A. Bekir, E. Aksoy and A. C. Cevikel, "Exact solutions of non-linear time fractional partial differential equations by sub-equation method", *Mathematical Methods in the Applied Sciences*, vol. **38**, no. 13, pp. 2779–2784, 2015.

[42] A. Bekir, E. Aksoy and Ö. Güner, "A generalized fractional sub-equation method for nonlinear fractional differential equations", *In AIP Conference Proceedings*, vol. **1611**, no. 1, pp. 78–83, 2014.

[43] A. Bekir and E. Aksoy, "Application of the subequation method to some differential equations of time-fractional order", *Journal of Computational and Nonlinear Dynamics*, vol. **10**, no. 5, 2015.

[44] A. Bekir and M. Kaplan, "Exponential rational function method for solving nonlinear equations arising in various physical models", *Chinese Journal of Physics*, vol. **54**, no. 3, pp. 365–370, 2016.

[45] Y. C. Hon and E. G. Fan, "Solitary wave and doubly periodic wave solutions for the Kersten–Krasil'shchik coupled KdV–mKdV system", *Chaos, Solitons and Fractals*, vol. **19**, no. 5, pp. 1141–1146, 2004.

[46] P. Kersten and J. Krasil'shchik, "Complete integrability of the coupled KdV–mKdV system", *Lie Groups, Geometric Structures and Differential Equations—One Hundred Years after Sophus Lie*, pp. 151–171, 2002.

[47] A. K. Kalkanli, S. Y. Sakovich and Í. Yurduşen, "Integrability of Kersten–Krasil'shchik coupled KdV–mKdV equations: Singularity analysis and Lax pair", *Journal of Mathematical Physics*, vol. **44**, no. 4, pp. 1703–1708, 2003.

[48] T. A. Abassy, M. A. El-Tawil and H. K. Saleh, "The solution of KdV and mKdV equations using Adomian Padé approximation", *International Journal of Nonlinear Sciences and Numerical Simulation*, vol. **5**, no. 4, pp. 327–339, 2004.

[49] W. Rui and X. Qi, "Bilinear approach to quasi-periodic wave solutions of the Kersten-Krasil'shchik coupled KdV-mKdV system", *Boundary Value Problems*, vol. **2016**, no. 130, 13 pages, 2016.

[50] N. Z. Petrovic and M. Bohra, "General Jacobi elliptic function expansion method applied to the generalized $(3 + 1)$-dimensional nonlinear Schrödinger equation", *Optical and Quantum Electronics*, vol. **48**, no. 268, 8 pages, 2016.

[51] H. Chen and H. Zhang, "Extended Jacobian elliptic function method and its applications", *Journal of Applied Mathematics and Computing*, vol. **10**, no. 1–2, pp. 119–130, 2002.

[52] C.-M. Wei, Z.-Q. Xia and N.-S. Tian, "Jacobian elliptic function expansion solutions of nonlinear stochastic equations", *Chaos, Solitons & Fractals*, vol. **26**, no. 2, pp. 551–558, 2005.

[53] H. A. Ghanny, "Analytical approach to exact solutions for the Wick-type stochastic space-time fractional KdV equation", *Chinese Physics Letters*, vol. **31**, no. 6, 10 pages, 2014.

[54] H. A. Ghanny and A. Hyder, "Exact solutions for the Wick-type stochastic time-fractional KdV equations", *Kuwait Journal of Science and Engineering*, vol. **41**, no. 1, pp. 75–84, 2014.

[55] Y. Chen and B. Li, "The stochastic soliton-like solutions of stochastic mKdV equations", *Czechoslovak Journal of Physics*, vol. **55**, no. 1, pp. 1–104, 2005.

[56] S. Saha Ray and S. Singh, "New exact solutions for the Wick-type stochastic Zakharov–Kuznetsov equation for modelling waves on shallow water surfaces", *Random Operators and Stochastic Equations*, vol. **25**, no. 2, pp. 107–116, 2017.

[57] S. Saha Ray and S. Singh, "New exact solutions for the Wick-type stochastic Kudryashov-Sinelshchikov equation", *Communications in Theoretical Physics*, vol. **67**, no. 2, pp. 197–206, 2017.

[58] S. Saha Ray and S. Singh, "New exact solutions for the Wick-type stochastic modified Boussinesq equation for describing wave propagation in nonlinear dispersive systems", *Chinese Journal of Physics*, vol. **55**, no. 4, pp. 1653–1662, 2017.

[59] N. Z. Petrović, M. R. Belić and W.-P. Zhong, "Exact traveling-wave and spatiotemporal soliton solutions to the generalized $(3+1)$-dimensional Schrödinger equation with polynomial nonlinearity of arbitrary order", *Physical Review E*, vol. **83**, article ID 026604, 4 pages.

[60] B. Chen and Y. Xie, "Periodic-like solutions of variable coefficient and wick-type stochastic NLS equations", *Journal of Computational and Applied Mathematics*, vol. **203**, no. 1, pp. 249–263, 2007.

[61] O. Bang, P. L. Christiansen, F. If2, K. Rasmussen and Y. B. Gaididei, "White noise in the two-dimensional nonlinear Schrodinger equation", *Applicable Analysis*, vol. **57**, no. 1–2, pp. 3–15, 1995.

[62] E. M. E. Zayed and K. A. E. Alurrfi, "New extended auxiliary equation method and its applications to nonlinear Schrödinger-type equations", *Optik*, vol. **127**, no. 20, pp. 9131–9151, 2016.

[63] E. M. E. Zayed and K. A. E. Alurrfi, "Extended auxiliary equation method and its applications for finding the exact solutions for a class of nonlinear Schrödinger-type equations", *Applied Mathematics and Computation*, vol. **289**, pp. 111–131, 2016.

[64] S. Singh and S. Saha Ray, "Exact solutions for the Wick-type stochastic Kersten-Krasil'shchik coupled KdV-mKdV equations", *The European Physical Journal Plus*, vol. **132**, no. 480, 12 pages, 2017.

[65] S. Ozgul, M. Turan and A. Yildirim, "Exact traveling wave solutions of perturbed nonlinear Schrodinger's equation (NLSE) with Kerr law nonlinearity", *Optik*, vol. **123**, no. 24, pp. 2250–2253, 2012.

[66] M. Panthee and M. Scialom, "Asymptotic behavior for a class of solutions to the critical modified Zakharov–Kuznetsov equation", *Studies in Applied Mathematics*, vol. **124**, no. 3, pp. 229–245, 2010.

[67] D. Lannes, F. Linares and J. C. Saut, *Studies in Phase Space Analysis with Applications to PDEs*, Birkhäuser Basel, New York, pp. 181–213.

[68] V. Y. Belashov and S. V. Vladimirov, "Solitary waves in dispersive complex media", Springer, Berlin, Heidelberg, vol. **149**, pp. 17–62, 2005.

[69] D. J. Benny, "Long nonlinear waves in fluid flows", *Journal of Mathematical Physics*, vol. **45**, pp. 52–63, 1966.

[70] Y. A. Berezin and V. I. Karpman, "Nonlinear evolution of disturbances in plasmas and other dispersive media", *Soviet Physics – Journal of Experimental and Theoretical Physics*, vol. **24**, pp. 1049–1056, 1967.

[71] M. Washimi and T. Taniuti, "Propagation of ion-acoustic solitary waves of small amplitude", *Physical Review Letters*, vol. **17**, no. 19, pp. 996–998, 1966.

[72] B. B. Kadomtsev and V. I. Petviashvili, "On the stability of solitary waves in weakly dispersive media", *Soviet Physics - Doklady*, vol. **15**, pp. 539–549, 1970.

[73] S. Monro and E. J. Parkes, "The derivation of a modified Zakharov-Kuznetsov equation and the stability of its solutions", *Journal of Plasma Physics*, vol. **62**, no. 3, pp. 305–317, 1999.

[74] S. Monro and E. J. Parkes, "Stability of solitary-wave solutions to a modified Zakharov-Kuznetsov equation", *Journal of Plasma Physics*, vol. **64**, no. 3, pp. 411–426, 2000.

[75] P. N. Ryabov, "Exact solutions of the Kudryashov-Sinelshchikov equation", *Applied Mathematics and Computation*, vol. **217**, pp. 3505–3590, 2010.

[76] P. N. Ryabov, D. I. Sinelshchikov and M. Kochanov, "Application of the Kudryashov method for finding exact solutions of the high order nonlinear evolution equations", *Applied Mathematics and Computation*, vol. **218**, no. 4, pp. 3965–3972, 2011.

[77] S. Saha Ray and S. Singh, "New exact solutions for the Wick-type stochastic Kudrashyov-Sinelshchikov equation", *Communications in Theoretical Physics*, vol. **67**, no. 2, pp. 197–206, 2017.

[78] S. T. Demiray, Y. Pandir and H. Bulut, "Generalized Kudryashov method for time-fractional differential equations", *New Trends on Fractional and Functional Differential Equations*, Hindawi Publishing Corporation, article ID 901540, pp. 1–13, 2014.

[79] S. M. Ege and E. Misirli, "The modified Kudryashov method for solving some fractional order nonlinear equations", *Advances in Difference Equations*, vol. **2014**, no. 135, 13 pages, 2014.

[80] S. Saha Ray, "New analytical exact solutions of time fractional KdV–KZK equation by Kudryashov methods", *Chinese Physics B*, vol. **25**, no. 4, 7 pages, 2016.

[81] S. Sahoo and S. Saha Ray, "A new method for exact solutions of variant types of time fractional Korteweg–de Vries equations in shallow water waves", *Mathematical Methods in the Applied Sciences*, vol. **40**, no. 1, pp. 106–114, 2017.

[82] A. Cesar and S. Gómez, "An improved sub-equation method for solving nonlinear fractional equations", *International Journal of Pure and Applied Mathematics*, vol. **101**, no. 2, pp. 133–140, 2015.

[83] J. F. Alzaidy, "Fractional sub-equation method and its applications to the space–time fractional differential equations in mathematical physics", *British Journal of Mathematics and Computer Science*, vol. **3**, no. 2, pp. 153–163, 2013.

[84] W. G. Wang and Z. T. Xu, "The improved fractional sub-equation method and its applications to nonlinear fractional partial differential equations", *Romanian Reports in Physics*, vol. **66**, no. 3, pp. 595–602, 2014.

[85] S. Guo, M. Liquan, Y. Li and Y. Sun, "The improved fractional sub-equation method and its applications to the space-time fractional differential equations in fluid mechanics", *Physics Letters A*, vol. **376**, no. 4, pp. 407–411, 2012.

[86] M. Ekici, E. M. E. Zayed and A. Sonmezoglu, "A new fractional sub-equation method for solving the space-time fractional differential equations in mathematical physics", *Computational Methods for Differential Equations*, vol. **2**, no. 3, pp. 153–170, 2014.

[87] S. Saha Ray and S. Sahoo, "New exact solutions of fractional Zakharov-Kuznestov equations using fractional sub equation

method", *Communications in Theoretical Physics*, vol. **63**, no. 1, pp. 25–30, 2015.

[88] S. Sahoo and S. Saha Ray, "Improved functional sub-equation method for (3+1) dimensional generalized fractional KdV-Zakharov-Kuznestov equations", *Computers and Mathematics with Applications*, **70**, no. 2, pp. 158–166, 2015.

[89] S. Zhang, A. Q. Zong, D. Liu and Q. Gao, "A generalized Exp-function method for fractional Riccati differential equations", *Communications in Fractional Calculus*, vol. **1**, no. 1, pp. 48–52, 2010.

[90] Q. Feng, "Jacobi elliptic function solutions for fractional partial differential equations", *IAENG International Journal of Applied Mathematics*, vol. **46**, no. 1, pp. 121–129, 2016.

[91] S. Liu, F. Fu, S. Liu and Q. Zhaoa, "Jacobi elliptic function expansion method and periodic wave solutions of nonlinear wave equations", *Physics Letters A*, vol. **289**, no. 2001, pp. 69–74, 2001.

[92] C. Huai-Tang and Z. Hong-Qing, "New double periodic and multiple soliton solutions of the generalized (2+1)-dimensional Boussinesq equation", *Chaos, Solitons & Fractals*, vol. **20**, no. 4, pp. 765–769, 2004.

[93] C.-M. Wei, Z.-Q. Xia and N.-S. Tian, "Jacobian elliptic function expansion solutions of nonlinear stochastic equations", *Chaos, Solitons & Fractals*, vol. **26**, no. 2, pp. 551–558, 2005.

[94] J. H. Choi, H. Kim and R. Sakthivel, "Exact solution of the Wick-type stochastic fractional coupled KdV equations", *Journal of Mathematical Chemistry*, vol. **52**, no. 10, pp. 2482–2493, 2014.

[95] G. Xu, "Extended auxiliary equation method and its applications to three generalized NLS equations", *Abstract and Applied Analysis*, vol. **2014**, 7 pages, 2014.

[96] K. Maleknejad, B. Basirat and E. Hashemizadeh, "Hybrid Legendre polynomials and Block-Pulse functions approach for nonlinear Volterra–Fredholm integro-differential equations", *Computers & Mathematics with Applications*, vol. **61**, no. 9, pp. 2821–2828, 2011.

[97] A. M. Wazwaz, *A First Course in Integral Equations*, World Scientific Publishing, London, 2015.

[98] P. E. Kloeden and E. Platen, *Numerical Solution of Stochastic Differential Equations*, Springer-Verlag, New York, USA, 1992.

[99] K. Maleknejad, M. Khodabin and M. Rostami, "Numerical solution of stochastic Volterra integral equations by a stochastic operational matrix based on block pulse functions", *Mathematical and Computer Modelling*, vol. **55**, no. 3–4, pp. 791–800, 2012.

[100] K. Maleknejad, E. Hashemizadeh and B. Basirat, "Numerical solvability of Hammerstein integral equations based on hybrid Legendre and Block-Pulse functions", *International Conference on Parallel*

and *Distributed Processing Techniques and Applications*, vol. **2010**, pp. 172–175, 2010.

[101] K. Maleknejad and M. T. Kajani, "Solving second kind integral equations by Galerkin methods with hybrid Legendre and Block-Pulse functions", *Applied Mathematics and Computation*, vol. **145**, no. 2–3, pp. 623–629, 2003.

[102] F. Mohammadi and P. Adhami, "Numerical study of stochastic Volterra-Fredholm integral equations by using second kind Chebyshev wavelets", *Random Operators and Stochastic Equations*, vol. **24**, no. 2, pp. 129–141, 2016.

[103] S. Saha Ray and A. K. Gupta, "Numerical solution of fractional partial differential equation of parabolic type with Dirichlet boundary conditions using two-dimensional Legendre wavelets method", *Journal of Computational and Nonlinear Dynamics*, vol. **11**, no. 1, 9 pages, 2016.

[104] M. Khodabin, K. Maleknejad, M. Rostami and M. Nouri, "Numerical approach for solving stochastic Volterra-Fredholm integral equations by stochastic operational matrix", *Computers and Mathematics with Applications*, vol. **64**, no. 6, pp. 1903–1913, 2012.

[105] B. Øksendal, *Stochastic Differential Equations: An Introduction with Applications*, 5th edition, Springer-Verlag, New York, 1998.

[106] J. C. Cortés, L. Jordan and L. Villafuerte, "Numerical solution of random differential equations: A mean square approach", *Mathematical and Computer Modelling*, vol. **45**, no. 7–8, pp. 757–765, 2007.

[107] D. J. Higham, "An algorithmic introduction to numerical simulation of stochastic differential equations", *SIAM Review*, vol. **43**, no. 3, pp. 525–546, 2001.

[108] A. Yildrim and Y. Gulkanat, "Numerical solutions of wave equations subject to an integral conservation condition by He's homotopy perturbation method", *International Journal of Modern Physics B*, vol. **25**, no. 32, pp. 4457–4469, 2011.

[109] E. Guariglia and S. Silvestrov, "Fractional-wavelet analysis of positive definite distributions and wavelets on $D'(C)$", *Engineering Mathematics II*, vol. **179**, pp. 337–353, 2016.

[110] E. Guariglia, "Riemann zeta fractional derivative—functional equation and link with primes", *Advances in Difference Equations*, vol. **261**, 15 pages, 2019.

[111] S. Yousefi and A. Banifatemi, "Numerical solution of Fredholm integral equations by using CAS wavelets", *Applied Mathematics and Computation*, vol. **183**, no. 1, pp. 458–463, 2006.

[112] S. Yousefi and M. Razzaghi, "Legendre wavelets method for the non-linear Volterra–Fredholm integral equations", *Mathematics and Computers in Simulation*, vol. **70**, no. 1, pp. 1–8, 2005.

[113] K. Maleknejad and S. Sohrabi, "Numerical solution of Fredholm integral equations of the first kind by using Legendre wavelets", *Applied Mathematics and Computation*, vol. **186**, no. 1, pp. 836–843, 2007.

[114] V. K. Singh, O. P. Singh and K. R. Pandey, "Numerical evaluation of the Hankel transform by using linear Legendre multi-wavelets", *Computer Physics Communications*, vol. **179**, no. 6, pp. 424–429, 2008.

[115] S. Saha Ray, "On Haar wavelet operational matrix of general order and its application for the numerical solution of fractional Bagley Torvik equation", *Applied Mathematics and Computation*, vol. **218**, no. 9, pp. 5239–5248, 2012.

[116] E. Babolian and F. Fattahzadeh, "Numerical solution of differential equations by using Chebyshev wavelet operational matrix of integration", *Applied Mathematics and Computation*, vol. **188**, no. 1, pp. 417–426, 2007.

[117] F. H. Shekarabi and T. Damercheli, "Numerical solution of stochastic mixed Volterra-Fredholm integral equations driven by space-time Brownian motion via two-dimensional block pulse functions", *AIP Conference Proceedings 1997*, vol. **020049**, 7 pages, 2018.

[118] K. Maleknejad, E. Hashemizadeh and B. Basirat, "Computational method based on Bernstein operational matrices for nonlinear Volterra–Fredholm–Hammerstein integral equations", *Communications in Nonlinear Science and Numerical Simulation*, vol. **17**, pp. 52–61, 2012.

[119] F. Mirzaee and N. Samadyar, "Numerical solution of nonlinear stochastic Itô-Volterra integral equations driven by fractional Brownian motion", *Mathematical Methods in the Applied Sciences*, vol. **41**, no. 4, pp. 1410–1423, 2017.

[120] P. K. Sahu and S. Saha Ray, "A new numerical approach for the solution of nonlinear Fredholm integral equations system of second kind by using Bernstein collocation method", *Mathematical Methods in the Applied Sciences*, vol. **38**, no. 2, pp. 274–280, 2015.

[121] S. K. Vanani, A. Yildrim, F. Soleymani and M. Khan, "A numerical algorithm for solving nonlinear delay Volterra integral equations by means of homotopy perturbation method", *International Journal of Nonlinear Sciences and Numerical Simulation*, vol. **12**, no. 1–8, pp. 15–21, 2011.

[122] A. Saeed and U. Saeed, "Generalized fractional order Chebyshev wavelets for solving nonlinear fractional delay-type equations", *International Journal of Wavelets, Multiresolution and Information Processing*, vol. **17**, no. 3, 14 pages, 2019.

[123] M. H. Heydari, "Chebyshev cardinal wavelets for nonlinear variable-order fractional quadratic integral equations", *Applied Numerical Mathematics*, vol. **144**, pp. 190–203, 2019.

[124] P. K. Sahu and S. Saha Ray, "A numerical approach for solving nonlinear fractional Volterra-Fredholm integro-differential equations with mixed boundary conditions", *International Journal of Wavelets, Multiresolution and Information Processing*, vol. **14**, no. 5, 1650036, 15 pages, 2005.

[125] K. Maleknejad, M. Khodabin and M. Rostami, "A numerical method for solving m-dimensional stochastic Itô–Volterra integral equations by stochastic operational matrix", *Computers & Mathematics with Applications*, vol. **63**, no. 1, pp. 133–143, 2012.

[126] C. Corduneanu, *Integral Equations and Applications*, Cambridge University Press, vol. 148, 1991.

[127] F. Mohammadi, "Haar wavelets approach for solving multidimensional stochastic Itô-Volterra integral equations", *Applied Mathematics E-Notes*, vol. **15**, pp. 80–95, 2015.

[128] P. K. Sahu and S. Saha Ray, "Hybrid Legendre Block-Pulse functions for the numerical solutions of system of nonlinear Fredholm–Hammerstein integral equations", *Applied Mathematics and Computation*, vol. **270**, pp. 871–878, 2015.

[129] A. N. Kolmogorov, "Wienersche Spiralen und einige andere interessante Kurven im Hilbertschen Raum", *The Academy of Sciences of the USSR (N.S.)*, vol. **26**, pp. 115–118, 1940.

[130] G. A. Hunt, "Random Fourier transforms", *Transactions of the American Mathematical Society*, vol. **71**, no. 1, pp. 38–69, 1951.

[131] A. M. Yaglom, "Correlation theory of processes with random stationary n^{th} increments", *Matematicheskii Sbornik*, vol. **37**, no. 79, pp. 141–196, 1955.

[132] F. Mirzaee and A. Hamzeh, "Stochastic operational matrix method for solving stochastic differential equation by a fractional Brownian motion", *International Journal of Applied and Computational Mathematics*, vol. **3**, no. 1, pp. 411–425, 2017.

[133] A. K. Gupta and S. Saha Ray, "Numerical treatment for the solution of fractional fifth order Swada-Kotera equation using second kind Chebyshev wavelet method", *Applied Mathematical Modelling*, vol. **39**, no. 17, pp. 5121–5130, 2015.

[134] P. K. Sahu and S. Saha Ray, "Chebyshev wavelet method for numerical solutions of integro-differential form of Lane-Emden type differential equations", *International Journal of Wavelets, Multiresolution and Information Processing*, vol. **15**, no. 2, article ID 1750015, 16 pages, 2017.

[135] M. Asgari, E. Hashemizadehm, M. Khodabin and K. Maleknejad, "Numerical solution of nonlinear stochastic integral equation by stochastic operational matrix based on Bernstein polynomials", *Bulletin mathématique de la Société des Sciences Mathématiques de Roumanie*, vol. **57**, no. 105, pp. 3–12, 2015.

[136] K. Maleknejad, E. Hashemizadeh and B. Basirat, "Computational method based on Bernstein operational matrices for nonlinear Volterra–Fredholm–Hammerstein integral equations", *Communications in Nonlinear Science and Numerical Simulation*, vol. **17**, pp. 52–61, 2012.

[137] H. Lisei and A. Soós, "Approximation of stochastic differential equations driven by fractional Brownian motion", *Seminar on Stochastic Analysis, Random Fields and Applications V*, vol. **59**, pp. 227–241, 2007.

[138] M. Zähle, "Integration with respect to fractal functions and stochastic calculus II", *Mathematische Nachrichten*, vol. **225**, pp. 145–183, 2001.

[139] S. Saha Ray and A. K. Gupta, *Wavelet Methods for Solving Partial Differential Equations and Fractional Differential Equations*, Chapman and Hall/CRC, New York, 2018.

[140] M. H. Heydari, M. R. Mahmoudi, A. Shakiba and Z. Avazzadeh, "Chebyshev cardinal wavelets and their application in solving nonlinear stochastic differential equations with fractional Brownian motion", *Communications in Nonlinear Science and Numerical Simulation*, vol. **64**, pp. 98–121, 2018.

[141] A. Atangana, "Convergence and stability analysis of a novel iteration method for fractional biological population equation", *Neural Computing & Applications*, vol. **25**, no. 5, pp. 1021–1030, 2014.

[142] A. Atangana, "On the new fractional derivative and application to nonlinear Fisher's reaction–diffusion equation", *Applied Mathematics and Computation*, vol. **273**, pp. 948–956, 2016.

[143] S. Petrovskii and N. Shigesada, "Some exact solutions of a generalized Fisher equation related to the problem of biological invasion", *Mathematical Biosciences*, vol. **172**, no. 2, pp. 93–94, 2001.

[144] A. Abdulle and G. A. Pavliotis, "Numerical methods for stochastic partial differential equations with multiple scales", *Journal of Computational Physics*, vol. **231**, no. 6, pp. 2482–2497, 2012.

[145] Q. Du and T. Zhang, "Numerical approximation of some linear stochastic partial differential equations driven by special additive noises", *SIAM Journal on Numerical Analysis*, vol. **40**, no. 4, pp. 1421–1445, 2002.

[146] A. Reardon and A. Novikov, "Front propagation in reaction-diffusion equations", Department of Mathematics, Pennsylvania State University, 2016.

[147] T. Mavoungou and Y. Cherruault, "Numerical study of Fisher's equation by Adomian's method", *Mathematical and Computer Modelling*, vol. **19**, no. 1, pp. 89–95, 1993.

[148] M. J. Ablowitz and A. Zeppetella, "Explicit solutions of Fisher's equation for a special wave speed", *Bulletin of Mathematical Biology*, vol. **41**, no. 6, pp. 835–840, 1979.

[149] S. E. A. Alhazmi, "Numerical solution of Fisher's equation using finite difference", *Bulletin of Mathematical Sciences and Applications*, vol. **12**, pp. 27–34, 2015.

[150] J. R. Branco, J. A. Ferreira and P. Oliveira, "Numerical methods for the generalized Fisher–Kolmogorov–Petrovskii–Piskunov equation", *Applied Numerical Mathematics*, vol. **57**, no. 1, pp. 89–102, 2007.

[151] A. L. Hodgkin and A. F. Huxley, "A quantitative description of membrane current and its application to conduction and excitation in nerve", *The Journal of Physiology*, **117**, no. 4, pp. 500–544, 1952.

[152] K. Wright, "Chebyshev collocation methods for ordinary differential equations", *The Computer Journal*, vol. **6**, no. 4, pp. 358–365, 1964.

[153] E. M. E. Elbarbary and M. El-Kady, "Chebyshev finite difference approximation for the boundary value problems", *Applied Mathematics and Computation*, vol. **139**, no. 2–3, pp. 513–523, 2003.

[154] S. Singh and S. Saha Ray, "Numerical solutions of stochastic Fisher equation to study migration and population behavior in biological invasion", *International Journal of Biomathematics*, vol. **10**, no. 7, article ID 1750103, 14 pages, 2017.

[155] Y. Zheng and J. Huang, "Stochastic stability of Fitzhugh-Nagumo systems perturbed by Gaussian white noise", *Applied Mathematics and Mechanics*, vol. **32**, no. 1, pp. 11–22, 2011.

[156] Y. Zheng and J. Huang, "Stochastic stability of FitzHugh-Nagumo systems in infinite lattice perturbed by Gaussian white noise", *Acta Mathematica Sinica*, vol. **27**, no. 11, pp. 2143–2152, 2011.

[157] H. C. Tuckwell and R. Rodriguez, "Analytical and simulation results for stochastic Fitzhugh-Nagumo neurons and neural networks", *Journal of Computational Neuroscience*, vol. **5**, no. 1, pp. 91–113, 1998.

[158] V. Chandraker, A. Awasthi and S. Jayaraj, "A numerical treatment of Fisher equation", *Procedia Engineering*, vol. **127**, pp. 1256–1262, 2015.

[159] A. H. Khater, R. S. Temsah and M. M. Hassan, "A Chebyshev spectral collocation method for solving Burgers'-type equations", *Journal of Computational and Applied Mathematics*, vol. **222**, no. 2008, pp. 333–350, 2008.

[160] S. Saha Ray, *Fractional Calculus with Applications for Nuclear Reactor Dynamics*, CRC Press, Boca Raton, 2015.

[161] A. E. Aboanber and A. A. Nahla, "Solution of the point kinetics equations in the presence of Newtonian temperature feedback by Padé approximations via the analytical inversion method", *Journal of Physics A: Mathematical and General*, vol. **35**, no. 45, pp. 9609–9627, 2002.

[162] B. D. Ganapol, "A highly accurate algorithm for the solution of the point kinetics equations", *Annals of Nuclear Energy*, vol. **62**, pp. 564–571, 2013.

[163] A. E. Aboanber and Y. M. Hamada, "Power series solution (PWS) of nuclear reactor dynamics with Newtonian temperature feedback", *Annals of Nuclear Energy*, vol. **30**, no. 10, pp. 1111–1122, 2003.

[164] J. J. A. Silva, A. C. M. Alvim, M. T. M. B. Vilhena, C. Z. Petersen and B. E. J. Bodmann, "On a closed form solution of the point kinetics equations with reactivity feedback of temperature", *International Journal of Nuclear Energy Science and Technology*, vol. **8**, no. 2, pp. 131–145, 2011.

[165] A. Henryk, *Nuclear Reactor Dynamics and Stability*, KTH Royal Institute of Technology, Stockholm, Sweden, 2011.

[166] Y. M. Hamada, "Trigonometric Fourier-series solutions of the point reactor kinetics equations systems", *Nuclear Engineering and Design*, vol. **281**, pp. 142–153, 2015.

[167] J. G. Hayes and E. J. Allen, "Stochastic point-kinetics equations in nuclear reactor dynamics", *Annals of Nuclear Energy*, vol. **32**, no. 6, pp. 572–587, 2005.

[168] C. Leonhard and A. Rößler, "An efficient derivative-free Milstein scheme for stochastic partial differential equations with commutative noise", *ArXiv e-prints*, 2015.

[169] S. Saha Ray, "Numerical simulation of stochastic point kinetic equation in the dynamical system of nuclear reactor", *Annals of Nuclear Energy*, vol. **49**, pp. 154–159, 2012.

[170] A. A. Nahla and A. M. Edress, "Efficient stochastic model for the point kinetics equations", *Stochastic Analysis and Applications*, vol. **34**, no. 4, pp. 598–609, 2016.

[171] A. Patra and S. Saha Ray, "On the solution of the nonlinear fractional neutron point-kinetics equation with Newtonian temperature feedback reactivity", *Nuclear Technology*, vol. **189**, no. 1, pp. 103–109, 2015.

[172] A. Patra and S. Saha Ray, "Numerical simulation based on Haar wavelet operational method to solve neutron point kinetics equation involving sinusoidal and pulse reactivity", *Annals of Nuclear Energy*, vol. **73**, pp. 408–412, 2014.

[173] G. Espinosa-Paredes, M.-A. Polo-Labarrios, E.-G. Espinosa-Martinez and E. Valle-Gallegos, "Fractional neutron point kinetics equations for nuclear reactor dynamics", *Annals of Nuclear Energy*, vol. **38**, no. 2–3, pp. 307–330, 2011.

[174] A. A. Nahla, "Stochastic model for the nonlinear point reactor kinetics equations in the presence Newtonian temperature feedback effects", *Journal of Difference Equations and Applications*, vol. **23**, no. 6, pp. 1003–1016, 2017.

[175] J. J. Duderstadt and L. J. Hamilton, *Nuclear Reactor Analysis*, John Wiley and Sons, New York, 1976.

[176] W. M. Stacey, *Nuclear Reactor Physics*, 2nd edition, Wiley-VCH Verlag GmbH & Co. KGaA, Weinheim, 2007.

[177] A. Patra and S. Saha Ray, "Solution of nonlinear neutron point kinetics equation with feedback reactivity in nuclear reactor dynamics", *International Journal of Nuclear Energy Science and Technology*, vol. **9**, no. 1, pp. 23 and 24, 2015.

[178] S. Singh and S. Saha Ray, "On the comparison of two split-step methods for the numerical simulation of stochastic point kinetics equations in presence of Newtonian temperature feedback effects", *Annals of Nuclear Energy*, vol. **110**, pp. 865–873, 2017.

[179] A. A. Nahla, "Taylor's series method for solving the nonlinear point kinetics equations", *Nuclear Engineering and Design*, vol. **241**, no. 5, pp. 1592–1595, 2011.

[180] W. Wagner and E. Platen, "Approximation of Itô integral equations", Preprint ZIMM, Akad. Wissenschaften, DDR, Berlin, 1978.

[181] S. Saha Ray and A. Patra, "Numerical solution for stochastic point-kinetics equations with sinusoidal reactivity in dynamical system of nuclear reactor", *International Journal of Nuclear Energy Science and Technology*, vol. **7**, no. 3, pp. 231–242, 2013.

[182] P. Wang and Y. Li, "Split-step forward methods for stochastic differential equations", *Journal of Computational and Applied Mathematics*, vol. **233**, no. 10, pp. 2641–2651, 2010.

[183] P. E. Kloeden, E. Platen and H. Schurz, *Numerical Solution of SDE through Computer Experiments*, Springer-Verlag, Berlin, Heidelberg, 1994.

[184] D. J. Higham, "An algorithm introduction to numerical solution of stochastic differential equations", *SIAM Review*, vol. **43**, no. 3, pp. 525–546, 2001.

[185] A. A. Nahla and A. M. Edress, "Analytical exponential model for stochastic point kinetics equations via eigenvalues and eigenvectors", *Nuclear Science and Techniques*, vol. **27**, no. 1, pp. 1–8, 2016.

[186] R.-I. Cázares-Ramírez, V. A. Vyawahare, G. Espinosa-Paredes and P. S. V. Nataraj, "On the feedback stability of linear FNPK equations", *Progress in Nuclear Energy*, vol. **98**, pp. 45–58, 2017.

[187] G. Espinosa-Paredes, M. A. Polo-Labarrios and J. Alvarez-Ramirez, "Anomalous diffusion processes in nuclear reactors", *Annals of Nuclear Energy*, vol. **54**, pp. 227–232, 2013.

[188] T. K. Nowak, K. Duzinkiewicz and R. Piotrowski, "Fractional neutron point kinetics equations for nuclear reactor dynamics – Numerical solution investigations", *Annals of Nuclear Energy*, vol. **73**, pp. 317–329, 2014.

[189] S. Saha Ray and A. Patra, "Numerical solution of fractional stochastic point kinetic equation for nuclear reactor dynamics", *Annals of Nuclear Energy*, vol. **54**, pp. 154–161, 2013.

[190] A. E. Aboanber, A. A. Nahla and A. M. Edress, "Developed mathematical technique for fractional stochastic point kinetics model in nuclear reactor dynamics", *Nuclear Science and Techniques*, vol. **29**, no. 9, article ID 132, 2018.

[191] C. Li, R. Wu and H. Ding, "High-order approximation to Caputo derivatives and Caputo-type advection-diffusion equations", *Communications in Applied and Industrial Mathematics*, vol. **6**, no. 2, 33 pages, 2015.

[192] H. Li, J. Cao and C. Li, "High-order approximation to Caputo derivatives and Caputo-type advection–diffusion equations (III)", *Journal of Computational and Applied Mathematics*, vol. **299**, pp. 159–175, 2016.

[193] J. X. Cao, C. P. Li and Y. Q. Chen, "High-order approximation to Caputo derivatives and Caputo-type advection-diffusion equations (II)", *Fractional Calculus and Applied Analysis*, vol. **18**, no. 3, pp. 735–761, 2015.

[194] M. G. Hafez, M. R. Talukder and R. Sakthivel, "Ion acoustic solitary waves in plasmas with nonextensive distributed electrons, positrons and relativistic thermal ions", *Indian Journal of Physics*, vol. **90**, pp. 603–611, 2016.

[195] H. Kim, J. H. Bae and R. Sakthivel, "Exact travelling wave solutions of two important nonlinear partial differential equations", *Zeitschrift für Naturforschung A*, vol. **69**, pp. 155–162, 2014.

[196] J. Lee and R. Sakthivel, "Exact travelling wave solutions of a variety of Boussinesq-like equations", *Chinese Journal of Physics*, vol. **52**, no. 3, pp. 939–957, 2014.

Index